T0235446

Specific Intermolecular Interactions of
Element-Organic Compounds

Alexei K. Baev

Specific Intermolecular Interactions of Element-Organic Compounds

 Springer

Alexei K. Baev
G.A. Krestov Institute of Solution Chemistry
Russian Academy of Sciences
Ivanovo
Russia

ISBN 978-3-319-36033-1   ISBN 978-3-319-08563-0 (eBook)
DOI 10.1007/978-3-319-08563-0
Springer Cham Heidelberg New York Dordrecht London

Printed on acid-free paper

Springer is part of Springer Science+Business Media (www.springer.com)

Preface

The nonstereotypical ideas concerning the participation of the carbon pentacoordinated atom in the formation of the specific intermolecular interaction with significantly undivided $2s^2(c)$ electron pairs of carbon atoms ($DAl \rightarrow CH_3-Al = 44.2$ kJ mol^{-1}) have been substantiated and the phenomenon of the intramolecular reverse dative bond between carbon atoms of the methyl group of alkyl ligand and the contacting central molecule atom (in the year 1995) led to the grounded refusal from the model of sp^3 hybridization of carbon atoms, proved by the quantum chemical calculations of the methyl compounds of group IV elements. The fourth nonstereotyped idea grounded in the author's works in the period 1969–1972 is in classifying the phase transformations to the chemical processes. The values of enthalpies and entropies of the processes are connected to the number and energies of the bursting specific intermolecular interactions. The theory of alternative character and the necessity of its further development led to the formulation of the grounding of different types of hydrogen bonds and specific intermolecular interactions, and also to the methodology of the thermodynamic calculation of their energies, building on the vaporization enthalpies applicable for all series of organic compounds.

In the monograph "Specific Intermolecular Interactions of Organic Compounds" (Springer, Heidelberg/New York/Dordrecht/London; 2012), a thermodynamic analysis of the different classes of organic compounds and their series, including the ethers, ketones, alcohols, carboxylic acids, aldehydes and complex ethers, saturated and unsaturated hydrocarbons, including isostructural alkyl groups, is performed. It settles the different types of specific intermolecular interactions and hydrogen bonds, forming the networks of interactions in the liquid and crystal conditions. It outlines the theoretical problems, establishes the natural numbers of stabilization of specific interactions and hydrogen bonds, and develops the concept of the extra stabilizing effect of the isostructural methyl group. It resulted in the creation of the united system of consistent, correlated specific interactions and hydrogen bonds reflecting their nature. The development of theoretical ideas of specific intermolecular interactions and hydrogen bonds, based on the four nonstereotypical ideas, continues in the monograph "Specific Intermolecular Interactions of Nitrogenated and Bioorganic Compounds" (Springer, Heidelberg/New York/Dordrecht/London; 2014).

Types of specific intermolecular interactions of compounds with nitrogenated and bioorganic compounds and including oxygenated hydrocarbon cycle, pentamerous heterocycle, oxazoles, isoxazoles and thyazoles, oxazolines, azetidimines, and aminoisoxazoles compounds with hexamerous heterocycles and their derivatives, as well as compounds with the cycle of 6-caprolactam with one and two carbonyl groups are discussed in 11 of the monograph's chapters. In the result of the implemented structural energy analysis it was created the unite system of a wide spectrum of the types of specific intermolecular interactions and hydrogen bonds, correctly reflecting their nature, and the energies were obtained and the natural numbers of their stabilization were grounded, which led to the significant development of the intermolecular interactions theory.

All preceding results of the implemented research in the field of the specificity of intermolecular interactions, including organic, nitrogenous, and compounds with two heteroatoms of nitrogen and oxygen in the molecules, provided the success in the problem solving of obtaining the energy value of the peptide hydrogen bond of all substantiated three types and the energy value of all realizing specific interactions in the formed networks in crystalline and liquid structure of peptides. Significant results in this direction are the grounding of the role of the methylene group and the fragment with the larger number of these groups, located between strong acceptors of electron density by the nitrogen atoms and carbonyl oxygen, expressing reduced ability in the formation of specific interaction in comparison with the similar fragment at the terminal part of the chain, and participation of the isostructural methyl group, connected with the marked methylene group, in the distribution of electron density in the molecule and its extra stabilizing effect to the fragment, fringing these isostructural groups, providing the introduction of additive energy contribution to the enthalpy characteristics independently from its number in peptide and protein. The obtained energy values of hydrogen bonds and specific interactions with the fifth coordinated carbon atom of peptides and deputies led to the determination of energy contribution by the conformer of chains and the value of sublimation enthalpies of dipeptide, tripeptide, and tetrapeptide chains of derivative aminoacetic acid, glycine, and L-(d)-α-alanine, aminopropionic acid. In the results, various protein secondary structures with parallel β–sheet and antiparallel β–sheet with all specific interactions and their energies are first presented.

The present book "Specific Intermolecular Interactions of Elementorganic Compounds" accomplishes a "trilogy" dedicated to the development of a theory of specific intermolecular interactions and to enrich the system of specific intermolecular interactions and hydrogen bonds and their energies.

As the result of the implemented thermodynamic analysis of hydrides, alkyls, and alkenes (Chap. 1) of the subgroup elements of boron, nitrogen, selenium, and zinc, the energies of hydrogen bonds were determined, grounded the types of specific interactions, and their energies were obtained. It was shown that these compounds are characterized by the formation of the intermolecular reverse dative bonds, which influence the change in stability of the specific interaction that is completed at the third carbon atom of the chain of the propyl ligand (Chap. 1). It is reasoned that only the molecule of diphenyl selenide from the series of methylphenyl–selenide–diphenyl selenide–diphenyl diselenide–diphenyl mercury compounds forms, by the two free electron pairs of selenium atom, the two specific interactions and forms one specific interaction with increased stability.

Properties of the silicon and carbon atoms analogous with essentially undivided $2s^2$(C) and $3s^2$(Si) electron air causes the similarity in the elements atoms in the manifestation of acceptor and donor properties, forming the hydrogen bonds and specific interactions (Chap. 2). In this connection, in the silane molecules like methane, the shifting of electron density from the hydrogen atom to the silicon atoms is accompanied by the change of differences in its positive charges, causing the formation of four hydrogen bonds Si–H•••H–Si while preserving the four-coordinated silicon atom. With a partial substitution of hydrogen atoms in the silane molecule to the methyl, ethyl, propyl, or to the differing alkyl groups in dependence of their location, the stabilizing effect of trans influence takes place. The compounds with propyl and butyl fragments in the molecules of dipropyldi-butylsilane and propyltributylsilane form specific interactions with reduced stability compared to liquid methylpropylsilanes, and the stabilizing effect of the trans influence is within experimental error. A distinctive feature of the molecules of trisilyl of nitrogen main group elements c SiH_3, an amino group with two or more silicone atoms of the chain, is the formation of a specific interaction with an essentially undivided $3s^2$ electron pair of silicone atom, located in the penta-coordinated condition. The specific interactions formed by the pentacoordinated silicone atom possess increased stability caused by the location of the shifted electron density at the more distant energy level in comparison with carbon. The maximum stability of the specific interaction is reached by the increasing number of silicone atoms of the chain, providing the completion of the influence of the reverse dative bond. According to the results of the implemented thermodynamic analysis of the linear silanes, oxygenated compounds, including methoxy, ethoxy, and propoxy groups, trimethoxy(methylthio)methyl)silane, silacyclobutanes, and 1,1-dimethylsilacyclopentane, with benzene ring, decamethylcyclopentasiloxane, and 1,3,7,5.4-pentametylcyclopentasiloxane, the nature of specific interactions was revealed, their energies determined, as well as that of hydrogen bonds, the natural numbers of stabilization were established, and the system of consistent energy properties was created, reflecting the nature of the organosilicon compounds.

The conducted thermodynamic analysis of the compounds with tetrahedral structure of the molecules of alkyls, alkylhydrides of germanium, tin and lead, and germanium alkyl compounds with the metal–metal bonds (Chap. 3) illustrate the analogue of the interaction with siliconorganic compounds with the corresponding alkyl ligand and practically unchanged energy value of the hydrogen bonds. It is shown that the energy value of hydrogen bonds formed by the hydrogen atom in the trans position has small changes following the substitution of four hydrogen atoms in the compounds of 1,1,2,2-tetramethyldigermane and 1,1,1,2.2-tetramethyldigermane to the methyl groups. The energies were determined of the specific interactions formed by pentacoordinated germanium atom in liquid triethyl (diethylamino)germane, methylthiogermane, hexamethyldigermoxane and methoxygermane, and ethoxygermanes, with the network of specific interactions and justified the numbers of their stabilization. Due to the obtained energy parameters of specific interactions formed by benzene ring of the crystalline compounds of tetraphenylgermane, tetraphenyltin, and tetraphenyllead, it was reasoned that the interaction of carbon atoms of benzene ring with the element atom with low coordinating ability and low donor properties is accompanied by the reduction of differences in the charges of the carbon atoms and the stability of the formed specific interactions.

It was reasoned that the intramolecular interaction of the reverse dative bond at the sulfur organic compounds with the completion of influence to the stabilization of the specific intermolecular interaction at the propyl ligand of the chain and at the six carbon atoms in the heterocycle. The types of specific interactions and hydrogen bonds were revealed (Chap. 4), forming the structure of liquid and crystalline conditions with the bonds network. The energies of the large number of specific interactions and hydrogen bonds of the various homological series were determined: sulfides and disulfides, with symmetrical and asymmetrical structure of the normal structure and with the isostructural methyl group, saturated and unsaturated ligands, and cyclic compounds including heteroatoms and thiols, cyclothiols, benzenethiols, and dithianes. The natural numbers of stabilization of the various types of specific interactions and hydrogen bonds were substantiated.

The result of the implemented thermodynamic analysis of sulfur, oxygen, and nitrogenated organic compounds is the reasoning of the types of specific interactions and determination of their energy parameters of the wide spectrum of the oxygenated and nitrogenated compounds: carbonyl sulfides and disulfides, dimethylsulfoxides and divinylsulfoxides, diacetylsulfides, 1-(methylthio)-2-(vinyloxy)ethanes, alkyl thiolactates and 1,3-dithiol-2-one, 1,3-dithiole-2-thione, sulfones with the linear, cyclic structure and benzene ring, p-tolyl sulfones, and diphenyl sulfone, dibenzyl sulfone (Chap. 5). The thermodynamic analysis of sulfur and nitrogenated compounds revealed information on the types and energies of specific intermolecular interactions of thiocyanates, isothicyanates and phenyl isothiocyanate and thioacetamide and dithiooxamide, thiosemicarbazide, thiocarbohydrazide and thiobenzamides, thiazole, its derivatives and benzothiazole, and (methylthio)pyridines, which results in the substantiation of the numbers of regularities change of the energies of specific interactions and the creation of a consistent system of energy properties of sulfonated compounds, correctly reflecting the nature of the intermolecular compounds.

The result of the performed theoretical studies based on the experimental thermodynamic parameters and structure data revealed a large number of hydrogen bonds with differing and identical stability of specific intermolecular interactions with the participation of pentacoordinated carbon atom and silicone atom, formed by the essentially undivided $2s^2(c)$ electron pair. A theory of hydrogen bonds and specific intermolecular interactions with the element atoms of the main groups was developed and the energies for a large number of compounds were determined. The result of the implemented studies is the creation of a system of corresponding energy values of hydrogen bonds and specific interactions of the organic, nitrogenated bioorganic and elementorganic, including siliconorganic and sulfonated, compounds with the participation of pentacoordinated atoms of carbon, silicon and nitrogen, sulfur, and other atoms of the main subgroup elements, adequately reflecting their nature.

Realizing that, over time, the volume of experimentally obtained vaporization enthalpies will increase, including enhanced accuracy at the standard conditions, the author considered it useful to suggest the calculations procedure and its use in the calculations of thermodynamic characteristics with a large number of experimental errors and significantly differing from the standard conditions.

Ivanovo, Russia   Alexei K. Baev

Synopsis

This book presents the development of the theory of hydrogen bonds and specific intermolecular interactions of nonstereotypic ideas and approaches, with the marking of motivated notions about five coordinated carbon atoms and silicone atoms of different elementorganic compounds of main group elements, including siliconorganic and sulfonated, sulfur and oxygenated, and sulfur and nitrogenated organic compounds. New types of specific interactions and hydrogen bonds are substantiated and their energies are determined, and the system of interconnected quantitative characteristics of stable specific interactions, reflecting identically their nature and law-governed changes, are created. A thermodynamic analysis of elementorganic compounds was fulfilled, which substantiated the new types of intramolecular interaction of the reverse dative bond, realized in organic, nitrogenated, and bioorganic compounds. This book is the part of trilogy of "Specific Intermolecular Interactions of Organic Compounds" and "Specific Intermolecular Interactions of Nitrogenated and Bioorganic Compounds", in combination with which was exposed the nature of specific interactions formed by organic, elementorganic compounds and a system of coordinated values of energies of different types of intermolecular interactions was created.

Abbreviations

PhE	Photoelectron
PhE spectroscopy	Photoelectron spectroscopy
(Iv)	Vertical ionization potential
eV	Electron volt
T m.p.	Melting point
T b.p.	Boiling point
DE-Me	Dissociation energies of bond
Ddis.dim	Dissociation energies of dimer molecule
Cn	Number of carbon atoms at chine

$M \longrightarrow CO$ -   Reverse dative bond on metal carbonyl

Reverse dative bond $Zn(CH_3)_2$

Reverse dative bond $Zn(C_2H_5)_2$

$Zn \longrightarrow CH_2 \longrightarrow CH_2 \longrightarrow CH_3$ Reverse dative bond $Zn(C_3H_7)_2$

$Zn \longrightarrow CH_2 \longrightarrow CH_2 \longrightarrow CH_2 - CH_3$ Reverse dative bond $Zn(C_4H_9)_2$

$\Delta_{vap}H^0(298\ K)$	Vaporization enthalpy at standard conditions
$\Delta_{vap}H^0(T\ K)$	Vaporization enthalpy at $T\ K$
$\Delta_{sub}H^0(298\ K)$	Sublimation enthalpy at standard conditions
$\Delta_{sub}H^0(T\ K)$	Sublimation enthalpy at $T\ K$
$\Delta_{evap}H^0(298\ K)$	Vaporization enthalpy at standard conditions

Energy of Specific Intermolecular Interactions of Zinc and Boron Subgroup Elements

$DM–CH_3 \rightarrow M$	Methyl group
$DM–CH_2–CH_3 \rightarrow M$	Ethyl group
$DM–CH_2–CH_2–CH_3 \rightarrow M$	Propyl group
$DH_3C \rightarrow H–CH_2$	Isomethyl group
$DC\equiv C \rightarrow C\equiv C$	Acetylene

Contributions' Energy at Enthalpy

$\sum DCH_2$
$DisoCH_3$
DH

Chapter 1

$\Delta_{vap}H^0(TK)etz$	Vaporization enthalpy of diethyl zinc
$\Delta_{vap}H^0(TK)pz$	Vaporization enthalpy of dipropyl zinc
$\Delta_{vap}H^0(TK)ipz$	Vaporization enthalpy of isopropyl zinc
$\Delta_{vap}H^0(TK)ibz$	Vaporization enthalpy of isobutyl zinc
$\Delta_{vap}H^0(TK)tbz$	Vaporization enthalpy of di-*tert*-butyl zinc
$\Delta_{vap}H^0(TK)tipb$	Vaporization enthalpy of triisopropylborane
$\Delta_{vap}H^0(TK)amph$	Vaporization enthalpy of allylmethylphosphine
$\Delta_{vap}H^0(TK)pph$	Vaporization enthalpy of 2-propylphosphine
$\Delta_{vap}H^0(TK)mps$	Vaporization enthalpy of methylphenyl selenide
$\Delta_{vap}H^0(TK)dps$	Vaporization enthalpy of diphenyl selenide
$\Delta_{vap}H^0(TK)dbds$	Vaporization enthalpy of dibenzyl diselenide

Energy of Hydrogen Bonds

$DB–H\bullet\bullet\bullet B$	Hydrogen bond
$DH_3E\bullet\bullet\bullet H–EH_2$	Hydrogen bond
$DS\bullet\bullet\bullet H–S$	Hydrogen bond
$DS \rightarrow S–H$	Specific interaction

Chapter 2

$\Delta_{vap}H^0(T\,\mathrm{K})$sn	Vaporization enthalpy of silane
$\Delta_{vap}H^0(T\,\mathrm{K})$mt	Vaporization enthalpy of methane
$\Delta_{vap}H^0(T\,\mathrm{K})$sn	Vaporization enthalpy of silane
$\Delta_{vap}H^0(T\,\mathrm{K})$mt	Vaporization enthalpy of methane
$\Delta_{vap}H^0(T\,\mathrm{K})$dmds	Vaporization enthalpy of 1,2-dimethyldisilane
$\Delta_{vap}H^0(T\,\mathrm{K})$xmds	Vaporization enthalpy of hexamethyldisilane
$\Delta_{vap}H^0(T\,\mathrm{K})$mphsi	Vaporization enthalpy of (2-methoxylphenyl) trimethylsilane
$\Delta_{vap}H^0(T\,\mathrm{K})$ephsi	Vaporization enthalpy of triethoxyphenylsilane
$\Delta_{vap}H^0(T\,\mathrm{K})$bz	Vaporization enthalpy of benzene
$\Delta_{vap}H^0(T\,\mathrm{K})$phtmsi	Vaporization enthalpy of phenoxytrimethylsilane
$\Delta_{vap}H^0(T\,\mathrm{K})$tmphmsi	Vaporization enthalpy of trimethoxy[(phenylthio)methyl] silane
$\Delta_{vap}H^0(T\,\mathrm{K})$mothms	Vaporization enthalpy of trimethoxy[(methylthio)methyl] silane
$\Delta_{vap}H^0(T\,\mathrm{K})$phxsi	Vaporization enthalpy of bis(2-methylphenoxy) dimethylsilane

Energy of Specific Intermolecular Interactions of Silicon of Organic Compounds

Energy of Specific Interactions

$D-SiH_2-H_3Si \rightarrow SiH_2-SiH_3$	Tetrasilane
$D-Si-H_3C \rightarrow H-CH_2-$	CH_3 group tetramethylsilane
$DSi-CH_2-CH_3 \rightarrow CH_2-CH_3,$	Tetraethylsilane
$DSi-CH_2-CH_2-CH_3 \rightarrow CH_3-CH_2-CH_2$	Tetrapropylsilane
$DSi-CH=CH_2 \rightarrow CH=CH_2$	Tetravinylsilane
$DSi-H_3C \rightarrow Si$	Methyl group
$DSi-CH_2-CH_3 \rightarrow Si$	Ethyl group
$DSi-CH_2-CH_2-CH_3 \rightarrow Si$	Propyl group
$D-Si-H_3C \rightarrow H-CH_2-$	H_3C
$DH_2C \rightarrow CH_2$	Silacyclobutane
$DS \rightarrow CH_2$	$Si-CH_2$ group

Energy of Hydrogen Bonds

DSi•••H–Si	Methyldisilane
D–Si–H → H–Si–	Hydrogen bond of silane
D–C–H → H–C–	Methane
DSi–O•••H–Si	Hydrogen bond of 1,3,7,5.4-pentametylcyclopentasiloxane

Energy of Specific Interactions

DN → SiH$_3$	Disilazane
DP → SiH$_3$	Trisilylphosphine
DAs → SiH$_3$	Trisilylarsine
DSb → SiH$_3$	Trisilylstibine
DN → CH$_2$–CH$_2$–Si	CH$_2$–CH$_2$– group
DisoCH$_2$–H$_3$C → CH$_2$–CH$_3$	isoC$_2$H$_5$ group
DH$_2$Si → SiH$_2$	SiH$_2$ group
DH$_3$C → SiH$_2$	Alkylsilanes

Energy of Specific Interaction

DO→SiH$_3$–SiH$_2$	bis(Disilanyl)ether
DSi–O•••H–O–Si	bis(Disilanyl)ether
DSi–O→CH$_3$–O–Si	–O–CH$_3$ group
DSi–O→CH$_3$–CH$_2$–O–Si	–O–CH$_3$–CH$_2$ group
DSi–O→R–O–Si	–O–R group
D DS→CH$_3$–S	S–CH$_3$ fragment
DS→CH$_3$–CH$_2$–S	S–CH$_2$–CH$_3$ fragment
DS→CH$_3$–CH$_2$–CH$_2$–S	S–CH$_2$–CH$_3$ fragment
DSi–O→CH$_3$–Si	Si–CH$_3$ and Si–O groups'

Chapter 3

$\Delta_{vap}H^0$(298 K)mg	Vaporization enthalpy of methylthiogermane
$\Delta_{vap}H^0$(T K)mtg	Vaporization enthalpy of methylthiogermane
$\Delta_{vap}H^0$(T K)mtg	Vaporization enthalpy of dimethylthiogermane
$\Delta_{vap}H^0$(T K)mdg	Vaporization enthalpy of methyldigermane
$\Delta_{vap}H^0$(T K)gsi	Vaporization enthalpy of germilsilane

Energy of Specific Intermolecular Interactions and Hydrogen Bond of Germanium Subgroup Elements

Energy of Specific Interactions

DGe–H$_3$C → H–CH$_2$–Ge	CH$_3$ group
DGe–CH$_2$–CH$_3$ → CH$_2$–CH$_3$	C$_2$H$_5$ group
DGe–CH$_2$–CH$_2$–CH$_3$ → CH$_2$–CH$_2$–CH$_3$	C$_3$H$_7$ group
DGe–S → CH$_3$–S–Ge	S–CH$_3$ group
DGe–O → CH$_3$–CH$_2$–O–Ge	O–CH$_2$–CH$_3$ fragment

Hydrogen Bonds

DGe–H•••H–Ge
DGe•••H–GeH$_2$

Chapter 4

$\Delta_{vap}H^0$ (298 K)das	Vaporization enthalpy of dialkyl sulfide
$\Delta_{vap}H^0$ (298 K)eips	Vaporization enthalpy of diethyl sulfide
$\Delta_{vap}H^0$ (298 K)dps	Vaporization enthalpy of dipropyl sulfide
$\Delta_{vap}H^0$ (298 K)ems	Vaporization enthalpy of ethyl methylsulfide
$\Delta_{vap}H^0$ (298 K)mps	Vaporization enthalpy of methyl propyl sulfide
$\Delta_{vap}H^0$ (298 K)eps	Vaporization enthalpy of ethyl propyl sulfide
$\Delta_{vap}H^0$ (298 K)ipps	Vaporization enthalpy of isopropyl propyl sulfide
$\Delta_{vap}H^0$ (298 K)eibs	Vaporization enthalpy of ethyl isobutyl sulfide
$\Delta_{vap}H^0$ (298 K)mtbs	Vaporization enthalpy of methyl *tert*-butyl sulfide
$\Delta_{vap}H^0$ (298 K)aes	Vaporization enthalpy of allyl ethyl sulfide
$\Delta_{vap}H^0$ (298 K)mphs	Vaporization enthalpy of methyl phenyl sulfide
$\Delta_{vap}H^0$ (298 K)eps	Vaporization enthalpy of ethyl phenyl sulfide
$\Delta_{vap}H^0$ (298 K)bms	Vaporization enthalpy of benzyl methyl sulfide
$\Delta_{vap}H^0$ (298 K)mtt	Vaporization enthalpy of (methylthio)toluene
$\Delta_{vap}H^0$ (298 K)mttp	Vaporization enthalpy of methyltetrahydrothiophene
$\Delta_{vap}H^0$ (298 K)thp	Vaporization enthalpy of tetrahydro-2H-thiopirane
$\Delta_{vap}H^0$ (298 K)mthp	Vaporization enthalpy of 4-methyltetrahydrothiophene
$\Delta_{vap}H^0$ (298 K)dth	Vaporization enthalpy of 3,4-dithiohexene
$\Delta_{vap}H^0$ (298 K)tmdth	Vaporization enthalpy of 2,2,5,5-tetramethyl-3,4-dithiohexene
$\Delta_{vap}H^0$ (298 K)mtph	Vaporization enthalpy of methylthiophene
$\Delta_{vap}H^0$ (298 K)tph	Vaporization enthalpy of Thiophene

$\Delta_{vap}H^0$(298 K)mc   Vaporization enthalpy of methyl mercaptan
$\Delta_{vap}H^0$(298 K)amc   Vaporization enthalpy of alkylthiols
$\Delta_{vap}H^0$(298 K)edto   Vaporization enthalpy of 1,2 ethanedithiols
$\Delta_{vap}H^0$(298 K)cpto   Vaporization enthalpy of cyclopentanethiol
$\Delta_{vap}H^0$(298 K)cp   Vaporization enthalpy of cyclopentane
$\Delta_{vap}H^0$(298 K)chto   Vaporization enthalpy of cyclohexanethiol
$\Delta_{vap}H^0$(298 K)ch   Vaporization enthalpy of cyclohexane
$\Delta_{vap}H^0$(298 K)pto   Vaporization enthalpy of 1-propanethiol
$\Delta_{vap}H^0$(298 K)btl   Vaporization enthalpy of benzenethiol
$\Delta_{vap}H^0$ (298 K)bmtl   Vaporization enthalpy of benzenemethanethiol
$\Delta_{vap}H^0$ (298 K)dtti   Vaporization enthalpy of 1,3-dithian-2-thyone
$\Delta_{sub}H^0$ (298 K)bht   Sublimation enthalpy of bicyclo[2,2,1]heptane-2-thione
$\Delta_{sub}H^0$ (298 K)bh   Sublimation enthalpy of bicyclo[2,2,1]heptane

Energies of Specific Intermolecular Interactions of Pentacoordinated Carbon Atom of Sulfonated Organic Compounds

Energies of Specific Interactions of Saturated Alkyl Series

DS → CH$_3$–S   Methyl
DS → CH$_2$–S   Methylene
DS → CH$_2$ Methylene
DS → CH$_3$–CH$_2$–S   Ethyl
DS → CH$_3$–CH$_2$–CH$_2$–S Propyl
D$^{-S-H_2C\cdots^{H}_{H}\cdots CH_2-S-}$ DisoCH3
DH$_2$C → CH$_2$ Methylene

Energies of Specific Interactions of Unsaturated Alkyl Series

DS → CH$_2$=CH–S   Vinyl
DS → CH$_2$=CH–CH$_2$–S   Allyl
=S → CH$_2$–CH$_2$–CH$_2$=S Propyl

Energies of Specific Interactions of Cyclic Compounds

$DC=S \rightarrow CH_2-S$	CH_2-S cycle
$DS \rightarrow CH$	CH of benzene
$DHC \rightarrow CH$	CH of benzene
$DHC \rightarrow C=$	$C=$ of benzene
$D=C \rightarrow C=$	$C=$ of benzene
$DS \rightarrow C\equiv C$	$C\equiv C-$ fragment
$DS \rightarrow C=$	$C=$ of benzene
$DS \rightarrow CH=$	$CH=$ of benzene
$DC=S \rightarrow C=S$	Thyone fragment
$DS \rightarrow C\equiv C$	Dithiin fragment
$DC=S \rightarrow CH_2-S$	CH_2-S fragment of cycle
$DS \cdots H-S$	Hydrogen bond

Chapter 5

$\Delta_{vap}H^0(T)$dms	Vaporization enthalpy of dimethylsulfoxide
$\Delta_{vap}H^0(T)$aa	Vaporization enthalpy of acetyl anhydride
$\Delta_{vap}H^0(T)$mtve	Vaporization enthalpy of 1-(methylthio)-2-(vinyloxy)ethane
$\Delta_{vap}H^0(T)$taa	Vaporization enthalpy of thioacetyl acid
$\Delta_{vap}H^0(350 \text{ K})$ot	Vaporization enthalpy of BY-oxythiane
$\Delta_{vap}H^0(350 \text{ K})$ttp	Vaporization enthalpy of tetrahydro-4H-thioperan-4-one
$\Delta_{vap}H^0(298 \text{ K})$tsdo	Vaporization enthalpy of thiete sulfone (2H-thiete)-1,1-dioxide
$\Delta_{vap}H^0(298 \text{ K})$thtdo	Vaporization enthalpy of tetrahydrothiophene-1,1-dioxide
$\Delta_{vap}H^0(298 \text{ K})$thmtp	Vaporization enthalpy of 3-methyltetrahydrothiophene
$\Delta_{vap}H^0(298 \text{ K})$phvs	Vaporization enthalpy of phenyl vinylsulfone
$\Delta_{sub}H^0(298 \text{ K})$mbs	Sublimation enthalpy of methylbenzyl sulfone
$\Delta_{sub}H^0(298 \text{ K})$dphs	Sublimation enthalpy of diphenyl sulfone
$\Delta_{vap}H^0(298 \text{ K})$tca	Vaporization enthalpy of thiocyanic acid
$\Delta_{vap}H^0 (298 \text{ K})$ ttca	Vaporization enthalpy of trithiocarbonic acid
$\Delta_{vap}H^0 (298 \text{ K})$thbt	Vaporization enthalpy of 4,5,6,7-tetrahydro-1,4-benzodithiol-2-tione

Energies of Specific Intermolecular Interactions of Sulfonated Organic Compounds with Oxygen Heteroatoms

DO → C	Carbon dioxide
DS → C	Carbonyl sulfide
DS = O → CH$_3$	Methyl group
DS=O → CH$_2$=CH–	Vinyl group
DC=O → C=O	Carbonyl group
DO → CH$_2$–C	Acetyl group
DS → CH$_2$–C	Acetyl group
DC=O•••H–CH$_2$	CH$_3$ group
DC=O → CH$_2$	CH$_2$ group
DS=O → C–C=	C–C=
DS=O → CH–CH	CH–CH
DC=S=O → C–C=	C–C=

Energies of Specific Interactions of Alkyl Sulfones

DS=O → CH$_2$	CH$_2$
DS=O → CH$_3$–CH$_2$	CH$_3$–CH$_2$
DS=O → CH$_3$–CH$_2$–CH$_2$	CH$_3$–CH$_2$–CH$_2$
DS=O → CH$_2$=CH	CH$_2$=CH
DS=O → CH=	CH=
DC=O → C=	C=

Energies of Specific Interactions of Thiazol and Trithiocarbonic Acid

DC=S → C=S	C=S
DS → N–N	Sulfur diimide
DC=N → C=N	C=N, pyridine's derivatives
DN → CH=	Thiazol and its derivatives
DS → C=	4,5,6,7-Tetrahydro-1,4-benzodithiol-2-tione

Contents

Chapter 1
Hydrogen Bonds of Hydrides and Specific Interactions of Alkyl Compounds of Main Group Elements

1.1 Hydrogen Bond Energies of Main Group Element Hydrides

Borane molecules and alkyl boron have an equilateral planar trigonal structure, with an angle of 120° between ligands. Based on the comparative analysis by Cvitaš et al. of the electron structure of isostructural molecules of diborane and ethylene molecules based on the photoelectron spectra and nonempirical calculations [1], Nefedov and Vovna [2] remark that significant differences are present in the nature of the MO for the $1b_{2u}$ orbital, the orbital of π-type ethylene, and bridgehead B–H → B in the diborane molecule. Observed changes in the binding type of the $1b_{2u}$ orbital are accompanied by its stabilization by 4.2 eV in relation to the $1(\pi)$ molecule C_2H_4. Shifts of the photoelectron band at the level of s-type $1a_g$ and $1b_{2u}$ at values of 3.0 and 2.1 eV, respectively, as well as the levels of p-type $1b_{2g}$, $2a_g$, and $1b_{1u}$ at 1.0, 1.5, and 2.1 eV, respectively, are correlated with the energy change due to ionization of the 2s and 2p electrons of carbon and boron atoms. As described by Bieri et al. [3], the electron population of atomic orbitals of s-3.137e, $2p_x$, $2p_y$-0.902, $2p_z$-0e, and hydrogen 1s-1.20e in molecules of BH_3, points to the insignificant negative charge at the hydrogen atom and its low donor properties in the formation of intermolecular hydrogen bonds in crystalline and liquid borane (Fig. 1.1). The planar structure of the borane molecule with six coordinated boron atoms forms a volumetric chain structure wherein, under the influence of the formed hydrogen bonds, the boron atoms are out of the plane and located at the apex of the mild trigonal pyramid. The chains are crosslinked by the hydrogen and boron atoms of the contacting chains due to hydrogen bond formation. Breakage of the seven hydrogen bonds in the vaporization process, with the transition to the vapor state of diborane, is accompanied by a reduction in the coordinated number of boron atoms from six to four, which is reflected by the stabilization of the two remaining intermolecular hydrogen bonds DB–H → B in the structure of the diborane molecule.

© Springer International Publishing Switzerland 2015
A.K. Baev, *Specific Intermolecular Interactions of Element-Organic Compounds*,
DOI 10.1007/978-3-319-08563-0_1

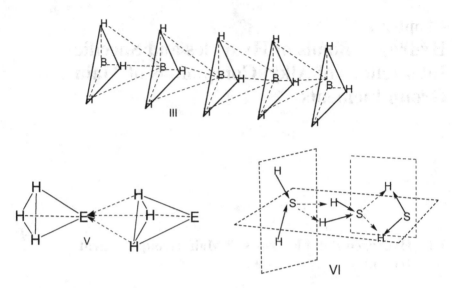

Fig. 1.1 Schematic of the crystalline and liquid structure of diborane, nitrogen, and sulfur subgroup elements with networks of hydrogen bonds

The average energy value of the ruptured hydrogen bonds, taking into account the simultaneous stabilization of the two hydrogen bonds in the diborane molecule, can be determined using the relationship of vaporization enthalpy and vaporization entropy, sublimation, and polymorphism with the number and energy of the bursting specific interactions [4–6] with the help of Eq. (1.1):

$$DB - H \bullet \bullet \bullet B = \Delta_{vap}H^0(T)/7 \qquad (1.1)$$

The measured energy value of the hydrogen bond of liquid borane (Table 1.1) at 210 K has a low value. The calculated energy value of the hydrogen bond $DB \rightarrow H_D$ of liquid perdeuterated diborane is equal to 2.2 kJ mol^{-1}. Given the difference in temperature of 46 K for definition of the vaporization enthalpy of borane and perdeuterated diborane and its temperature dependence, one can assume that the limited expression of deuterium atoms is due to the stabilization of the hydrogen bonds at any given temperature.

Table 1.1 Energies (kJ mol^{-1}) of the hydrogen bonds of liquid diborane and nitrogen and sulfur subgroup elements

Compounds	Formula	Structure	$\Delta_{vap}H^0$ (298 K)	T (K)	$DH_3B\cdots H\text{-}BH_2$ / $DH_3E\cdots H\text{-}EH_2$ / $DH_2E\cdots H\text{-}EH$	$DS \to S\text{-}H$
Diborane	B_2H_6		12.8	210	1.83	–
Perdeuterated diborane	B_2D_6		15.3	164	2.2	–
Ammonia	NH_3		22.7 / 23.5 / 23.4	308 / 239 / 239	3.8 / 3.93	– / –
Phosphine	PH_3		14.61 ± 0.02	185.73	2.43	–
Arsine	AsH_3		18.12 ± 0.04	210.68	3.02	–
Stibine	SbH_3		21.1 ± 2.1	254.8	3.51	–
Hydrogen sulfide	H_2S		23.2 ± 0.01 / 23.3	186 / 190	5.8	– / –
Hydrogen selenide	H_2Se		19.92 ± 0.21	231.8	4.98	–
Hydrogen selenide-d_2	SeD_2		22.2	217	5.55	–
Hydrogen telluride	H_2Te		23.4 ± 1.7	271	5.68	–
Dihydrogen disulfide	H_2S_2	H–S–S–H	33.8 ± 0.1	293	8.45	–
Dihydrogen trisulfide	H_2S_3	H–S–S–S–H	45.5 ± 0.2	293	8.45	11.5 ± 0.2
Dihydrogen tetrasulfide	H_2S_4	H–S–S–S–S–H	56.8 ± 0.3	293	8.45	11.5 ± 0.2
Dihydrogen pentasulfide	H_2S_5	H–S–S–S–S–S–H	68.4 ± 0.6	293	8.45	11.53 ± 0.2

The low energy values of hydrogen bonds of diborane and perdeuterated diborane point to the fact that the given compounds are unstable at higher temperatures and that they should dissociate in pairs, with permanent shifting of the balance to the side of the partial pressure reduction of diborane and its absence at standard conditions. Therefore, reliable studies of the measurement of partial components of the diborane molecule and the borane molecule are possible only with the use of static methods.

The ammonia molecule and its derivatives phosphine, arsine, and stibine, with a trigonal pyramid and an equilateral triangle at the bottom, form a volumetric chain structure in the crystalline and liquid condition by hydrogen bonds. The chains are crosslinked with the participation of hydrogen and nitrogen atoms, or the analog of the contacting chains forming hydrogen bonds (Fig. 1.1).

The process of vaporizing liquid compounds is accompanied by the transition to vapor of monomer molecules with the rupture of six hydrogen bonds; therefore, the energies of these bonds can be determined with the help of Eq. (1.1a) [8]:

$$DN \bullet \bullet \bullet H - N = \Delta_{vap}H^0(T)/6 \tag{1.1a}$$

The energies of hydrogen bonds of hydrides of the nitrogen subgroup elements obtained at different temperatures point to the general regularity of its stabilization:

Phosphine (2.43) 185.73 K < Arsine (3.02) 210.68 K < Stibine (3.51) 254.8 K ≤ Ammonia (3.93 kJ mol^{-1}) 239 K

According to the work of Nefedov and Vovna [2], the nature of the bonds in EH_2 molecules is similar to the interaction of water molecular orbitals. The angled molecular structure of these compounds is caused by p-electrons and the electron populations of atom orbitals N s-3.137e, $2p_x$ and $2p_y$-0.902e, $2p_z$-1.861e, and hydrogen 1s-0.734e with great deficit provides it with high acceptor properties and, from the formation of hydrogen bonds, with increased stability compared with the nitrogen subgroup elements. The monomeric nature in the pairs of hydrides of sulfur subgroup elements causes rupture of the four hydrogen bonds during the transition of the molecules to vapor. It follows that the hydrogen bond energy formed by the hydride molecules of sulfur subgroup elements, as for water hydrogen bonds [9], is obtained by Eq. (1.1b):

$$DS \bullet \bullet \bullet H - S = \Delta_{vap}H^0(T\,K)/4 \qquad (1.1b)$$

The obtained energies of the hydrogen bonds of hydrides of this type of compound are described by the following equation:

Hydrogen selenide (4.98) 231.8 K < Hydrogen selenide-d_2 (5.55) 217 K < Hydrogen telluride (5.68) 271 K < Hydrogen sulfide (5.8 kJ mol^{-1}) 186 K,

where the increased value of the hydrogen bond in H_2S, in comparison with H_2Te, reflects the influence of the location of valence electrons at the energy level nearest the core. A more striking influence is displayed by the location of valence electrons at the second energy level of hydrogen atoms, forming hydrogen bonds with an energy of 10.99 kJ mol^{-1} [9].

Molecules of dihydrogen disulfide form the structure of the liquid condition with a network of hydrogen bonds (Fig. 1.2a). The presence in the molecule of two sulfur atoms and two hydrogen atoms causes the increased shifting of electron density from the hydrogen atoms, which provides it with an increased positive charge and acceptor properties. In turn, the sulfur free electron pairs create the possibility to shift the increased electron density to the hydrogen atom, which is accompanied by stabilization of the formed hydrogen bonds. The energies of these interactions are determined by the division of the vaporization enthalpy by the number of formed hydrogen bonds with the help of Eq. (1.1b). The obtained energy of the hydrogen bonds is given in Table 1.1.

The molecules of dihydrogen trisulfide, dihydrogen tetrasulfide, and dihydrogen pentasulfide, with one to three sulfur atoms, form four hydrogen bonds and, by the sulfur atoms, one, two, and three specific interactions D–S → S–, respectively (Fig. 1.2b). This allows the energy of specific interactions to be calculated from Eq. (1.1c), taking into account the energy contribution of the four hydrogen bonds and the corresponding energies of the D–S → S– interactions contributed by the

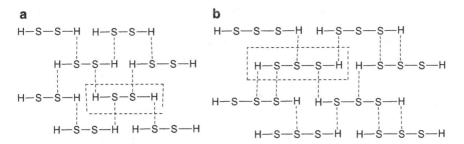

Fig. 1.2 Schematic of the liquid structure of dihydrogen disulfide (**a**) and dihydrogen trisulfide (**b**) with networks of hydrogen bonds

appropriate number of sulfur atoms of the molecule to the enthalpy characteristic, where n is the number of sulfur atoms in the molecule forming specific interactions:

$$D - S \rightarrow S- = \left(\Delta_{vap}H^0(TK) - 4DS \bullet \bullet \bullet H - S\right)/n \qquad (1.1c)$$

Vacant electron pairs of the sulfur atoms in the molecule of dihydrogen disulfide (8.45 kJ mol^{-1}) provide increased stability to the hydrogen bonds in comparison with the formed molecule of hydrogen sulfide (5.8 kJ mol^{-1}, 190 K). Thus, the energy value of the hydrogen bonds can be used to determine the energy of the specific interaction D–S $\rightarrow$ S– with the help of Eq. (1.1c). The obtained constant energy value of this type of specific interaction (11.5 kJ mol^{-1}) with dihydrogen trisulfide, dihydrogen tetrasulfide, and dihydrogen pentasulfide reflects the real behavior and the correctness of the concept of specific interactions.

1.2 Specific Interactions of Alkyls of Zinc Subgroup Elements

1.2.1 Zinc Alkyls

The electron configuration of MMe$_2$ molecules of zinc subgroup elements is rather complex. Two lower MOs are C 2s electrons, splitting the energy level, which, in the case of Cd atoms, is equal to 0.53 eV. Two C–H connecting pseudo-orbitals $1\pi_u$ and $2\pi_g$ give a single broad band at 12–14 eV according to the photoelectron spectra [10–14]. Based on the same spectra, it is shown that the molecular orbitals $3\sigma_g$ cause a binding sequence of C $2p_z$ with M-ns orbitals and a $2\sigma_u$ sequence C $2p_z$ with M-np_z orbitals. The information obtained in the photoelectron spectra allow establishment of the sequence of $d_{s/2}$ subshell levels, which proved the electrostatic nature of splitting and the small contribution of a covalent component.

The electron structures of atoms of the sulfur subgroup elements and the carbon atom in alkyl molecules have distinct differences in the donor and acceptor properties, providing shifting of the electron density from the element atom of this group to the essentially undivided $2s^2$ electron pair of the carbon atom of the alkyl chain [9]. An additional excess of the negative charge of carbon atoms is a result of the shifting of electron density from the related hydrogen atoms.

Realization of the reverse dative bond, caused by the partial transfer of electron density from the carbon atom to the element atom of the sulfur subgroup, reduces its positive charge. The shifting of electron density from the element atom of the sulfur subgroup to the essentially undivided $2s^2$ electron pair of the carbon atom of the methyl group in the dimethyl compound of the elements of this group provide it with donor properties and the ability to form a specific intermolecular interaction $DZn–CH_3 \rightarrow Zn$ or, in the general case, $DM–CH_2–CH_3 \rightarrow M$, $DM–CH_2–CH_2–CH_3 \rightarrow M$. The completion of the influence of the reverse dative bond at the third or fourth carbon atom provides the maximum difference in the corresponding positive and negative charge at the carbon atom of the terminal methyl group of the propyl or butyl compound and the metal atom. Thus, stabilization of the specific interaction due to an increase in the length of the alkyl chain occurs:

$$DM - CH_3 \rightarrow M < DM - CH_2 - CH_3 \rightarrow M < DM - CH_2 - CH_2 - CH_3 \rightarrow M$$

The relationship between the vaporization enthalpy and vaporization entropy, sublimation, polymorphic transformation, and melting and the number and energy of the bursting specific interactions causes the expressed dependence of the corresponding thermodynamic property on the number of carbon atoms of the alkyl chain [9]. A clear distinction in the dependence of the vaporization enthalpy of zinc alkyls (Fig. 1.3) on the number of carbon atoms for compounds $Zn(CH_3)_2$ to $Zn(C_3H_7)_2$ and $Zn(C_4H_9)_2$ to $Zn(C_7H_{15})_2$ is caused by the completion of the influence of the reverse dative bond at the third carbon atom of the chain. Further increase in the enthalpy characteristic reflects the contribution of the increasing number of methylene groups. This dependence allows us to refine the value of the vaporization enthalpy of dipentylzinc (53.5 kJ mol^{-1}), given in the literature without information on temperature and error.

Zinc alkyls with a linear molecular structure form a network in the crystalline and liquid conditions that is formed by specific intermolecular interactions with the

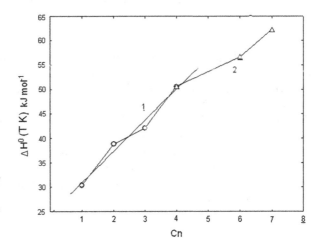

Fig. 1.3 Dependence of vaporization enthalpy of zinc alkyls on the number of carbon atoms (*Cn*) in the chain: *1* influence of reverse dative bond, *2* contribution of energy of CH$_2$ groups

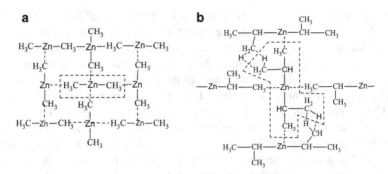

Fig. 1.4 Schematic of the liquid structure of dimethylzinc (**a**) and diisopropylzinc (**b**) with networks of specific interactions

participation of the pentacoordinated carbon atom of the methyl, propyl, butyl, or alkyl ligand and the four-coordinated zinc atom (Fig. 1.4a). Consequently, the vaporization process is accompanied by the transition to vapor of the monomer molecules with the rupture of four specific interactions, whose energy is determined using Eq. (1.2):

$$DZn - (CH_2)n - CH_3 \rightarrow Zn = \Delta_{vap}H^0(T)/4 \tag{1.2}$$

In calculating the energy of specific interactions, the most correct values of the vaporization enthalpies of the compounds were used, obtained with the help of a static measuring method of vapor pressure in vacuum with a low error level. Table 1.2 gives the energy values of specific interactions, which are described by the sequence of stabilization:

$$DZn - CH_3 \rightarrow Zn(7.6) < DZn - CH_2 - CH_3 \rightarrow Zn(9.70)$$

$$< DZn - CH_2 - CH_2 - CH_3 \rightarrow Zn(10.68\,kJ\,mol^{-1})$$

The energy of specific interaction of dipropylzinc is close to the energy of a hydrogen bond of liquid water (10.99 kJ mol^{-1}). The energy contribution of methylene groups to the vaporization enthalpy of alkyls with the largest number of carbon atoms in the chain is the difference between the vaporization enthalpy of the neighboring alkyl compound in the series or the value obtained as the difference between the vaporization enthalpy of the compound and the enthalpy of dipropylzinc (Table 1.2). The isostructural methyl group of zinc alkyls participates in the formation of the network of specific interactions (Fig. 1.4b) and forms two specific interactions of low stability with a similar group of contacting molecules via the hydrogen atom in the *trans* position to the chain carbon atom.

Table 1.2 Energies (kJ mol^{-1}) of specific interactions of liquid zinc alkyls

Compounds	Formula	Structure	$\Delta_{vap}H^0$ (298 K)	References	T (K)	DZn $\rightarrow$ CH$_3$-(CH$_2$)$_n$	DH$_3$C $\rightarrow$ H-CH$_2$
Dimethylzinc	ZnC$_2$H$_6$	H$_3$C–Zn–CH$_3$	30.4 ± 0.2	[15]	298	$n = 0$ 7,6	–
Diethylzinc	ZnC$_4$H$_{10}$	C$_2$H$_5$–Zn–C$_2$H$_5$	38.8 ± 0.4	[15]	298	$n = 1$ 9.70	–
Dipropylzinc	ZnC$_6$H$_{14}$	C$_3$H$_7$–Zn–C$_3$H$_7$	42.7 ± 0.4	[16]	345	$n = 2$ 10.68	–
Diisopropylzinc	ZnC$_6$H$_{14}$		41.8 ± 0.5	[16]	324	$n = 1$ 9.70	1.6:4 = 0.4
Dibutylzinc	ZnC$_8$H$_{18}$	C$_4$H$_9$–Zn–C$_4$H$_9$	50.7 ± 0.3	[17]	342	$n = 2$ 10.68	$\sum$DCH$_2$ = 4.0 × 2
Di-sec-butylzinc	ZnC$_8$H$_{18}$		40.9 ± 0.2	[17]	330	9.82	1.6:4 = 0.4
Diisobutylzinc	ZnC$_{8\rightarrow}$H$_{18}$		44.6 ± 0.2	[17]	330	$n = 2$ 10.68	1.9:4 = 0.5
Di-$tert$-butylzinc	ZnC$_8$H$_{18}$		44.3	[17, 18]	–	$n = 1$ 9.45	6.5:8 = 0.8
Dipentylzinc	ZnC$_{10}$H$_{22}$	C$_5$H$_{11}$–Zn–C$_5$H$_{11}$	53.5	[7]	342	12.7	$\sum$DCH$_2$ = 2.7 × 4
Dihexylzinc	ZnC$_{12}$H$_{26}$	C$_6$H$_{13}$–Zn–C$_6$H$_{13}$	56.2	[7]		10.68	$\sum$DCH$_2$ = 2.25 × 6
Diheptylzinc	ZnC$_{14}$H$_{30}$	C$_7$H$_{15}$–Zn–C$_7$H$_{15}$	62.3	[7]	–	10.68	$\sum$DCH$_2$ = 2.45 × 8

$$-\text{H}_2\text{C} \underset{\text{H}}{\overset{\text{H}}{<>}} \text{CH}_2 -$$

Taking into account that the ethyl group of diisopropylzinc contributes equivalently to the similar group of diethylzinc, the energy contribution to the vaporization enthalpy of four specific interactions of low stability formed by the two isostructural methyl groups can be obtained from the difference in the vaporization enthalpies of these compounds:

$$\text{DH}_3\text{C} \rightarrow \text{H} - \text{CH}_2 = \left(\Delta_{\text{vap}}H^0(T\,\text{K})\text{ipz} - \Delta_{\text{vap}}H^0(T\,\text{K})\text{et}\right)/4 \qquad (1.2a)$$

Similarly, we define the energies of specific interactions of this type in diisobutyl zinc by the difference between the vaporization enthalpies of diisopropylzinc and dipropylzinc:

$$\text{DH}_3\text{C} \rightarrow \text{H} - \text{CH}_2 = \left(\Delta_{\text{vap}}H^0(T\,\text{K})\text{ipz} - \Delta_{\text{vap}}H^0(T\,\text{K})\text{pz}\right)/4 \qquad (1.2b)$$

Correspondingly, double the number of formed specific interactions of the same type, formed in liquid di-*tert*-butylzinc, can be obtained from the difference in the vaporization enthalpies of diethylzinc and di-*tert*-butylzinc:

$$\text{DH}_3\text{C} \rightarrow \text{H} - \text{CH}_2 = \left(\Delta_{\text{vap}}H^0(T\,\text{K})\text{tbz} - \Delta_{\text{vap}}H^0(T\,\text{K})\text{etz}\right)/8 \qquad (1.2c)$$

$\text{DH}_3\text{C}{\rightarrow}\text{H}{-}\text{CH}_2$: Diisopropyl zinc (0.4) = Diisobutyl zinc (0.5) < Di-tert-butyl zinc (0.8 kJ mol^{-1})

The low values for the energies of the specific interactions reflect some stabilization of liquid di-*tert*-butylzinc. The energy contribution of the formed specific interactions of low stability of the isostructural methyl groups of di-*sec*-butylzinc can be taken as equal to that for the similar diisopropylzinc. This allowed the specific interaction energies of the formed propyl group of compounds to be obtained (Table 1.2).

1.2.2 Alkyls of Cadmium and Mercury

In the work by Nefedov and Vovna [2], it is noted that the values of exciton binding (Eb) for the internal Hg 4f 7/2 level with an energy of 107.14 eV and for the C 1s level with a value of 289.6 eV in comparison with the values of 107.1 eV for the mercury atom and 290.7 eV for C_2H_6 point to the insignificant positive charge of the Hg atom in the molecule of $Hg(CH_3)_2$. In this connection, the small difference in the vaporization enthalpies (Fig. 1.5) and, consequently, the low energy

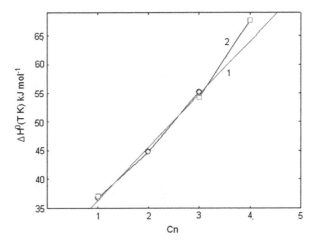

Fig. 1.5 Dependence of the vaporization enthalpy of liquid zinc alkyls on the number of carbon atoms in the chain: *1* mercury alkyls, *2* cadmium alkyls

difference of specific interactions of $Cd(CH_3)_2$ and $Hg(C_2H_3)_2$ point to the low positive charge at the cadmium atom. Interrelation of the vaporization enthalpy and sublimation enthalpy with the energy and number of bursting specific interactions is equal for cadmium and mercury alkyls with the same number of carbon atoms and the sharp increase in the values of the enthalpy characteristic of the alkyl elements (Fig. 1.5) indicates the significant shifting of electron density from the metal atoms to the end methyl group with an increasing number of carbon atoms in the chain. The measured charges reported in the literature [11–14] for the mercury atoms in $Hg(Alk)_2$, changing from +0.023 in $Hg(CH_3)_2$ to −0.070e in molecules of $Hg(i-Bu)_2$, are caused by the participation of isostructural methyl groups in the redistribution of electron density in the molecules of isobutylmercury. As shown in Fig. 1.5, the dependency of the alkyl vaporization enthalpy on the number of carbon atoms in the chain was used to clarify the value of the enthalpy characteristic of diethylcadmium (Table 1.3), used in the determination of the energy of specific interactions. The analogy of linear molecules of cadmium alkyls and mercury alkyls with the structure of zinc alkyl molecules, provides a similarity in the structure of their liquid and crystal conditions with the network of specific interactions, respectively, a normal structure and with the isostructural group (Fig. 1.5). The energy calculations of specific interactions of liquid cadmium and mercury alkyls of a normal structure, using Eq. (1.2), illustrate their stabilization (Table 1.3) with an increasing number of carbon atoms in the chain. High energies of the compounds with propyl ligands indicate significant differences in the positive charge of the metal atom and the negative charge of the carbon atom of the end methyl group. Note that the maximum stability of specific interactions occurs for the compounds with butyl ligands. The calculated energy contribution of the isostructural methyl group of diisopropylmercury and the formed specific interaction of low stability $DH_3C \rightarrow H–CH_2$ (1.9 kJ mol^{-1}) using Eq. (1.2b) points to its

Table 1.3 Energies (kJ mol^{-1}) of specific interactions of liquid cadmium alkyls

Compounds	Formula	Structure	$\Delta_{vap}H^0$ (298 K) [7]	References	T (K)	DM → CH$_3$–(CH$_2$)$_n$	DH$_3$C → H–CH$_2$
Dimethylcadmium	CdC$_2$H$_6$	H$_3$C – Cd – CH$_3$	37.10 ± 0.07	[18, 19]	324	9.27	–
Diethylcadmium	CdC$_2$H$_{10}$	C$_2$H$_5$ – Cd – C$_2$H$_5$	46.0 ± 0.4	[18, 19]	324	11.5	–
Dipropylcadmium	CdC$_6$H$_{14}$	C$_3$H$_7$ – Cd – C$_3$H$_7$	54.2 ± 0.4	[19]	342	13.6	–
Dibutylcadmium	CdC$_8$H$_{18}$	C$_4$H$_9$ – Cd – C$_4$H$_9$	67.7 ± 9.2	[18, 19]	356	16.75	–
Dimethylmercury	HgC$_2$H$_6$	H$_3$C – Hg – CH$_3$	36.7 ± 0.1	[18]	321	9.18	–
Diethylmercury	HgC$_4$H$_{10}$	C$_2$H$_5$ – Hg – C$_2$H$_5$	44.8 ± 1.7	[18]	–	11.2	–
Dipropylmercury	HgC$_6$H$_{14}$	C$_3$H$_7$ – Hg – C$_3$H$_7$	55.2 ± 1.3	[7, 18]	–	13.8	–
Diisopropylmercury	HgC$_6$H$_{14}$	Hg(CH–CH$_3$)$_2$ with CH$_3$	53.6 ± 1.7	[7]	–	11.2	7.6:4 = 1.9

significant stabilization in comparison with that in liquid diisopropylzinc $(0.4 \text{ kJ mol}^{-1})$.

1.3   Specific Interactions of Alkyls of Boron Subgroup Elements

The calculation results [20] for the electron density distribution in molecules of B $(CH_3)_3$, performed using an improved version of the method of Huckel [21] with various approximations, indicate shifting of the electron density from the boron and hydrogen atoms to the carbon atom of the methyl group.

A part of the charge ($\sim$0.3e) is transferred from the carbon atom to the $3p_z$ orbital of the boron atom as the reverse dative bond, stabilizing the intermolecular interactions. The result of the electron density redistribution are the charges on the boron atom and the methyl group [22], forming specific interactions in liquid and crystal conditions.

$$+0.704 \quad B \begin{array}{l} CH_3 \quad -0.234 \\ CH_3 \quad -0.234 \\ CH_3 \quad -0.234 \end{array}$$

Alkyl boranes with a planar molecular structure in the crystal and liquid conditions show similarity to boranes with respect to the volume of the network of specific interactions with the six-coordinated boron atom (Fig. 1.1). The vaporization process is accompanied by the transition to vapor in practically the monomer condition, which allows the energy calculation to be performed using Eq. (1.3):

$$DB - (H_2)n - CH_3 \rightarrow B = \Delta_{vap}H^0(T\,K)/6 \qquad (1.3)$$

Taking into account the equivalent energy contribution by the ethyl fragments of triethylborane and triisopropylborane, the energy contribution of the three isostructural methyl groups to the vaporization enthalpy of triisopropylborane or the contribution of the six specific interactions formed by the isostructural methyl groups, can be determined from the difference between the vaporization enthalpies of the compound and triethylborane (1.3a):

Table 1.4 Energies (kJ mol^{-1}) of specific interactions of liquid borane alkyls

Compounds	Formula	Structure	$\Delta_{vap}H^0$ (298 K) [7]	T (K)	DB → CH$_3$– (CH$_2$)$_n$–B	DH$_3$C → H– CH$_2$
Trimethylborane	BC$_3$H$_9$	CH$_3$ B—CH$_3$ CH$_3$	20.9 ± 0.1	239	3.5	–
Triethylborane	BC$_6$H$_{15}$	C$_2$H$_5$ B—C$_2$H$_5$ C$_2$H$_5$	36.8 ± 0.4	298	6.13	–
Tripropylborane	BC$_9$H$_{21}$	C$_3$H$_7$ B—C$_3$H$_7$ C$_3$H$_7$	41.8 ± 1.3	–	7.0	–
Triisopropylborane	BC$_9$H$_{21}$	CH$_3$ CH—CH$_3$ H$_3$C—CH—B CH$_3$ CH—CH$_3$ CH$_3$	41.8 ± 1.3 40.0	–	6.13	5.9:6 = 0.8

$$DH_3C \rightarrow H - CH_2 = \left(\Delta_{vap}H^0(K)tipb - \Delta_{vap}H^0(K)teb\right)/6 \qquad (1.3a)$$

The calculation results (Table 1.4) indicate that the general pattern of stabilization of specific interaction energies formed by the alkyl ligand with an increasing number of chain carbon atoms is caused by the relative increase in electron density shift and also by reduction in the influence of the reverse dative bond:

DB–(CH$_2$)n–CH$_3$→B: Trimethylborane   (3.5) < Triethylborane   (6.13) < Tripropylborane (7.0 kJ mol^{-1})

This draws attention to the low stability of the hydrogen bonds of liquid diborane in comparison with the energy of specific interactions of liquid trimethylborane and dimethylzinc, formed by the pentacoordinated carbon atom, reflecting the ability of the carbon atom of the methyl group to shift the electron density:

DB–H→B, Diborane (1.83 kJ mol^{-1}, 210 K) < DB → CH$_3$–(CH$_2$)n–B (239 K), Trimethylborane < DZn → CH$_3$–(CH$_2$)n (7.6 kJ mol^{-1}, 298 K), Dimethyl zinc

Aluminum alkyl compounds are the only representatives of alkyl elements of the third to sixth group of the periodic system that are characterized by the association in pairs. However, inconsistency in the results of thermodynamic studies on them demanded significant changes in the type and quality of the experimental approaches. Research [18, 23–25] using the static method using zero gauge glass with a substance vacuum distillation system and filling of the membrane chamber, measuring of the vapor pressure with an error of ±0.1 torr, the possibility of the determination of molar mass, and the study of the related processes in the vapor phase has ensured accurate qualitative measurement. The obtained data on the vapor composition, substantiated by the presence of vapor molecules in dimeric form over the liquid trimethylaluminum and triethylaluminum with proportions of 88.8 and 30.1%, respectively, at around 320 K and decreasing with increasing temperature, provided their thermodynamic properties [26, 27].

Detailed studies of the thermal behavior of trialkylaluminums showed that the mechanism of thermal decomposition of Al(C$_2$H$_5$)$_3$, including the generation stage and the open circuit, is a sequence of free-radical processes. Thus, the process of decomposition belongs to the radical and olefin–hydride mechanism. The obtained information on the thermal behavior of $AlEt_3$, $AlBu_3$ [24, 25], and $AlOkt_3$ [23] substantiated the tendency to the olefin–hydride process in a series of trialkylaluminums with a normal structure and an increasing number of atoms in the ligand. It complies with the results given by Korneev [28]: $AlMe_3 > AlEt_3 > AlPr_3 > AlBu_3$.

As a result, there was an appropriate opportunity to choose the conditions for an accurate study of $AlPr_3$. The vapor pressure measured in independent experiments and the molecular weights of compounds as measured in the vapor phase (165.0 c. u.) and as calculated according to the formula (156.0 c.u.), clearly indicated the presence of dimeric molecules in the vapor. The low partial pressure of dimeric molecules in the vapor of $AlPr_3$, not exceeding 5.8 % at low temperatures, is connected with the reduction in the tendency to dimerization in the series $AlMe_3 > AlEt_3 > AlPr_3 > AlBu_3$.

For unsaturated vapor up to the compound decomposition temperature (413 K), the change in vapor pressure with temperature corresponds to Gay–Lussac's law; thus, for saturated vapor, the dependence $P = f(T\,K)$ was processed according to the appropriate equations and resulted in a vaporization enthalpy and vaporization entropy of $43.0 \pm 2.0\ kJ\,mol^{-1}$ and $88 \pm 6\ kJ\,mol^{-1}\,K^{-1}$, respectively. These values correspond to those given by Chickos and Acree [7]. According to the research results, at temperatures below the commencement of decomposition processes at 320 K, vapor up to 5 mm participates in the vaporization process with the transition to vapor of the monomer and dimer molecules of $AlBu_3$. Similar results for trialkylaluminum [23] reflect the balance between the starting material and the components formed in the liquid and vapor phases:

$$\{AlOkt_3\}_{lig} \leftrightarrow (AlOkt_3)_{vap} \tag{1.4}$$

$$\{AlOkt_3 + AlOkt_3 \bullet Okt_2AlH + C_8H_{10}\}_{lig} \leftrightarrow (Okt_2AlH)_{vap} + (C_8H_{10})_{vap} \tag{1.5}$$

$$(AlOkt_3)_{vap} \leftrightarrow (Okt_2AlH)_{vap} + (C_8H_{10})_{vap} \tag{1.6}$$

After the complete evaporation of the starting material in the system, equilibrium of the trialkylaluminum is established and $AlOkt_3 \bullet Okt_2AlH$ (Eq. 1.5) is formed in the solution. Above 430 K in the unsaturated vapor, molecules of the two compounds are present and the vapor molecular mass calculated from the total pressure and the quantity of the starting material (in the two independent experiments) are in good agreement with each other (254 c.u.) and are consistent with Eq. (1.4) [23].

The subsequent increase in the vapor pressure at increasing temperature is caused by the balance shown in Eq. (1.4) shifting towards the more complete transformation of trialkylaluminum to dialkylaluminum hydride. At 483 K, trialkylaluminum is totally dissociated and the system presents equimolar amounts of dialkylaluminum hydride and olefin. The average molecular weight is equal to 183 c.u., and that calculated according to the Mendeleev–Clapeyron equation is 193.7 c.u. At temperatures above that specified, the vapor pressure changes in accordance with Gay–Lussac's law, i.e., olefin present in the system is in an unsaturated state. This allowed the dependence $P = f(T)$ for Okt_2AlH and thermodynamic vaporization parameters $\Delta H^0(TK) = 51.9$ and $\Delta S^0(TK) = 95.5$ kJ mol^{-1} K^{-1} at 388–433 K to be obtained. The obtained value of the vaporization enthalpy at 11.2 kJ mol^{-1} is lower than that established (63.1 kJ mol^{-1}) for the unbalanced conditions at 333–393 K [29]. Therefore, it is necessary to state that uncounted overlay processes related to the vaporization of the basic substance of aluminum alkyls cause significant differences in the vaporization enthalpy values of alkyl compounds, which are irrelevant with respect to the general laws of their change with an increasing number of carbon atoms in the alkyl chain. It is not possible to determine the energy values of the formed specific interactions in the liquid aluminum alkyls. The correct values of the vaporization enthalpies of a number of compounds in the literature [24–26] were obtained under conditions that were significantly different from and incomparable with the standard conditions and, therefore, do not provide comparable values for the energy characteristics. These peculiarities take place in hydrogen and oxygen derivatives of the aluminum alkyls [18, 24, 25, 30]. However, of particular interest is tetraethylaluminoxan, in which dimeric molecular forms are stable in the vapors and undergo dissociation [30]:

$$((C_2H_5)_2Al - O - Al(C_2H_5)_2)_2 \leftrightarrow 2((C_2H_5)_2Al - O - Al(C_2H_5)_2)$$

Tripropylaluminum is the only trialkylaluminum with a low ability to form dimeric molecules, which provides it with properties similar to gallium, indium, and thallium alkyls. X-ray diffraction studies of trimethylindium crystals showed the presence of four indium atoms in its structure, forming square and trigonal trimethylindium units located in the plane and perpendicular to the plane of In_4 [31]. The tetrameric unit is held by the groups located in the directions of lines connecting indium atoms, and the distance $(H_3C)_3In–CH_3$ (3.1 Å) indicates the formation of weak covalent bonds.

Research studies on the structure of trimethylindium and trimethylgallium crystals [32, 33] showed the formation of a distorted trigonal bipyramidal structure with coordination of the five methyl groups with In–C interatomic distances of 2.06, 2.12, and 2.16 Å (Fig. 1.6). In the solid state, it has the structure of a distorted trigonal bipyramid with interatomic distances of Tl–Tl equal to 5.46 ± 0.01 and 5.63 ± 0.01 Å and interatomic distances between carbon atoms of 3.76, 3.91, 4.25, and 4.43 Å. The distances Tl–C of the alkyl groups are equal to 2.22–2.30 and 2.34 Å [33]. Methyl groups of the coordinated molecules are located almost perpendicular to the distorted planar structure of $Tl(CH_3)_3$, with distances of 3.16 and 3.31 Å, with the formation of the intermolecular bond $(H_3C)_3Tl–CH_3Tl$. The

Fig. 1.6 The structure of
trimethylindium in solid
and liquid states

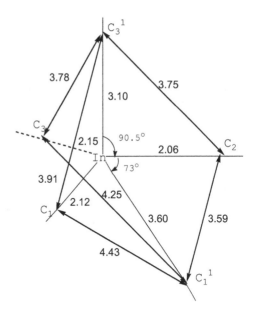

data allow assumption of a similar structure with five intermolecular bonds
$R_3M \to R$ for the liquid and crystal gallium, indium, and thallium alkyls. The
energies of specific interactions formed by the pentacoordinated carbon atom can
be calculated using Eq. (1.7):

$$DM - (H_2)n - CH_3 \to M = \Delta_{vap}H^0(TK)/5 \qquad (1.7)$$

The calculation results given in Tables 1.5 and 1.6 are described by the natural
sequence of stabilization of the specific interactions in dependence on the increasing number of carbon atoms in the ligand chain:

$$\text{Trimethylgallium}(6.52) < \text{Triethylgallium}(8.62) < \text{Tripropylgallium}(9.32)$$
$$< \text{Tributylgallium}\left(10.32\,\text{kJ}\,\text{mol}^{-1}\right)$$

The stabilization is caused by the increasing electron density shift, with the
maximum value at the fourth carbon atom of the ligand and completion of the
reverse dative bond influence. This correlates with the transfer of part of the
received electron density from the carbon atom to the p_z orbital of the gallium
atom. This draws attention to the lack of monotonicity in the stabilization at the
third carbon atom of the ligand of tripropylgallium (Table 1.6) and manifestation of
the stabilizing effect at the location of electron density at the fourth energy level of
the gallium atom $DM-CH_2-CH_2-CH_3 \to M$: tripropylaluminum (8.6) <
tripropylgallium (9.32 kJ mol^{-1}) and, at the same time, at the location of electron
density at the fifth energy level of the thallium atom:

Table 1.5 Energies (kJ mol^{-1}) of specific interactions of liquid trialkylaluminums

Compounds	Formula	Structure	$\Delta_{vap}H^0$ (298 K)	References	T (K)	DM–(CH$_2$)$_n$–CH$_3$ → M	DH$_3$C → H–CH$_2$
Tripropylaluminum	AlC$_9$H$_{21}$	C$_3$H$_7$–Al–C$_3$H$_7$ / C$_3$H$_7$	42.5 ± 1.2 / 43.0 ± 2.0	[7] / [34]	–	$n=2$, 8.5 / 8.6	–
Trimethylgallium	GaC$_3$H$_9$	H$_3$C–Ga–CH$_3$ / H$_3$C	33.1 ± 0.8 / 32.6 ± 0.1	[35]	266–305	$n=0$, 6.52	–
Triethylgallium	GaC$_6$H$_{15}$	C$_2$H$_5$–Ga–C$_2$H$_5$ / C$_2$H$_5$	43.1 ± 1.6	[35]	299–387	$n=1$, 8.62	–
Tripropylgallium	GaC$_9$H$_{21}$	C$_3$H$_7$–Ga–C$_3$H$_7$ / C$_3$H$_7$	46.6 ± 0.5	[35]	316–385	$n=2$, 9.32	–
Triisopropylgallium	GaC$_9$H$_{21}$	H$_3$C–CH–Ga–CH–CH$_3$ / CH$_3$ CH$_3$	46	[35]	–	$n=1$, 8.62	0.5
Tributylgallium	GaC$_{12}$H$_{27}$	C$_4$H$_9$–Ga–C$_4$H$_9$ / C$_2$H$_9$	51.6 ± 1.3	[35]	330–378	$n=3$, 10.32	–
Trimethylindium	InC$_3$H$_9$	H$_3$C–In–CH$_3$ / H$_3$C	42.1 ± 4.1	[36]	361–408	$n=0$, 8.42	–
Trimethylindium mp	InC$_3$H$_9$ mp	H$_3$C–In–CH$_3$ / H$_3$C	15.8	[37]	361	–	–
Triethylindium	InC$_6$H$_{15}$	C$_2$H$_5$–In–C$_2$H$_5$ / C$_2$H$_5$	44.95 ± 0.7	[38]	326–376	$n=1$, 8.99	–
Triisopropylindium	InC$_9$H$_{21}$	H$_3$C–CH–In–CH–CH$_3$ / CH$_3$ CH$_3$	52.3 ± 0.7	[18]	318–366	$n=1$, 8.99	1.23
Triethylthallium	TlC$_6$H$_{15}$	C$_2$H$_5$–Tl–Cl / C$_2$H$_5$	52.3 ± 0.7	[7]	297	$n=1$, 10.46	–

Table 1.6 Energies (kJ mol^{-1}) of specific interactions of crystalline trimethylgallium, trimethylindium, and trimethylthallium

Compounds	Formula	Structure	$\Delta_{sub}H^0$ (298 K)	References	T (K)	DM– $CH_3 \rightarrow M$
Trimethylgallium	GaC$_3$H$_9$ sub	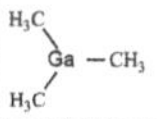	45.2	[35]	252	9.04
Trimethylindium	InC$_3$H$_9$ sub	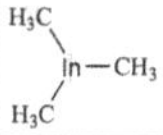	48.5 $\pm$ 2.5	[38]	298	9.7
Trimethylthallium	C$_3$H$_9$Tl sub	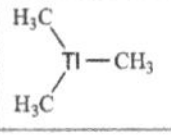	57.3	[39]	285	11.46

$$DM - CH_2 - CH_3 \rightarrow M : \text{Triethylgallium}(8.62) < \text{Triethylindium}(8.99)$$
$$< \text{Triethylthallium}\left(10.46 \, kJ \, mol^{-1}\right)$$

A similar pattern in the stabilization of specific interactions takes place in the crystal methyl compounds of the elements of the gallium subgroup:

$$DM - CH_3 \rightarrow M : \text{Trimethylgallium}(9.04) < \text{Trimethylindium}(9.7)$$
$$< \text{Trimethylthallium}\left(11.46 \, kJ \, mol^{-1}\right)$$

This indicates the increased stabilization of the specific interaction due to the shifted electron density from the carbon atom to the fifth energy level of the indium atom, and an even more advantageous location at the thallium atom in comparison with the location at the fourth gallium atom or at the third aluminum atom. This is possible if: first, by shifting from the carbon atom of the terminal methyl group there is ever increasing electron density at the more distant energy level of the metal atom of the third group; and second, the metal atom transfers more electron density from the more distant energy level to the carbon atom of the alkyl ligand.

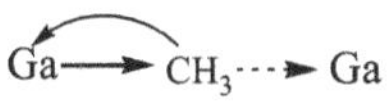

Therefore, the third energy level of the aluminum atom is less able to shift the electron density to the carbon atom and has an even lower ability to take it from the same carbon atom in the shape of the reverse dative bond. As a result, the increase in negative charge remains significantly reduced at the carbon atom, with the formation of the specific interaction of the methyl compound or AlR$_3$ with the aluminum atom:

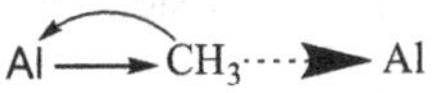

As a consequence, the trialkylaluminums form dimeric molecules (Fig. 1.7), and the dissociation energies are described by the natural sequence of their reduction (Table 1.7):

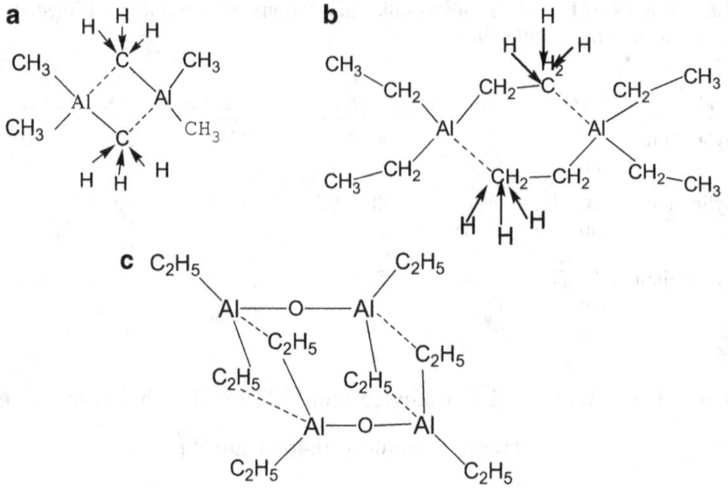

Fig. 1.7 Dimer molecule of trimethylaluminum (**a**), triethylaluminum (**b**), and tetraethylalumoxan (**c**)

Table 1.7 Dissociation energies (kJ mol^{-1}) of dimer molecules of alkylaluminum and its derivatives

Compounds	Formula	Structure	T (K)	$\Delta_{vap}H^0$ (TK)	$\Delta_{dis.dim}H^0$ (TK)	DAl–(CH$_2$)$_n$–CH$_3 \rightarrow$ Al
Trimethylaluminum	AlC$_3$H$_9$	H$_3$C\ Al–CH$_3$ / H$_3$C	297–367	63.2 ± 1.7	88.4 ± 0.9	$n = 0$
						44.2
Triethylaluminum	AlC$_6$H$_{15}$	C$_2$H$_5$\ Al–H / C$_2$H$_5$	323–423	63.7 ± 37	66.2 ± 1.1	$n = 1$
						33.1
Tetraethylalumoxan	(C$_2$H$_5$)Al–O–Al(C$_2$H$_5$)$_2$	((C$_2$H$_5$)$_2$Al–O–Al(C$_2$H$_5$)$_2$)$_2$	410–432	69.0 ± 3.3	89.3 ± 5.8	$n = 1$
						22.4

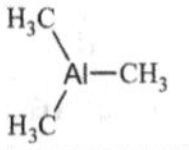

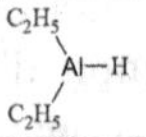

$$Al_2(CH_3)_6 \; < \; Al_2(C_2H_5)_6 \; < \; Al_2(C_3H_7)_6$$

The substantiation of dimerization in the couples trimethylaluminum and triethylaluminum [26, 27] and the obtained energy values of their dissociation became the key for new ideas and concepts. These include: the idea of the pentacoordinated condition of the carbon atom, the existence of a new type of interaction – the intermolecular reverse dative bond and its influence on the energies of the specific intermolecular interactions formed by the carbon atoms, substantiation of the changes in the enthalpy characteristics of the vaporization of alkyls of main group elements, and the failure of the model of sp^3-hybridization of the carbon atom [9].

The value of the enthalpy of dissociation of the dimeric molecule of tetraethy-lalumoxan obtained later [30] (Table 1.7) indicated the possibility of the formation of dimeric molecules with a large number of ethyl fragments by the oxygen compounds of aluminum. The location of the oxygen atom between the aluminum atoms in the molecule of tetraethylalumoxan points to the shifting of the reduced electron density to the carbon atom of the ethyl group, in comparison with the molecules of triethylaluminum. Nevertheless, the formation of dimeric molecules of tetraethylalumoxan points to the fact that the excessive electron density at the terminal methyl group of the ethyl fragment of this compound is sufficient for the formation of the stable bond in its dimeric molecule. With the dissociation energy value of the dimeric molecule of tetraethylalumoxan being $89.3 \pm 5.8 \, kJ \, mol^{-1}$, and four bonds in the molecule, the value of a single bond is equivalent to $22.4 \, kJ \, mol^{-1}$. This energy value is significantly higher than the energy value of the specific interaction $DAl-CH_2-CH_2-CH_3 \rightarrow Al$ ($8.6 \, kJ \, mol^{-1}$) of liquid tripropylaluminum.

1.4   Specific Intermolecular Interactions of Alkyls of the Nitrogen Subgroup Elements

1.4.1   Specific Interactions of Trialkyls and Trivinyls

The molecules of phosphine and alkyl compounds of phosphorus are characterized by a trigonal pyramidal structure with an equilateral triangle at the bottom and a phosphine atom at the top of the pyramid. Nevertheless, following the replacement of the hydrogen atom in the methyl group, one can observe the reduction in the ionization potential $I(n)$ by 1.03, 0.55, and 0.44 eV, the reduced stability in methylamines [2] being 1.26, 0.72, and 0.56 eV, respectively, and the difference in the effective charges at the phosphine atoms being -0.04 and $+0.07e$. The charges at the nitrogen atom and the methyl group in the molecule of trimethylamine are $+0.024$ and $-0.008e$ [22] as a result of the manifestation of the reverse dative bond, accompanied by the shifting of electron density from the p_z orbital of the nitrogen atom to the significantly undivided $2s^2$ electron pair of the carbon atom. Therefore, in analogy with phosphine, the energies of the formed specific interactions of trimethylphosphate and phosphorus alkyls should have relatively low stability. In this connection, we used the correlation dependency of the vaporization enthalpy of trialkylphosphines and trialkylamines and specified the value of the enthalpy characteristic of tripropylphosphine (Table 1.8).

Presented in Fig. 1.8 is the dependence of the vaporization enthalpy of the arsenic alkyls (line 1) and phosphorus alkyls (line 2) on the number of carbon atoms in the alkyl ligand and, on the other hand, on the number of methyl groups in the molecules of the methylstibine to trimethylstibine series (line 3). This illustrates the stabilization of specific interactions by the increasing number of carbon atoms in the chain and, for the example of methylstibines, dimethylstibines, and

Table 1.8 Energies (kJ mol^{-1}) of specific interactions of liquid trialkyls of main group elements

Compounds	Formula	Structure	$\Delta_{vap}H^0$ (298 K)	T (K)	References	DE→CH$_3$–(CH$_2$)$_n$–E	DH$_3$C→H–CH$_2$
Trimethylphosphine	C$_3$H$_9$P		28.8	–	[40]	$n=0$; 4.8	–
Triethylphosphine	C$_6$H$_{15}$P		38.3	306	–	$n=1$; 6.4	–
Tripropylphosphine	C$_9$H$_{21}$P		39.4; 43.5^a	340	[18]	$n=2$; 7.21	–
Tributylphosphine	C$_{12}$H$_{27}$P		51.72 ± 0.45; 53.94	390	[18]	$n=3$; 7.21	$\sum$DCH$_2$ = 8.2
Trivinylphosphine	PC$_6$H$_9$		32.7	304	[7]	5.45	–
Trimethylarsine	AsC$_3$H$_9$		27.7 ± 0.2; 28.9 ± 1.3	240–280	[41]	$n=0$; 4.61; 4.81	–
Triethylarsine	AsC$_6$H$_{15}$		38.10 ± 1.45; 38.52 ± 0.74	273–339	[42]	$n=1$; 6.35; 6.41	–
Tripropylarsine	AsC$_9$H$_{21}$		44,0 ± 0.7	314–420	[43]	$n=2$; 7.3	–
Triisopropylarsine	AsC$_9$H$_{21}$		45.22 ± 0.46	346–405	[18]	$n=1$; 6.35; 6.41	1.1
Trivinylarsine	AsC$_6$H$_9$		35.6	310	[7]	$n=1$; 5.9	–

							$DC{\equiv}N \to C{\equiv}N$
Trimethylstibine	SbC_3H_9		32.2 31.15	308	[38]	$n=0$; 5.38 5.2	—
Triethylstibine	SbC_6H_{15}		39.9 ± 1.3 41.8	306	[44]	$n=1$;6.65 7.0	—
Trivinylstibine	SbC_6H_9		38.7	308	[7]	$n=1$; 6.45	—
Trimethylbismutine	BiC_3H_9		36.0 ± 1.3 34.75	298	[7]	$n=0$; 6.0 5.8	—
Triethylbismutine	BiC_6H_{15}		43.9	322	[7]	$n=1$; 7.31	—
Trivinylbismutine	BiC_6H_9		48.6	306	[7]	$n=1$; 8.1	—
Cyano(ethyl)propylarsine	$AsNC_6H_{12}$	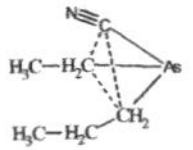	38.5 ± 0.7	303–334	[7]	6.41	8.1
Tricyanophosphine	PC_3N_3 sol	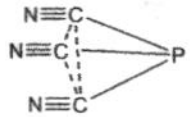	75.3 ± 2.9^b	298	[7]	—	12.6^b

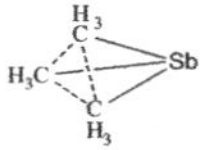

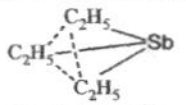

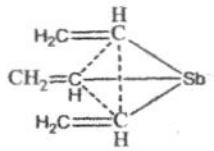

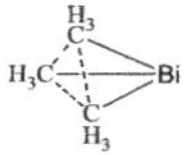

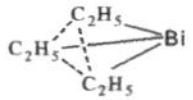

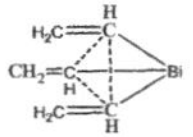

[a]Estimated using the correlation of vaporization enthalpies of trialkylphosphines and trialkylamines
[b]Sublimation

Fig. 1.8 Dependence of the vaporization enthalpy on the number of carbon atoms in alkyl chain for *1* trialkylarsines, *2* trialkylphosphines, and *3* methylstibine to trimethylstibine series. *Lines 1–3* show the influence of reverse dative bonds; *line 2* shows the contribution to the energy by CH_2 groups

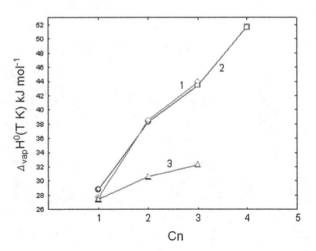

trimethylstibines (line 3), the energy contribution to the enthalpy characteristic following the replacement of the hydrogen atom in the molecule by a methyl group. The tendency of energy reduction due to the increased contribution to the vaporization enthalpy during the transition from triethylphosphine to tripropylphosphine reflects the reduction of the reverse dative bond with the increase in the number of carbon atoms of the alkyl ligand, with completion of the influence in the compound with the propyl fragment. The further increase in the enthalpy characteristic of tributylphosphine by 8.2 kJ mol^{-1} is determined by the energy contribution of the increasing number of methylene groups. It follows that it is possible to estimate the vaporization enthalpies of tripentylphosphine and trihexylphosphine as being equal to 59.9 and 68.1 kJ mol^{-1}, respectively.

The alkyl molecules of the elements of the nitrogen subgroup with three alkyl groups at the bottom of a trigonal pyramid form liquid and crystal structures with the six-coordinated element atom, wherein the bond vacancies form six specific intermolecular interactions with the pentacoordinated carbon atom, DE → CH_3–E, and the propyl ligand, DE → CH_3–CH_2–CH_2–E:

$$DE \rightarrow CH_3 - (H_2)n - E = \Delta_{vap}H^0(298\,K)/6 \qquad (1.8)$$

The influence of the reverse dative bond ends at the third carbon atom of the chain (Fig. 1.1). As a result, the energies of the specific interactions are determined from the vaporization enthalpy or sublimation enthalpy, interconnected with the number of bursting bonds and their energies [4, 5], using Eq. (1.8). The results of the calculations are described by the natural sequence of its stabilization with an increasing number of carbon atoms in the alkyl ligand, with the maximum energy value of the specific interaction in the compound with a butyl ligand:

Trimethylphosphine (4.8) < Triethylphosphine (6.4, 308 K) = Tripropylphosphine
    (17.21, 340 K) < Tributylphosphine (7.21 kJ mol^{-1}, 360 K)

Trimethylarsine (4.81) < Triethylarsine (6.41) < Tripropylarsine $(7.3 \, \text{kJ} \, \text{mol}^{-1})$
  Trimethylstibine (5.2) < Triethylstibine $(7.0 \, \text{kJ} \, \text{mol}^{-1})$
  Trimethylbismutine (6.0) < Triethylbismutine $(7.67 \, \text{kJ} \, \text{mol}^{-1})$

This is caused by the completion of the influence of the reverse dative bond at the fourth carbon atom of the chain. Attention needs to be paid to the fact that the changes in the stability of specific interactions were obtained using the vaporization enthalpies measured at temperatures that were higher than the standard condition. Therefore, the actual energy values of the specific interactions for, for example, tripropylphosphine, tributylphosphine, triethylstibine, and triethylbismutine, should have increased values in comparison with those given in Table 1.8:

Trimethylphosphine (4.8) ≈ Trimethylarsine (4.81) < Trimethylstibine (5.2) <
  Trimethylbismutine $(6.0 \, \text{kJ} \, \text{mol}^{-1})$
Triethylphosphine $(6.4, \, 306 \, \text{K})$ ≈ Triethylarsine $(6.41, \, 306 \, \text{K})$ < Triethylstibine
  $(7.0, \, 306 \, \text{K})$ < Triethylbismutine $(7.67 \, \text{kJ} \, \text{mol}^{-1})$

  Tripropylphosphine $(7.21, 306 \, \text{K})$ < Tripropylarsine $(7.3 \, \text{kJ} \, \text{mol}^{-1}, 322 \, \text{K})$

According to the specific interaction energy of compounds with a similar alkyl ligand, from the increasing sequence number of the methyl and ethyl compounds of the group element atom, the location of the shifting electron density from the carbon atom to the third and fourth energy level of the phosphorus atom and the arsine atom is not practically expressed by the energy of the specific intermolecular interaction. The influence of the shifting electron density is expressed at its location on the fourth and fifth energy level in trimethylstibine, triethylstibine, trimethyl-bismutine, and triethylbismutine. The increase in the number of carbon atoms in the ligand chain causes the shifting electron density, with its location at the third and fourth energy level in tripropylarsine and trimethylbismutine.

   The energy of the specific interaction of low stability, formed by the isostructural methyl group of triisopropylarsine, can be determined from the difference between the vaporization enthalpies of the compounds with a similar ethyl fragment of triethylarsine, obtained using Eq. (1.8a) and given in Table 1.8:

$$DH_3C \rightarrow H - CH_2 = \Delta_{vap}H^0(TK)ip - \Delta_{vap}H^0(TK)et/2 \qquad (1.8a)$$

A clearly defined pattern of the stabilization of the specific interaction formed by the unsaturated vinyl fragment $DE \rightarrow CH_2{=}CH{-}E$ in a number of compounds, as presented in the literature without errors (Table 1.8), is caused by an increase in the shift of electron density towards the more distant energy level of atoms of the elements P (third) < As (fourth) < Sb (fifth) < Bi (sixth):

Fig. 1.9 Schematic of the fragment of the volume chain of the network specific interaction of liquid cyano(ethyl)propylarsine

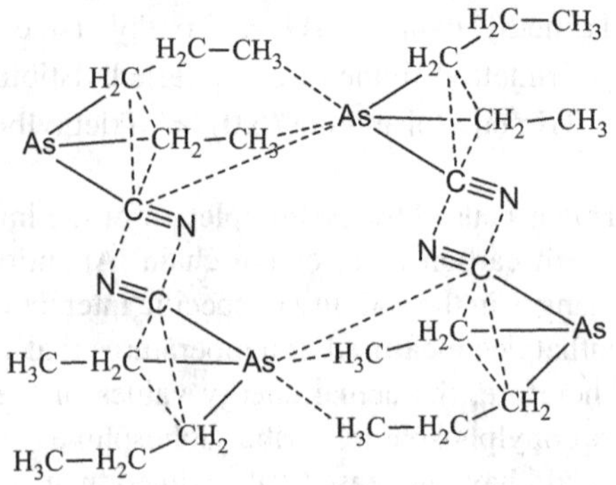

Trivinylphosphine   (5.45) < Trivinylarsine   (5.9) < Trivinylstibine   (6.45) <
    Trivinylbismutine (8.1 kJ mol^{-1}),

The increased energy values of the specific interactions formed by the ethyl ligand DE $\rightarrow$ CH$_3$–CH$_2$–E in comparison with the formed vinyl fragment DE $\rightarrow$ CH$_2$=CH–E in liquid compounds is natural because the reduced number of hydrogen atoms of the vinyl ligand provides the reduced electron density at the carbon atoms and, relatively, at the atoms of arsine and stibine:

$$\text{Triethylarsine } (6.35) \; > \; \text{Trivinylarsine } \left(5.9 \, \text{kJ mol}^{-1}\right);$$
$$\text{Triethylstibine } (7.0) \; > \; \text{Trivinylstibine } \left(6.45 \, \text{kJ mol}^{-1}\right)$$

In this connection, there are doubts about the correctness of the values given in the literature for the vaporization enthalpies of triethylbismutine and trivinylstibine, without any indication of the error:

$$\text{Triethylbismutine } (7.67) \; < \; \text{Trivinylbismutine } \left(8.1 \, \text{kJ mol}^{-1}\right)$$

The carbon atom of the cyanide group of molecules of cyano(ethyl)propylarsine, with the cyano, ethyl, and propyl fragments, participates in the formation of the specific interaction with the arsenic atom and the nitrogen atom, and is in a pentacoordinated condition (Fig. 1.9). An energy contribution to the vaporization enthalpy of cyano(ethyl)propylarsine is given by the two formed propyl fragments, two ethyl fragments, and two specific interactions of the nitrile group. Taking the energies of the specific interactions DAs $\rightarrow$ CH$_3$–CH$_2$–As and DAs $\rightarrow$ CH$_3$–CH$_2$–CH$_2$–As equal to 6.4 and 7.3 kJ mol^{-1}, respectively, which are the energies of the same interactions in liquid triethylarsine and tripropylarsine, we determined the energy of the specific interaction of DC$\equiv$N $\rightarrow$ C$\equiv$N of liquid cyano(ethyl) propylarsine, with the help of Eq. (1.9):

$DC\equiv N\rightarrow C\equiv N$

$$= \left(\Delta_{vap}H^0(TK) - 2DAs\rightarrow CH_3 - CH_3 - As - DAs\rightarrow CH_3 - CH_2 - CH_2 - As\right)/2$$
$$(1.9)$$

The energy of the specific interaction $DC\equiv N\rightarrow C\equiv N$ of crystal tricyanophosphine, formed by the three cyanide groups, is determined by dividing the sublimation enthalpy by the six bond vacancies. The obtained energy values of the interaction of the two compounds are presented in Table 1.8 and illustrate its high stability for the compounds.

1.4.2 Alkyl Hydrogens and Alkene Hydrogens of the Phosphorus Subgroup Elements

Shown in Table 1.9 are the vaporization enthalpies of trialkyl hydrogens of the phosphorus subgroup elements, which point to an increase of 3.2 and 1.6 kJ mol^{-1}in the vaporization enthalpy following replacement of a hydrogen atom by a methyl group.

Methylstibine (27.4, 248 K) < Dimethylstibine (30.6, 254 K) < Trimethylstibine (32.2 kJ mol^{-1}, 308 K)

Similar changes take place in the compounds of the bismuth derivatives. Based on these data, it can be concluded that the formed specific interactions are more stable than the formed hydrogen bonds. The low energy values of the specific interactions of liquid methyl, ethyl, and butyl compounds of the phosphorus subgroup elements (Table 1.9) with one to four carbon atoms of the alkyl chain are little susceptible to the change in energy of the formed specific intermolecular interactions following replacement of the alkyl ligand by the hydrogen atom. This allows estimation of the energy value of the hydrogen bond, formed by the hydrogen atom in liquid methylstibine, dimethylstibine, and in the corresponding compounds of arsenic and bismuth. The energies of the formed hydrogen bonds of the compounds are obtained from Eqs. (1.10) and (1.11):

$$DSb \bullet \bullet \bullet H - Sb = \left(\Delta_{vap}H^0(TK) - 2DSb \rightarrow CH_3 - Sb\right)/4 \qquad (1.10)$$

$$DSb \bullet \bullet \bullet H - Sb = \left(\Delta_{vap}H^0(TK) - 4DSb \rightarrow CH_3 - Sb\right)/2 \qquad (1.11)$$

The calculation results for the hydrogen bonds (Table 1.9) indicate the stabilization of the hydrogen bonds DSb•••H–Sb following the replacement of the hydrogen atom in stibine by the methyl group, showing the influence of the increasing shift of electron density from the hydrogen atom to the Sb atom:

Table 1.9 Energies (kJ mol^{-1}) of specific interactions of liquid methyl- and dimethylphosphine, arsine, stibine, and bismutine

Compounds	Formula	Structure	$\Delta_{vap}H^0$ (298 K) [7]	T (K)	DE $\to$ CH$_3$–(CH$_2$)$_n$–E	DE$\cdot\cdot$H–E
Methylstibine	SbCH$_5$		27.4	248	5.38	4.16
Dimethylstibine	SbC$_2$H$_7$		30.6	254	5.38	4.55
Methylbismutine	BiCH$_5$		29.9	224	6.0 5.8	4.42
Dimethylbismutine	BiC$_2$H$_7$		32.7	228	6.0	4.47
Diethylarsin	AsC$_4$H$_{11}$		34.2	298	6.41	4.3
Allylmethylphosphine	C$_4$H$_9$P		34.4	276	4.8 DP $\to$ CH$_3$–CH=CH–P = 8.3	4.0
2-Propylphosphine	C$_3$H$_7$P		32.7	241	4.8 DP $\to$ CH$_2$=CH–CH$_2$–P = 8.3	4.0
3-Butenylphosphine	C$_4$H$_9$P		34.5	273	DPCH$_2$=CH–CH–CH$_2$–P = 8.3 DCH$_2$ = 1.8	4.0
Phospholane	C$_4$H$_9$P		37.4	272	10.7	4.0
Propynylphosphine	C$_3$H$_5$P		36.8	298	4.8 D–C≡C– = 5.5	4.0

DSb•••H–Sb: Stibine (3.51, 254.8 K) < Methylstibine (4.16, 248 K) < Dimethylstibine (4.55 kJ mol^{-1}, 254 K)

The obtained increased negative charge of the Sb atom transferred partly from the p_z orbital to the carbon atom, in turn, provides the carbon atom with the ability to form more stable specific interactions DSb → CH$_3$–Sb > DSb•••H–Sb by the methyl group in comparison with the hydrogen bond. At the same time, by the method of photoelectron spectroscopy [45, 46], it was shown that the replacement of the hydrogen atom by a methyl group in amines is accompanied by an insignificant reduction in the electron density at the nitrogen atom and reduced stability of the formed specific interaction DN → CH$_3$–N < DN•••H–N, which is proved by the obtained energy values [9]. The same sequence in the stabilization of the hydrogen bond following replacement of the hydrogen atom in the molecule of bismutine by methyl groups is expressed in methylbismutine and dimethylbismutine, with a disparity in the energies of the hydrogen bonds and the specific interaction DBi•••H–Bi < DBi → CH$_3$–Bi:

$$DBi \bullet \bullet \bullet H - Bi : Methylbismutine(4.42, 224\,K) < Dimethylbismutine$$
$$\left(4.47\,kJ\,mol^{-1}, 228\,K\right) < DBi \rightarrow CH_3 - Bi : Trimethylbismutine\left(6.0\,kJ\,mol^{-1}\right)$$

The replacement of the hydrogen atom in arsine by ethyl groups is accompanied by an increased difference in the stabilization of the hydrogen bond DAs•••H–As and with the energy of specific interactions formed by the ethyl ligand:

$$DAs \bullet \bullet \bullet H - As : Arsine(3.02, 210.68\,K) < Diethylarsine\left(4.3\,kJ\,mol^{-1}, 298\,K\right)$$
$$< DAs \rightarrow CH_3 - CH_2 - As : Triethylarsine\left(6.41\ kJ\ mol^{-1}\right)$$

Based on the obtained data, it can be concluded that the low negative charge of the phosphorus atom in alkyl and trialkyl hydrides increases in the analogs of arsenic, antimony, and bismuth, and that reduced stability of the hydrogen bonds can be seen in comparison with the energies of specific interactions formed by the methyl and ethyl ligands.

The reduced energy value of the formed specific interaction of the vinyl ligand in comparison with the ethyl ligand is caused by the reduced shifting of the electron density, for example, at the phosphorus atom. This indicates that allyl and 2-propyl ligands with the same number of carbon atoms give the phosphorus atom practically adequate electron densities. This means that the difference in the vaporization enthalpies of allylmethylphosphine and 2-propylphosphine (1.7 kJ mol^{-1}) is equal to the difference between the energies of specific interaction formed by the methyl group and the energy of the hydrogen bonds formed by hydrogen atoms with the phosphorus of these compounds:

$$\Delta_{vap}H^0(T\,K)amph - \Delta_{vap}H^0(T\,K)pph = 2(DP \rightarrow CH_3 - P - 2DP \bullet \bullet \bullet H - P)$$

Taking into account the energy value of the specific interaction DP → CH$_3$–P

(4.8 kJ mol^{-1}), the energy of the hydrogen bond can be obtained as being equal to 4.0 kJ mol^{-1}. This makes it possible to calculate the energy value of the specific interaction formed by the allyl ligand of allylmethylphosphine, with the help of Eq. (1.12):

$$DP \rightarrow CH_3 - CH = CH - P = DP \rightarrow CH_2 = CH - CH_2 - P$$

$$= \left(\Delta_{vap}H^0(TK)amph - 2DP \rightarrow CH_3 - P - 2DP \bullet \bullet \bullet H - P\right)/2 \qquad (1.12)$$

The obtained energy values of the hydrogen bonds (Table 1.9) allows the determination of the energies of the specific interactions of liquid 2-propylphosphine, 3-butenylphosphine, and phospholane from the difference in vaporization enthalpies of the compounds using Eq. (1.12a), taking into account the energy contribution of the four bonds formed by the two hydrogen atoms of the contacting molecules:

$$DP \rightarrow CH_2 = CH - CH_2 - P, DP \rightarrow CH_2 = CH - (CH_2)_2 - P$$

$$= \left(\Delta_{vap}H^0(TK) - 4DP \bullet \bullet \bullet H - P\right)/2 \qquad (1.12a)$$

Taking into account the completion of the reverse dative bond influence at the third carbon atom of the chain, the energy contribution of the methylene group to the vaporization enthalpy is subtracted, equal to the difference with the vaporization enthalpy of 2-propylphosphine. The calculated energy values of the specific interactions of the liquid compounds of the phosphines with unsaturated ligands are described by the natural sequence of the energy stabilization with an increase in the number of carbon atoms in the chain (up to three), caused by the reduction of the influence of the reverse dative bond on the formation of intermolecular interactions:

Trivinylphosphine   (5.45) < Allylmethylphosphine   (8.3) = 2-Propylphosphine
    (8.3) < 3-Butenyl-Phosphine (8.3 kJ mol^{-1})

The molecule of propynylphosphine with the triple bond –C≡C– gains two additional bond vacancies, expressed by the carbon atoms with the triple bond [9]. The formed specific interactions participate in the crosslinking of the volumetric chains of the liquid structure (Fig. 1.10). The methylene group of the propynyl fragment forms specific interactions with energy values close to the energy of the interaction formed by the methyl group (4.8 kJ mol^{-1}). The known energy values of hydrogen bonds, formed by the hydrogen atoms of the propynylphosphine molecule, allow the energy of the specific interaction D–C≡C– to be calculated with the help of Eq. (1.13):

$$DC \equiv \rightarrow C \equiv$$

$$= \left(\Delta_{vap}H^0(TK) - 2DP \rightarrow CH_3 - P - 4DP \bullet \bullet \bullet H - P\right)/2 \qquad (1.13)$$

The obtained energy value of this interaction type is equal to 5.5 kJ mol^{-1} and

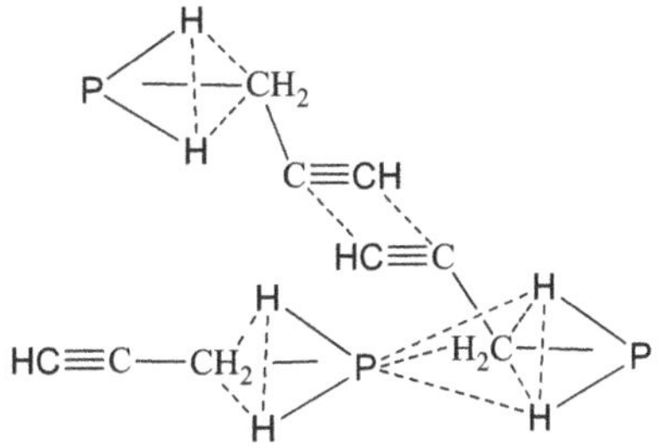

Fig. 1.10 Schematic of a
fragment of the volume–
chain liquid structure of
propynylphosphine

agrees with the result for liquid organic compounds in the literature
(5.0 kJ mol^{-1}) [9].

1.5   Energies of Specific Interactions of Selenium and Tellurium Alkyls and Their Unsaturated Analogs

1.5.1   Selenium and Tellurium Alkyls

Research into the electron structure of dimethyloxygen and dimethylsulfur point to
a match in the sequence of the four upper orbitals and to the contribution increase of
the *p*-orbitals in a number of elements, the increased shifting of electron density
with its location at the more distant energy level [2]. This may cause the stabilization of specific interactions formed by the element atom of the sixth group, shifting
the electron density from the more distant energy level. The increasing value of the
vaporization enthalpies of dialkyl elements of the group (Table 1.10) is related to
the number and energy of the bursting specific interactions, combined with the
conclusion drawn from the results of spectral studies.

Presented in Fig. 1.11 is the dependency of the vaporization enthalpies of
selenium alkyls and telluride alkyls on the number of carbon atoms in the alkyl
chain. There is a distinctly inadequate energy contribution because of the reduction
of the reverse dative intramolecular bond influence on the shifting of the electron
density on the chain, ending at the propyl ligand. The further sharp increase in the
vaporization enthalpy of the butyl compounds of selenium and telluride is caused
by the contribution of the fourth carbon atom, free from the influence of the reverse
dative bond. Therefore, the contribution of each following chain methylene group is
additive for this group of compounds, and the value is equal to the difference in the
vaporization enthalpies of neighboring compounds in the series or to the vaporization enthalpy of the group of compounds divided by the corresponding number of
methylene groups [9]. The molecule of the dimethyl, diethyl, and dipropyl compound of the sixth group elements, with two bond vacancies of the two free electron
pairs and two terminal methyl groups with essentially undivided $2s^2$ electron pairs
of carbon atom of the terminal methyl group, form four specific interactions

Table 1.10 Energies (kJ mol⁻¹) of specific interactions of liquid dialkyl selenides and tellurides

Compounds	Formula	Structure	$\Delta_{vap}H^0$ (298 K)	T (K)	References	$DE \rightarrow CH_3-(CH_2)_2-E$	$DH_3C \rightarrow H-CH_2$
Dimethyl selenide	C_2H_6Se	CH₃–Se–CH₃	31.2 ± 0.1	273–333	[18, 47, 48]	7.8	–
Diethyl selenide	$C_4H_{10}Se$	CH₂–CH₃ / Se / CH₂–CH₃	38.9 / 37.95 ± 0.1	298	[39] / [48, 49]	9.49	–
Diisopropyl selenide	$C_6H_{14}Se$		43.1 ± 1.0	298	[7]	9.49	5.2:4 = 1.3
Dipropyl selenide	$C_6H_{14}Se$	CH₂–CH₂–CH₃ / Se / CH₂–CH₂–CH₃	41.1ᵃ	298	–	10.3	–
Dibutyl selenide	$C_8H_{18}Se$	CH₂–CH₂–CH₂–CH₃ / Se / CH₂–CH₂–CH₂–CH₃	47.3	298	[7]	10.3 $\sum DCH_2 = 6.2$	–
Dipentyl selenide	$C_{10}H_{22}Se$	CH₂–(CH₂)₃–CH₃ / Se / CH₂–(CH₂)₃–CH₃	51.9 ± 1.0	298	[7]	10.3 $\sum DCH_2 = 10.8$	–
Dimethyl telluride	C_2H_6Te	CH₃ / Te / CH₃	37.4 ± 0.7 / 35.7 ± 0.1	298 / 273–372	[47, 48]	8.93	–
Diethyl telluride	$C_4H_{10}Te$	CH₂–CH₃ / Te / CH₂–CH₃	41.6 ± 1.0 / 41.6 ± 0.2Б	298 / 273–341	[48, 49]	10.4	–
Diethyl telluride sub	$C_4H_{10}Te$ sub	CH₂–CH₃ / Te / CH₂–CH₃	53.4 ± 2.3		[49, 50]	13.35ᵃ	–
Dipropyl telluride	$C_6H_{14}Te$	CH₂–CH₂–CH₃ / Te / CH₂–CH₂–CH₃	46.5 ± 0.7 / 45.5 ± 0.3Б	298 / 298–434	[51]	11.4	–
Diisopropyl telluride	$C_6H_{14}Te$		47.6 ± 0.1 / 40.4 ± 0.1Б	358 / 298–399	[48, 51]	10.4	–
Dibutyl telluride	$C_8H_{18}Te$	CH₂–CH₂–CH₂–CH₃ / Te / CH₂–CH₂–CH₂–CH₃	53.4 ± 0.1	298	[48, 51]	11.4 $DCH_2 = 7.9$	–

Diisobutyl telluride	C$_8$H$_{18}$Te	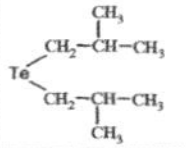 [structure]	47.6 ± 0.1Б	303–409	[48, 51]	11.4	2.1:4 = 0.5
Di-sec-butyl telluride	C$_8$H$_{18}$Te	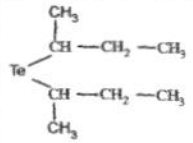 [structure]	49.6 ± 0.9 303–372	338	[51, 52]	11.4	2.0:4 = 0.5
Dipentyl telluride	C$_{10}$H$_{22}$Te	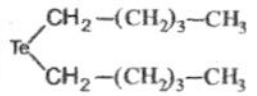 [structure]	59.5 ± 0.8	343–403	[52]	11.4 ΣDCH$_2$ = 7.9	–
Diisopentyl telluride	C$_{10}$H$_{22}$Te	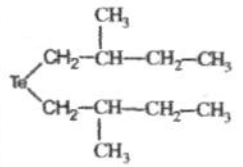 [structure]	51.9 ± 0.7	343–403	[52]	11.4 DCH$_2$ = 6.4	0.9:4 = 0.2

[a]Estimated on the basis of the correlation $\Delta_{vap}H^0(T\,K)SeR_2 = f(\Delta_{vap}H^0(T\,K)TeR_2)$

Fig. 1.11 Dependence of vaporization enthalpy on the number of carbon atoms in the alkyl chain of dialkyl tellurides (*1*) and dialkyl selenides (*2*)

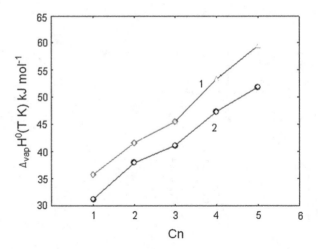

DSe → CH$_3$–Se with the molecules of the near environment. The angled molecule of the alkyl of the sulfur subgroup, with the ligand at an angle of 90° or more, causes the formation of the volumetric chain structure of the liquid condition, for which the energies can be derived by the division of the vaporization enthalpy by the number of bond vacancies. The number of formed bonds can be found using Eq. (1.14):

$$DE \rightarrow CH_3 - (CH_2)n - E = \Delta_{vap}H^0(T\,K)/4 \qquad (1.14)$$

For calculation of the energy of the hydrogen bonds formed in liquid hydrogen sulfide, taking into account that the reverse dative bond completes its influence at the third carbon atom of the chain in the molecule with the propyl ligand, there is reason to assume that the energy value is constant in the compound with the butyl ligand and the ligand with a large number of carbon atoms in the alkyl chain. The energies of the specific interactions formed by the butyl fragments are determined using Eq. (1.15):

$$DE \rightarrow CH_3 - CH_2 - CH_2 - E = \left(\Delta_{vap}H^0(T\,K) - DqCH_2\right)/4 \qquad (1.15)$$

where *q* is the number of methylene groups of the compound with butyl or bigger ligands.

Taking into account that the energy contribution of the propyl fragment of diisobutyl selenide and diethyl selenide to the vaporization enthalpy of these compounds is similar, the value of the energy contribution of isostructural methyl groups can be obtained from the difference in the enthalpy characteristics of these compounds by division by the number (*n*) of methyl groups:

$$DisoCH_3 = 2DH_3C \rightarrow H - CH_2$$
$$= \left(\Delta_{vap}H^0(T\,K)ip\right) - \left(\Delta_{vap}H^0(T\,K)et\right)/n \qquad (1.16)$$

The calculated energy results of the formed liquid specific interactions of the considered compounds (Table 1.10) are described by the natural sequence of stabilization of tellurides, with increased stability of the specific interactions for the liquid dialkyl tellurides:

Dimethyl selenide $(7.8) <$ Diethyl selenide $(9.45) <$ Dipropyl selenide $(10.) <$ Dibutyl selenide $(10.3$ kJ mol$^{-1})$.

Dimethyl telluride $(8.93) <$ Diethyl telluride $(10.4) <$ Dipropyl telluride $(11.40 =$ Dibutyl

It follows that the electron pair at the more distant energy level of the tellurium atom provides the increased shifting of the electron density from the alkyl chains in comparison with selenium and especially the sulfur atoms. This observation is confirmed by the research results on the electron structures using photoelectron spectroscopy of the methyl compounds of the basic group VI elements [53, 54] and the calculation of the electron structure by an ab initio method [54]. This work shows that the sequence of the four upper orbitals of $O(CH_3)_2$ and $S(CH_3)_2$ coincide: $1b_1 < 4a_1 < 3b_2 < 1a_2$. Following the replacement of sulfur by selenium and tellurium, the contribution of the p-orbitals of the sixth group to the three upper MOs increases. According to the literature [55], using the example of compounds of $E(CH_2)_4$ of the oxygen group elements, it is shown that, following the increase in the size of the heteroatom, the oscillation circuit of the first band significantly narrows and, therefore, the level of localization of the n-electrons at atom E^{VI} of the group significantly increases. The obtained energy values of the specific interaction $DTe \rightarrow CH_3-CH_2-Te$ in liquid $(10.4$ kJ mol$^{-1})$ and crystal $(13.35$ kJ mol$^{-1})$ diethyl telluride allows a conclusion to be made regarding the significant influence of the crystalline field on the stabilization of the bonds of tellurium alkyls, caused by the participation of the electrons located at the sixth energy level in the formation of the bonds. As shown in Table 1.10, the energies of the specific interactions (except for diisopropyl selenide) of isostructural compounds of tellurium indicate their low stability, which lies within the rather high error values of the experimentally measured vaporization enthalpies.

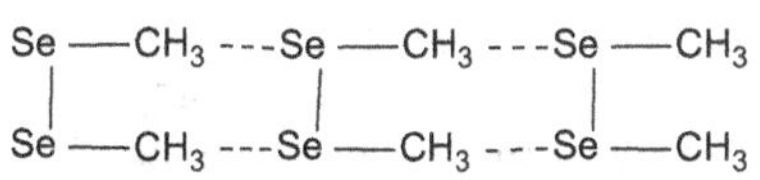

Se —CH$_3$ --- Se —CH$_3$ --- Se —CH$_3$
| | |
Se —CH$_3$ --- Se —CH$_3$ --- Se —CH$_3$

The selenium and tellurium atoms of the molecules of dialkyldiselenide and dibutyl ditelluride with six bond vacancies retain one free electron pair at the selenium and tellurium atoms not engaged during formation of the specific interactions between the selenium atom or the tellurium atom by crosslinking chains or during shifting of the electron density to the carbon atom of the methyl group of

Table 1.11 Energies (kJ mol^{-1}) of specific interactions of liquid dialkyl diselenides and dibutyl ditelluride at 298 K

Compounds	Formula	Structure	$\Delta_{vap}H^0$ (298 K) [7]	DE → CH$_3$–(CH$_2$)$_n$
Dimethyl diselenide	C$_2$H$_6$Se$_2$	Se—CH$_3$ Se—CH$_3$	42.0 ± 1.0	$n = 0$, 10.5
Diethyl diselenide	C$_4$H$_{10}$Se$_2$	Se—CH$_2$—CH$_3$ Se—CH$_2$—CH$_3$	47.1 ± 0.9	$n = 1$, 11.8
Dibutyl ditelluride	C$_8$H$_{18}$Te$_2$	Te-CH$_2$-(CH$_2$)$_2$-CH$_3$ Te-CH$_2$-(CH$_2$)$_2$-CH$_3$	57.3 ± 1.0	$n = 3$, 14.32

dimethyl diselenide, the end methyl group of diethyl diselenide, or dibutyl ditelluride. The shifting of the electron density should be expressed by the stabilization of the specific interaction of DE → CH$_3$–(CH$_2$)$_n$, where n is the number of methylene groups in the chain. As a result, the selenium and tellurium atoms of the molecules of dialkyl diselenide and ditelluride in their liquid condition are in the three-coordinated condition and the molecule forms four specific interactions, whose energies can be calculated using Eq. (1.14).

The obtained energy values of the bonds (Table 1.11) point to its stabilization, completing at the three carbon atoms of the alkyl chain. Consequently, the real value of the specific interaction energy of dibutyl ditelluride is overstated by approximately 1.2 kJ mol^{-1}, which is equal to the energy contribution of the methylene group.

1.5.2 Dialkenes of Selenium and Tellurium

The vaporization enthalpies of alkyls of selenium and tellurium have been determined for a limited number of compounds. The value of the vaporization enthalpy of divinyl selenide (Table 1.12) raises doubts about its correctness because it is greater than the enthalpy characteristic of diethyl selenide and dipropyl selenide (Table 1.10). The similarity in the molecular structures with the saturated and unsaturated ligands determines the analogous structure of the liquid and crystal conditions and the formed network of specific interactions. Thus, the energy values of the formed specific interactions are calculated using Eq. (1.15). Cyclic molecule of selenophene with the planar structure has four bond vacancies two of which display the end of the CH group and the two free electron pairs of tellurium atom.

Table 1.12 Energies (kJ mol^{-1}) of specific interactions of liquid dialkene tellurides and selenophenes

Compounds	Formula	Structure	$\Delta_{vap}H^0$(298 K) [7]	T (K)	DE → CH$_2$=CH-
Divinyl selenide	C$_4$H$_6$Se	Se(HC=H$_2$C)(HC=CH$_2$)	42.0 ± 1.0	298	13.0
Divinyl telluride	C$_4$H$_6$Te	Te(HC=CH$_2$)(HC=CH$_2$)	38.1 ± 2.1	–	9.52
Diallyl telluride	TeAll$_2$	Te(CH=CH–CH$_3$)(CH=CH–CH$_3$)	44.3 ± 0.5 [18]	314–397	DTe → CH$_3$–CH=CH– = 11.1
Selenophene	C$_4$H$_4$Se	Se(CH=CH)(CH=CH)	38.1 ± 1.0	298	DSe → CH=CH– = 9.52
Selenophene	C$_4$H$_4$Se	Se(CH=CH)(CH=CH)	$\Delta_{sub}H^0$(225) [39] 47.1	225	DSe → CH=CH– = 11.8

The two free electron pairs of the tellurium atom in the network of the liquid and crystal conditions participate in four specific interactions $DTe \rightarrow CH=CH-Te$, whose energies are determined from Eq. (1.15). Comparison of the data (Table 1.12) indicates invariance in the energy of the formed $DTe \rightarrow CH_2=CH-Te$ (9.52 kJ mol^{-1}) in the liquid divinyl telluride and $DSe \rightarrow CH=CH-Se$ (9.52 kJ mol^{-1}) in the liquid selenophene with the cyclic ligand and, on the other hand, in the liquid diethyl telluride (10.4 kJ mol^{-1}) with the saturated ethyl ligand. A similar tendency in the energies of the specific interactions takes place at the replacement of the unsaturated allyl ligand of liquid diallyl telluride by the saturated propyl ligand of dipropyl telluride: $DTe \rightarrow CH_2=CH-CH_2-Te$ (11.1 kJ mol^{-1}) and $DTe \rightarrow CH_3-CH_2-CH_2-Te$ (11.4 kJ mol^{-1}). Note that the energy values of the specific interactions with allyl and propyl ligand are the highest and remain practically unchanged in the corresponding series of compounds. The obtained energy values of the specific interactions of selenophene illustrate the stabilization of $DSe \rightarrow CH=CH-Se$ (9.52 kJ mol^{-1}) by 2.3 kJ mol^{-1}during interaction by the crystal field (11.8 kJ mol^{-1}). In accordance with the given data, the energies of the formed specific intermolecular interactions of the unsaturated ligand have a reduced value and its stabilization of the crystal field in diethyl telluride is lower by 2.0 kJ mol^{-1}.

1.5.3   Energies of Specific Interactions of Phenyl Selenides and Diphenylmercury

In the literature [9, 56], a detailed analysis is presented of the specificity of the interactions of the benzene molecule, caused by the alternating charge difference of the carbon atom of the CH groups and the energy contribution of the saturated hydrogen atom to the vaporization enthalpy and sublimation enthalpy, equal to 0.60 and 1.00 kJ mol^{-1}, respectively. This provides the energies of the three types of specific interactions formed in liquid and crystal benzene in dependence on the number of saturated ring hydrogen atoms: $DHC \rightarrow CH$ (5.63), $DHC \rightarrow C$ (5.35), and $DC \rightarrow C$ (5.0 5 kJ mol^{-1}) in liquid conditions and $DHC \rightarrow CH$ (7.43), $DHC \rightarrow C$ (6.90), and $DC \rightarrow C$ (6.40 kJ mol^{-1}) in crystals.

Molecules of methylphenyl selenide with five bond vacancies of the CH group, one of the carbon atom with a saturated hydrogen atom of the benzene ring, the carbon atom of the methyl group (expressed by the essentially undivided $2s^2$ electron pair), and two free electron pairs of the selenium atoms is able to form nine specific interactions. There are five interactions $4DHC \rightarrow CH$ and $DHC \rightarrow C$ (Fig. 1.12a) with known energy values, $DSe \rightarrow CH$ and $DSe \rightarrow C$ interactions of similar energies, and two $DSe \rightarrow CH_3-Se$ interactions. Taking the energy value of the specific interaction of $DSe \rightarrow CH_3-Se$ to be equal to the value of the same type of interaction of the liquid dimethyl selenide (7.8 kJ mol^{-1}) and using the known energy values of the formed specific interactions of the benzene ring, the total

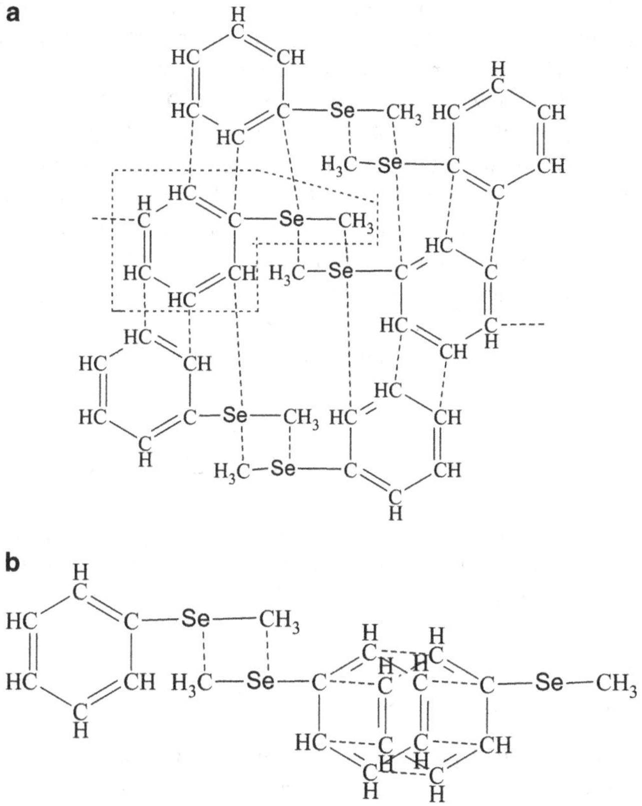

Fig. 1.12 (**a,b**) Schematics of the liquid methylphenyl selenide with the chain network of specific intermolecular interactions

energy value of the two interactions $DSe \rightarrow CH$ and $DSe \rightarrow C$ can be obtained using Eq. (1.17):

$$DSe \rightarrow CH + DSe \rightarrow C$$
$$= \left(\Delta_{vap}H^0(T\,K)\text{mps} - 4DHC \rightarrow CH - DHC \rightarrow C - DSe \rightarrow CH_3 - Se\right)$$

$$(1.17)$$

The obtained value of 8.75 kJ mol^{-1} for $DSe \rightarrow CH > DSe \rightarrow C$ plus the value of the energy contribution of the saturated hydrogen atom (0.60 kJ mol^{-1}) allows the estimation of their values as 4.70 and 4.10 kJ mol^{-1}, respectively. Taking into account the reduced energy values of the two types of interactions in comparison with those of the formed benzene ring, $DHC \rightarrow CH$ (5.63) and $DHC \rightarrow C$ (5.35 kJ mol^{-1}), it can be concluded that the specific interactions $DSe \rightarrow CH$ and $DSe \rightarrow C$ are realized in the liquid and crystal methylphenyl selenide.

The rigidity of the benzene ring and the increased ability of the carbon atom of the methyl group to shift the electron density to the remote energy level of the selenium atom provide the increased difference in charge of the carbon atom and the increased stability of the formed specific interaction $DSe \rightarrow CH_3-Se$. Therefore, the selenium atom of the methylphenyl selenide molecule participates in the formation of the specific interaction by one free electron pair, and the second pair provides it with the increased stabilization. Given in Fig. 1.12b is a schematic of the structure of liquid methylphenyl selenide with the network of eight specific interactions: six interactions of two types, $4DHC \rightarrow CH$ and $2DHC \rightarrow C$ formed by the benzene ring, and two interactions of $DSe \rightarrow CH_3-Se$ with unknown energy calculated using Eq. (1.18), summarizing the energy contributions of all types of interactions to the vaporization enthalpy [9]:

$$DSe \rightarrow CH_3 - Se$$

$$= \left(\Delta_{vap}H^0(T\,K)mps - 4DHC \rightarrow CH - 2DHC \rightarrow C\right)/2 \qquad (1.18)$$

Note that the CH groups of benzene rings are able to form specific interactions with the neighboring chains, matching them with stable interactions. The calculated energy value of the interaction $DSe \rightarrow CH_3-Se$ (Table 1.13) is described by the natural sequence:

$DSe \rightarrow CH_3-Se$: Dimethyl selenide (7.8) < Methylphenyl selenide (9.65) < Dimethyl diselenide (10.5 kJ mol^{-1})

This reflects the correct position of the high electron density shift from the carbon atom of the CH_3 group to the selenium atom of methylphenyl selenide in comparison with dimethyl selenide, providing an increased difference in the charges of the selenium and carbon atoms and the stability of the formed specific interaction. It follows that the rigid benzene ring should be quite clearly expressed by the stability reduction of the formed specific interaction of the selenium atom of diphenyl selenide with the carbon atom of two rings.

Presented in Fig. 1.13a is a schematic of liquid diphenyl selenide with the network of 14 formed specific interactions of two types, $10DHC \rightarrow CH$ and $4DSe \rightarrow CH_3-Se$, which allows the unknown energy value of the second type to be obtained with the help of Eq. (1.19):

$$DSe \rightarrow CH_3 - Se = \left(\Delta_{vap}H^0(T\,K)dps - 10DHC \rightarrow CH\right)/4 \qquad (1.19)$$

The calculated energy value of the specific interaction $DSe \rightarrow CH_3-Se$ is 2.5 kJ mol^{-1}. Table 1.13 illustrates its low stability, which is significantly reduced in comparison with the energy of the same type of interaction of liquid diphenyl selenide < dimethyl selenide (7.8 kJ mol^{-1}).

The low ability of CH groups of the benzene ring to shift the electron density is also expressed by diphenyl diselenide, whose selenium atoms participate in the formation of specific interactions by one free electron pair. Figure 1.13b presents a

Table 1.13 Energies (kJ mol^{-1}) of specific interactions of diphenyl diselenide and diphenyl mercury

Compounds	Formula	Structure	$\Delta_{evap}H^0$ (298 K) [7]	T (K)	DHC → CH	DHC → C=	DSe → CH$_3$–Se / DSe → C–Se
Liquid							
Methylphenylselenide	C$_7$H$_8$Se		52.5	282	5.63 × 4	5.33 × 2	9.4
Diphenyl selenide	C$_{12}$H$_{10}$Se		63.6 ± 2.5	294	5.63 × 10	–	2.5
			63.4	395			
Crystalline [39]							
Diphenyl diselenide	C$_{12}$H$_{10}$Se$_2$		116.7 ± 2.5	313	7.43 × 10	–	10.6
Diphenylmercury	C$_{12}$H$_{10}$Hg		112.8 ± 0.8	–	7.43 × 10	–	9.6
Dibenzyl diselenide	C$_{14}$H$_{14}$Se$_2$		130.5	–	7.43 × 8	6.93 × 4	DSe → CH$_2$–Se = 10.8

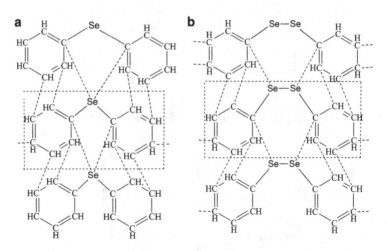

Fig. 1.13 Schematic of the liquid diphenyl selenide (**a**) and crystal diphenyl diselenide (**b**) with the chain network of specific intermolecular interactions

schematic of the crystal diphenyl diselenide structure, which reflects the specificity of the interaction and points to the formation, by CH groups, of ten specific interactions $10DHC \rightarrow CH$ and the specific interactions of $DSe \rightarrow C\text{-}Se$ by the two selenium atoms and the two carbon atoms of the two benzene rings, whose energies are derived from Eq. (1.19). The obtained energy value of the specific interaction (10.6 kJ mol^{-1}) of the crystal diphenyl diselenide is unusually low, despite the stabilizing effect of the crystal field. Thus, the Se–Se bond in the molecule of diphenyl diselenide does not significantly influence the stabilization of the specific interaction $DSe \rightarrow C\text{-}Se$.

The molecule of diphenylmercury has a similar structure to the diphenyl selenide molecule, the same number of bond vacancies, and forms the same types of specific interactions (Fig. 1.13a). Therefore, the energy of the specific interaction $DSe \rightarrow C\text{-}Se$ can be obtained using Eq. (1.19). Table 1.13 shows the calculated energy value of the diphenylmercury interaction, which reflects the manifestation of the stabilizing influence of the crystal field and, on the other hand, the electron configuration of the mercury atom. The molecule of dibenzyl diselenide with CH_2 groups contains two more bond vacancies and forms 16 specific interactions (Fig. 1.14): eight $8DHC \rightarrow CH$ and four $4DHC \rightarrow C$ formed by the benzene cycle, and four interactions $DSe \rightarrow CH_2\text{-}Se$ with unknown energy, which can be obtained using Eq. (1.20):

$$DSe \rightarrow CH_2 - Se$$

$$= \left(\Delta_{vap} H^0 (T\,K) dbds - 8DHC \rightarrow CH - 4DHC \rightarrow C \right) /4 \qquad (1.20)$$

The calculated energy value of the specific interaction is within the range of experimental error:

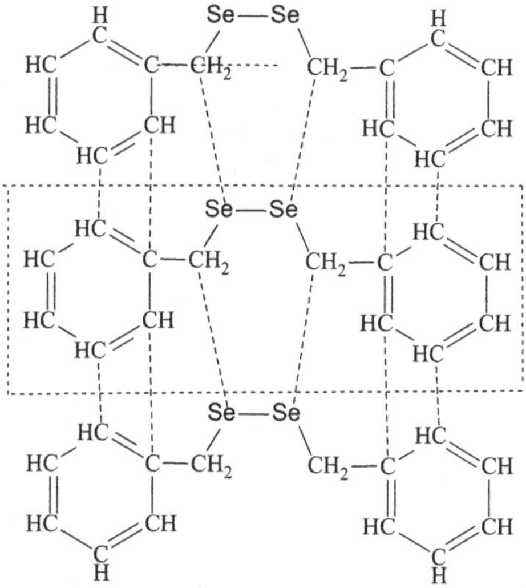

Fig. 1.14 Schematic of the crystalline dibenzyl diselenide with the chain network of specific intermolecular interactions

$$DSe \rightarrow CH_2 - Se, \text{ dibenzyl diselenide}(10.8) \approx DSe \rightarrow C - Se\left(10.6\,kJ\,mol^{-1}\right)$$

Based on the established fact that there are small differences in the energies of liquid [9] esters, ketones, alcohols, and acids formed by the methylene group at its location with the benzene ring (5.5–5.7 kJ mol^{-1}), the obtained energy value should be accepted as being constant when calculating the energy properties of dibenzyl analogs of the sulfur subgroup elements.

It should be noted that the experimentally measured sublimation enthalpies of triphenylphosphine, N-ethyl triphenylphosphine imine, triphenylarsine, triphenylantimony, and triphenylboron with values in the range 113.2 ± 3.0 to 92.1 ± 2.5 kJ mol^{-1} are not correct because of their significant reduction compared with the total energy value contributed by the three benzene rings (130.8 kJ mol^{-1}), with the saturated hydrogen atom in the case of bis(benzyl)mercury, matching the energy value (88.7 kJ mol^{-1}) of two benzene rings.

1.5.4   Energies of Specific Interactions of (Dimethyl-amino) methylarsine, Dimethylstibineborane, and Dimethyl (methylthio)borane

The molecules of EH$_3$– for the series N–P–As–Sb are characterized by a decline in the binding character $1a_1$ MO and $2a_1$ MO. At the same time, the reduction in the destabilizing B3MO n–σ C–H interaction in the increasing series of nitrogen subgroups elements is accompanied by an increase in the ionization potential: N (CH$_3$)$_3$ < P(CH$_3$)$_3$ < As(CH$_3$)$_3$ and by the stabilization of intermolecular bonds

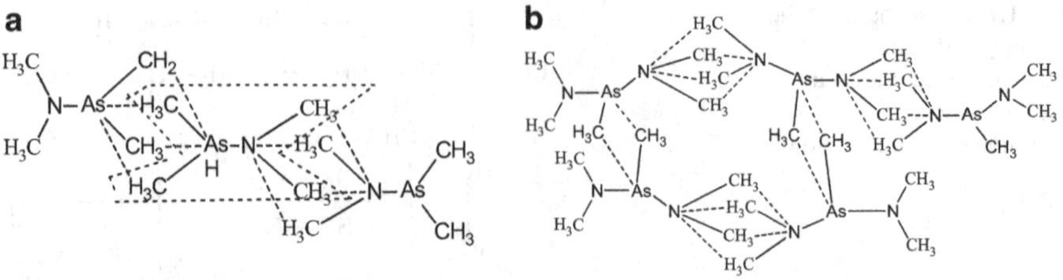

a

b

Fig. 1.15 Schematic of the network of specific interactions of crystalline (dimethylamino) dimethylarsine (**a**) and bis(dimethylamino)methylarsine (**b**)

[2]. It can be concluded that there is stabilization of the specific intermolecular interactions formed by the liquid and crystal methyl compounds of the series. This corresponds to the data in Table 1.8. Thus, it can be supposed that the shifting of the electron density in the molecule of (dimethylamino)dimethylarsine from the arsenic atom to the nitrogen atom causes the increase in the negative charge of the carbon atom of the methyl group and the difference in charge compared with the As atom. As a result, there occurs stabilization of the formed specific interaction DN $\rightarrow$ CH$_3$– N < DAs $\rightarrow$ CH$_3$–As, formed by the molecules of (dimethylamino)dimethylarsine and bis(dimethylamino)methylarsine. Since the amine and arsenide groups of (dimethylamino)dimethylarsine molecules do not form *trans* isomerism, the eight bond vacancies of the molecule form the four specific interactions DN $\rightarrow$ CH$_3$–N and DAs $\rightarrow$ CH$_3$–As (Fig. 1.15). In the literature [9], it was shown that the nitrogen atom at the maximum number of methyl groups in the molecules of N(CH$_3$)$_3$, NH (CH$_3$)$_2$, and NH$_2$CH$_3$ forms the specific interaction DN $\rightarrow$ CH$_3$–N with reduced stability in the liquid condition (4.25, 4.8, and 5.35 kJ mol^{-1}). Taking this value of the adequate energy of the compound in liquid (dimethylamino)dimethylarsine, we obtained the energy value of the specific interaction of the second type as being equal to 4.9 kJ mol^{-1} using Eq. (1.21):

$$DAs \rightarrow CH_3 - As = \left(\Delta_{vap}H^0(TK) - 4DN \rightarrow CH_3 - N\right)/4 \qquad (1.21)$$

This value fits the calculated energy value of the same type of interaction (4.81 kJ mol^{-1}) in liquid trimethylarsine (Table 1.8).

The molecule of bis(dimethylamino)methylarsine with ten bond vacancies forms eight specific interactions of DN $\rightarrow$ CH$_3$–N with known energy and two interactions of the second type DAs $\rightarrow$ CH$_3$–As, whose energy can be derived from Eq. (1.21a):

$$DAs \rightarrow CH_3 - As = \left(\Delta_{vap}H^0(TK) - 8DN \rightarrow CH_3 - N\right)/2 \qquad (1.21a)$$

As shown in Table 1.14, the energy value corresponds to the energy of the same type of interaction for (dimethylamino)dimethylarsine, lying within the error of the

Table 1.14 Energies (kJ mol^{-1}) of specific interactions and hydrogen bonds of (dimethylamino)methylarsine, dimethylstibineborane, and dimethyl (methylthio)borane

Compounds	Formula	Structure	$\Delta_{vap}H^0$ (298 K) [7]	T (K)	DE•••H–E	DN → CH$_3$–N	DAs → CH$_3$–As
(Dimethylamino)dimethylarsine	AsNC$_4$H$_{12}$		36.7	274–342	–	4.25	4.9
bis(Dimethylamino)methylarsine	AsN$_2$C$_5$H$_{15}$		39.2	–	–	3.7 (4.25)	4.8
Dimethylstibineborane	C$_2$H8BSb		32.1	254	2.65	DSb → CH3– Sb = 5.2	–
Dimethyl(methylthio)borane	BSC$_3$H$_9$		31.6	263	–	DB–CH$_3$ → B = 3.5	DS → CH$_3$–S = 8.8

experimentally measured vaporization enthalpy (± 0.2 kJ mol^{-1}) of trimethylarsine, according to the work of Baev et al. [41].

The calculated energy values of the specific interaction for the two compounds, differing by the number of amine groups, are almost equal. This is promising as it allows the evaluation of the energies of other similar compounds. The molecule of dimethylstibineborane with the same number of bond vacancies and the formed specific interactions DSb $\rightarrow$ CH$_3$–Sb (with energy of 5.2 kJ mol^{-1}), and a similar network structure of specific interactions of liquid (dimethylamino)dimethylarsine (Fig. 1.15a), allows estimation of the energy value of the hydrogen bond of borane with the help of Eq. (1.22). Using the value of enthalpy sublimation of dimethyl-stibineborane (32.1 kJ mol^{-1}, measured at 254 K), a somewhat high value of the hydrogen bond energy DB•••H–B (2.65 kJ mol^{-1}) is obtained in comparison with the energy of the same type of liquid diborane (1.83 kJ mol^{-1}) (Table 1.14). The molecule of dimethyl(methylthio)borane, with six bond vacancies and a vaporization enthalpy of 31.6 kJ mol^{-1} measured at 264 K, forms in the liquid condition four specific interactions DB–CH$_3$ $\rightarrow$ B with an energy of 3.5 kJ mol^{-1}, as calculated from the enthalpy characteristic measured at 239 K (Table 1.4). From these data the energy of the specific interaction can be calculated using Eq. (1.21b):

$$DS \rightarrow CH_3 - S = \left(\Delta_{vap}H^0(TK) - 4DB - CH_3 \rightarrow B\right)/2 \qquad (1.21b)$$

$$DB \bullet \bullet \bullet H - B = \left(\Delta_{vap}H^0(T\ K) - 4DSb \rightarrow CH_3 - Sb\right)/4 \qquad (1.22)$$

The energy is equal to 8.8 kJ mol^{-1} at 239 K. The energy of the same type of interaction of liquid sulfur dimethyl at 303 K has a lower value (7.8 kJ mol^{-1}). It should be noted that the vaporization enthalpies available in the literature for a great number of alkyl complexes of the third, fifth, and sixth main group elements were obtained under conditions of the joint transition to vapor of the molecules of individual compounds. This is the reason why thermodynamic analysis is not possible.

References

1. Cvitaš T, Klasinc L, McDiarmid R (1984) Vibrational structures in the excited states of the ethylene radical cation. Int J Quant Chem 18:537–546
2. Nefedov VI, Vovna VI (1989) Electronic structure of organic and elementorganic compounds. Nauka, Moscow, 198 p
3. Bieri G, Åsbrink L, von Niessen WJ (1982) 30.4-nm He (II) photoelectron spectra of organic molecules: Part VII. Miscellaneous compounds. J Electron Spectrosc 27(2):129–178
4. Baev AK (1969) Problems of chemical nature of phase transformation, vol 1, General and applied chemistry. Vaisheshika Educating, Minsk, pp 197–206
5. Baev AK (1969) Phase condition and complex formation ability of galogenide metals, vol 1, General and applied chemistry. Vaisheshika Educating, Minsk, pp 207–218
6. Baev AK (1972) Complex formation ability Halogenides of second–six groups periodical system, vol 5, General and applied chemistry. Vaisheshika Educating, Minsk, pp 35–51

7. Chickens JS, William EA (2002) Enthalpies of vaporization of organic and organometallic compounds, 1880–2002. J Phys Chem Rev 31(2):537

8. Lebedev YA, Miroschnichenko EA (1981) Thermochemistry of the vaporization of organic compounds. The vaporization and sublimation enthalpy and saturated pressure. Nauka, Moscow, p 285

9. Baev AK (2012) Specific intermolecular interactions of organic compounds. Springer, Heidelberg/Dordrecht/London/New York, 434 p

10. Bancroft GM, Creber DK, Basch H (1977) Valence and core electron energies in alkyl cadmium compounds from ab initio calculations and photoelectron spectra: electric field gradients in gas phase Cd compounds. J Chem Phys 67(11):4891–4897

11. Bice JE, Tan KH, Bancroft GM, Yates BW (1987) A variable energy photoelectron study of the valence and Hg 5d levels of $Hg(CH3)2$. J Chem Phys 87(2):821–829

12. Creber DK, Bancroft GM (1980) Photoelectron studies of dialkyl group 2B compounds: ligand field splittings and intensity variations with photon energy. Inorg Chem 19(13):643–648

13. Eland JHD (1970) Photoelectron spectra and chemical bonding of mercury(II) compounds. Int J Mass Spectrom Ion Phys 4(1):37–49

14. Fehiner NP, Ulman G, Nugent WA, Kochi JK (1976) Effect of alkyl subtends on the first ionization potential and on $5d^{10}$ ionization in dialkyl mercury compounds. Inorg Chem 15 (8):2544

15. Sokolovskii AE, Baev AK (1984) Thermodynamic study of the evaporation process of dimethyl and diethyl zinc. Russ J Gen Chem 54(1):103–106

16. Sokolovskii AE, Baev AK (1984) Thermodynamics of vaporization di-n-propyl- and diisoropylzinc. Vesty A N BSSR Ser Chem Navuk 1:115–117

17. Sokolovskii AE, Baev AK (1984) Thermodynamic study of vaporisation process of butyl zinc compounds. Russ J Phys Chem 58(11):2692–2694

18. Baev AK (2001) Thermodynamic evaporation of alkyl compounds of elements second-six groups. Russ J Izvestiya Vischich Ucheb Zaved Chem Chemo Technnol 44(1):3–13

19. Sokolovskii AE, Baev AK (1985) Evaporation thermodynamics of organic cadmium compounds. Vesty A N BSSR Ser Chem Navuk 5:112–114

20. Ehricson S (1968) Environmental effects in the MO treatment of the trimethylamine-trimethylboron addition reaction. Theor Chimica Acta 10(193):209

21. Boer FP, Newton MD (1966) Molecular orbitals for boron hydrides parameterized from SCF model calculations. J Am Chem Soc 88:2361–2366

22. Gurianova EN, Goldstein IG, Romm IP (1973) Donor-acceptor bonds. Chemistry, Moscow, 365 p

23. Baev AK, Schischko MA (2002) Thrioctylaluminium as equilibrium system of interactions dioctylaluminumhydride. Russ J Org Comp 3(8):1157–1160

24. Baev AK, Schischko MA (2002) Thermical conduct trialkylaluminium. Термическое поведение триалкилов алюминия. Russ J Appl Chem 74(1):26–29

25. Baev AK, Schischko MA, Korneev NN (2002) Thermodynamics and thermochemistry of aluminium organic compounds. Russ J Gen Chem 72(9):1474–1480

26. Baev AK, Vitalev SE (1989) Thermodynamic vaporization and association of triethylaluminium. Vesti Akademii Navuk Belarus 2:58–63

27. Vitalev SE, Baev AK (1993) Thermodynamic vaporization and association of trimethylaluminum. Russ J Izvestiya Vischich Ucheb Zaved Chem Chemo Technol 36(6):36–40

28. Korneev NN (1979) Chemistry and technology of aluminumorganic compounds. Nauka, Moscow, 247 p

29. Korneev NN, Govorov RN, Tomashevskie MV (1988) Aluminiumorganic compounds. Review information, Series of aluminiumorganic compounds. NIITEOChem, Moscow, 48 p

30. Baev AK (2004) Physic-chemical parameters and thermodecomposition of derivatives of aluminiumalkyls. Russ J Phys Chem 78(8):1519–1521

31. Posson M (1970) Chemistry of metalorganical compounds. Mir, Moscow, 238 p

32. Pauling L, Laubengayer AW (1941) The alkyls of the third group elements. II. The electron diffraction study of indium trimethyl. J Am Chem Soc 63(2):480–481
33. Sheldrick GM, Sheldrick WS (1970) Crystal structure of trimethylthallium. J Chem Soc Sec A Inorg Phys Theor Chem: 28–30
34. Zarov VV, Gubar YuL, Gribov VG, Baev AK (1973) Thermodynamic investigation of vaporization process and dissociation metallalkyl coordination compounds. Abstracts. I All-Union conference on chemistry of vaporous condition. USSR, Minsk, pp 84–85
35. Sokolovskii AE, Baev AK (1994) Thermodynamics of gallium alkyl compounds and systems of trimethyl–triethylgallium. Russ J Applauded Chem 67(9):1476–1481
36. Baev AK, Sokolovskii AE (1993) Thermodynamics of vaporization of trimethylindium and binary system of trimethylindium-triethylindium. Russ J Thermodyn Thermochem 2(2):173–178
37. Krans CF, Noonder FF (1933) Enthalpy of melting of trimethylindium. Proc Natl Acad Sel 19:292
38. Baev AK, Mikhailov VE, Podcovshov AI, Chernyak IN (1995) Investigation of equilibrium liquid-vapor ethyl compounds of indium, bismuth and binary systems of In(C2H5)3-Sb(C2H5)3, Zn(C2H5)2-Te(C2H5)3. Russ J Applauded Chem 68(6):918–922
39. Chickos JS, Acree WE Jr (2003) Enthalpies of sublimation of organic and organometallic compounds. 1910–2001. J Phys Chem Rev 31(2):537
40. Dennis LM, Patnode WJ (1932) Thermodynamics of vaporization of trimethylphosphine. J Am Chem Soc 54:183–184
41. Baev AK, Mikhailov VE, Chernyak IN (1995) Thermodynamics of the trimethylgallium-trimethylarsine system. Russ J Applauded Chem 68(6):923–926
42. Baev AK, Silivanchik IP, Kozirkin BI (1980) Thermodynamic investigation of system triethylarsenicum with diethyl ether. Russ J Gen Chem 50(9):1931–1937
43. Baev AK, Mikhailov VE, Chernyak IN (1995) Thermodynamics of system trimethylgallium – tri-n-propylarsin. Russ J Gen Chem 65(7):1073–1076
44. Baev AK, Silidanchik IP (1980) Thermodynamic investigation of system triethylstibium with diethylether. Russ J Phys Chem 55(9):2242–2245
45. Karlson TA (1981) Photo electronic and OGE-spectroscopy. Mechanical Engineering, Leningrad, 431 p
46. Nefedov VI (1984) Infrared spectroscopy of chemical compounds. Chemistry, Moscow, 255 p
47. Baev AK, Gubar YL, Kozirkin BI, Silidanchik IP (1975) Thermodynamic investigation of dimethylselenium and dimethyltellurium. Russ J Phys Chem 49(10):2651–2654
48. Baev AK, Gubar YuL, Gaidim IL, Kozirkin BI (1976) Vapor pressure of some alkyl compounds of selenium and tellurium. Abstracts. II All-Union conference on thermodynamics of organic compounds. Gorkie, Russia, p 22
49. Podkovirov AI, Baev AK (1994) Investigation of equilibrium of liquid-vapor at system Zn(CH3)2-Zn(C2H5)2, Te(CH3)3-Te(C2H5)3. Russ J Applauded Chem 67(3):430–432
50. Baev AK, Podkovirov AI (1994) Thermodynamics and associated interactions at system of dimethylzinc – dimethylsulfur. Russ J Thermodyn Thermochem 2(1):80–87
51. Sokolovski AE, Baev AK, Cherneak IN, Kozirkin BI (1996) Temperature dependence of pressure saturated vapor of alky derivative selenium and tellurium. Russ J Applauded Chem 69(1):160–161
52. Baev AK (2001) Thermodynamic vaporization of alkyl compounds of II–VI main group elements. Russ J Izvestiya Vischich Ucheb Zaved Chem Chemo Technol 44(1):3–13
53. Chang FC, Young VY, Prather JW, Cheng KL (1986) Study of methyl chalcogen compounds with ultraviolet photoelectron spectroscopy. J Electron Spectrosc 40(4):363–383
54. Auc DN, Welts HM, Davidson WR et al (1980) Proton affinities and photoelectron spectra of three-membered – ring hetecycles. J Am Chem Soc 102(16):5151–5157
55. Pignataro S, Distefano G (1974) n-σ Mixing in pentatomic heterocyclic compounds of sixth group by photoelectron spectroscopy. Phys Lett 26(3):356–360
56. Baev AK (2014) Specific intermolecular interactions of nitrogenated and bioorganic compounds. Springer, Heidelberg/Dordrecht/London/New York, 579 p

Chapter 2
Specific Intermolecular Interactions of Silanes and Their Derivatives

2.1 Types of Specific Interactions of Silanes and Compounds with a Tetrahedral Molecular Structure

2.1.1 Silanes n = 4 + 9 and Compounds with a Tetrahedral Molecular Structure

Research on the electron structures of the hydrides of the silicon element subgroup by photoelectron spectroscopy showed that, in the case of their ions, the minimum energy is accounted for by the C_{2v} configuration, similarly to CH_4 [1–3]. The ionization fields of silanes Si_nH_{2n+2} with two to five silicon atoms is thought to divide into the ionization fields Si–Si and Si–H binding electrons [4, 5]. The increasing number of chain silicon atoms is accompanied by broadening of the ionization field of $\sigma(SiSi)$ electrons, similar to the levels of s- and p-type in the alkane molecules, which reflects the similarity in their properties. However, the bandwidth responsible for the Si–H binding level varies little. With increasing chain length, the ionization field of $\sigma(SiSi)$ electrons is broadens, similar to the levels of s- and p-type in the alkane molecules [2]. In the case of pentasilane, this field is spreads from 9 to 11.5 eV. For the compounds Si_nH_{2n+2} where $n = 4$ and 5 in the field of ionization of Si–Si binding levels, it was discovered that the contributions corresponded to 2–4 rotamers [2, 3]. These compounds are characterized by the shifting of electron density from the hydrogen atoms to the essentially undivided $3s^2$ electron pair of the silicon atom and by the difference in the charges of the chain silicon atoms.

It should be noted that the thermodynamic characteristics of the vaporization processes of silanes and methylsilanes are extremely limited (Table 2.1), and the data available from the literature were obtained using differing standard conditions [6]. Taking into account the similarity of the spectral properties of the saturated hydrocarbons and silanes, we assume that the compounds of these two related

A.K. Baev, *Specific Intermolecular Interactions of Element-Organic Compounds*,
DOI 10.1007/978-3-319-08563-0_2

Table 2.1 Energies (kJ mol^{-1}) of specific interactions of liquid tetrasilane and trialkylsilanes at 298 K

Compounds	Formula	Structure	$\Delta_{vap}H^0$ (298 K) [6]	DSi–H$_3$C → H–CH$_2$–	DSi–CH$_2$–CH$_3$ → CH$_2$–CH$_3$	DSi–CH$_2$–CH$_2$–CH$_3$ → CH$_2$–CH$_2$–CH$_3$
Tetrasilane	H$_{10}$Si$_4$		35.6	–	–	D–SiH$_2$–H$_3$Si → SiH$_2$–SiH$_3$ = 8.9
2,2-Dimethylpropane	C$_5$H$_{12}$		21.8	DC–H$_3$C → H–CH$_2$– = 2.73	–	–
Tetramethylsilane	C$_4$H$_{12}$Si		26.0 ± 0,6	3.25	–	–
Tetraethylsilane	C$_8$H$_{20}$Si		39.0 ± 0.7	–	4.9	–
Tetrapropylsilane	C$_{12}$H$_{28}$Si		42.2 ± 0.7	–	–	5.27
Tetravinylsilane	C$_8$H$_{12}$Si		42.7 ± 0.7	–	5.34	–

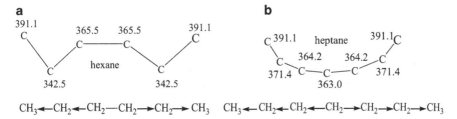

Fig. 2.1 Distribution of dissociation energies (kJ mol^{-1}) of carbon atoms of hexane (**a**) and heptane (**b**) molecules

element atoms with similar electron configuration are similar as regards the peculiarities of electron density shifting in the molecules. At the same time, the sequence in the distribution of electron density at the silicon atoms and hydrogen atoms should be preserved at the first chain atoms, even and odd chain atoms in general, as well as even and odd numbers of chain atoms. The energies of the bond dissociation of C–C [7] in the molecules of the saturated hydrocarbon compound are illustrated in Fig. 2.1 [8]. It is clear from the data shown in Fig. 2.1 that, independently of the number of carbon atoms in the molecules of hexane and heptane, the separation of the methyl groups is accompanied by the maximum energy value. The separation energy of the second and fifth carbon atoms of the hexane chain with an even number of carbon atoms has a significantly lower value (342.5 kJ mol^{-1}) compared with the separation energy of the second and sixth carbon atoms of heptane (371.4 kJ mol^{-1}). At the same time, the separation of the central carbon atoms is accompanied by approximately equal energy expenditure. This distribution pattern of the energy separation of carbon atoms is caused by the distribution of electron density in the molecules of the compounds, providing the high energy separation of the end methyl groups and the differing separation energy values for carbon atoms connected to the end methyl groups. These data also show that the small difference in the separation energies of all further central chain carbon atoms is caused by the small difference in the negative charges at the even-numbered and odd-numbered carbon atoms. The analogy in the nature of changes of the energy separation of the Si–Si bonds and separation into SiH, SiH$_2$, SiCH$_3$, and Si(CH$_3$)$_3$ should manifest in the silanes.

In heptane molecules with an even number of carbon atoms, the shifting of the electron density occurs from the three central carbon atoms to the second atom of each side chain, which, in turn, shifts the electron density to the carbon atom of the end methyl group, providing it with an increased negative charge and, as a result, the formation of a more stable specific interaction. At the same time, at even-numbered carbon atoms in the chain, an equal energy value of the bond separation of C–C occurs in each of the three average carbon atoms and, consequently, each of them possesses an equal negative charge, reflecting the same stability of the formed intermolecular interactions.

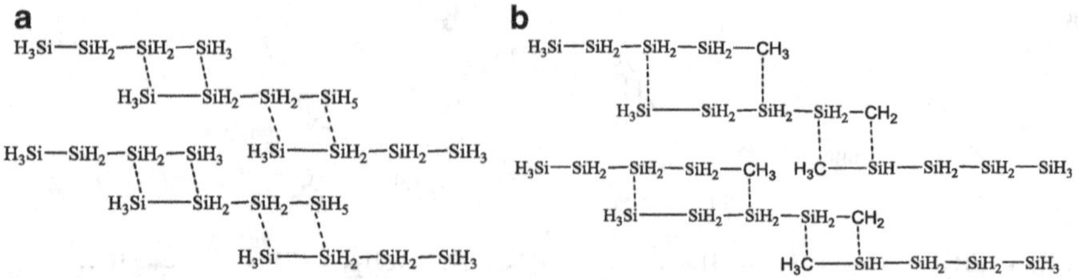

a

H$_3$Si—SiH$_2$—SiH$_2$—SiH$_3$

　　　　H$_3$Si——SiH$_2$—SiH$_2$—SiH$_5$

H$_3$Si—SiH$_2$—SiH$_2$—SiH$_3$　H$_3$Si——SiH$_2$—SiH$_2$—SiH$_3$

　　　　H$_3$Si——SiH$_2$—SiH$_2$—SiH$_5$

　　　　　　　H$_3$Si——SiH$_2$—SiH$_2$—SiH$_3$

b

H$_3$Si—SiH$_2$—SiH$_2$—SiH$_2$—CH$_3$

　　　　H$_3$Si——SiH$_2$—SiH$_2$—SiH$_2$-CH$_2$

H$_3$Si—SiH$_2$—SiH$_2$—SiH$_2$—CH$_3$　H$_3$C——SiH—SiH$_2$—SiH$_2$—SiH$_3$

　　　　H$_3$Si——SiH$_2$—SiH$_2$—SiH$_2$-CH$_2$

　　　　　　　H$_3$C——SiH—SiH$_2$—SiH$_2$—SiH$_3$

Fig. 2.2 Schematic of the network of specific interactions of liquid tetrasilane (**a**) and pentasilane (**b**)

$$D CH_2—CH_2—CH_3 \cdots \blacktriangleright H_2C—CH_2—CH_3$$

The maximum high-energy separation of the first C–C bond of hexane and heptane points to the fact that their atoms have an increased difference between their charges. Therefore, in the liquid and solid conditions, the alkane molecules have the possibility to coordinate mutually and, always, the carbon atoms with increased negative charge lie near the carbon atom with the reduced charge. The carbon atoms of the end methyl groups with the increased negative charge form the specific interaction because of the donor–acceptor mechanism, with the transfer of electron density of the essentially undivided $2s^2$ electron pair of the carbon atom with the biggest negative charge of $DH_2C-H_3C \rightarrow CH_2-CH_3$ to the undivided $2s^2$ electron pair of the carbon atom of the methyl group with the reduced negative charge [8].

$$D SiH_2—SiH_2—SiH_3 \cdots \blacktriangleright H_2Si—SiH_2—SiH_3$$

Similar specific interactions are formed in the liquid and crystal tetrasilane, pentasilane, and hexasilane, with the network of these bonds (Fig. 2.2) being crisscrossed by the less stable interactions formed by the hydrogen atoms of SiH$_2$ groups with low stability.

　　Such distribution of the structural energy parameters with the four interactions of the same type $D-SiH_2-SiH_3 \rightarrow SiH_2-SiH_3$, formed in the liquid tetrasilane, allows the energy of the vaporization enthalpy of the compound to be calculated using Eq. (2.1):

$$D - SiH_2 - H_3Si \rightarrow SiH_2 - SiH_3 = \Delta_{vap}H^0(TK)/4 \qquad (2.1)$$

The minimum energy separation of the C–C bond at the fourth carbon atom of the heptane chain indicates that the central methylene group in the compound with nine chain carbon atoms remains unshifted, as its electron density is insensitive to the influence of the reverse dative bond of the end methyl groups. Consequently, the carbon atom of this group does not shift its electron density by the chain and the given methylene group contributes energy to the vaporization enthalpy (sublimation), which remains practically unchanged for the definite number of compounds of the given homologous series. A further increase in the number of

Fig. 2.3 Dependence of vaporization enthalpy on the number of carbon atoms in the alkane chain [8]

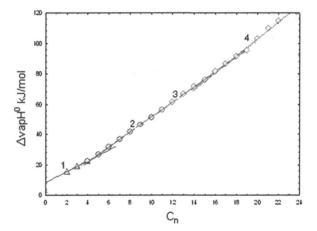

methylene groups in the chain can be accompanied by some reduction or increase in a similar energy contribution to the enthalpy of vaporization [8].

This rule is confirmed by the homogeneous series of all the considered organic compounds and the dependence of the enthalpy characteristics of a large number of compounds of the homogenous series on the number of chain carbon atoms. The accuracy of the experiment is observed by the number of series of compounds with a stable energy contribution from the increasing number of CH_2 groups [8]. This rule occurs in alkanes (Fig. 2.3), which are characterized by the equal energy contribution of CH_2 groups of the series of hexane compounds, dodecane, dodecane–octadecane, and octadecane–dodecosane. A similar dependence is used for the description of the vaporization enthalpies of the saturated and unsaturated silanes, alkylsilanes, and alkenesilanes. As presented in Fig. 2.4, the vaporization enthalpies of trialkylsilanes are dependent on the number of carbon atoms of the alkyne chain and illustrate the description of the three character compounds by sections with increasing values of the enthalpy characteristics. The first section reflects the stabilization of specific interactions $DSi–CH_2–CH_3 \rightarrow Si < DSi–CH_2–CH_2–CH_3 \rightarrow Si < DSi–CH_2–CH_2–CH_2–CH_3 \rightarrow Si$ with an increasing number of carbon atoms in the alkyl ligand, accompanied by a reduction in the influence of the reverse dative bond. The completion of the tributylsilane influence is accompanied by a change in the nature of the considered dependence in the compounds with five to seven methylene groups in the alkyl chain, each group adding an energy contribution to the vaporization energy. Further increase in the number of methylene groups leads to a reduction in the energy contribution to the vaporization enthalpy of trioctylsilane–tris(decyl)silane.

Based on a previous work [2], one should notice that there is a reduction in the value of the vertical ionization potential $I(n_o)$ from 9.19 to 8.96 eV with an increasing number of methylene groups in the hydrocarbons $O=C(CH_2)_n$ from five to ten. The weak effect manifests in a similar dependence $I(n_o)$ of the organosilicon compounds, caused by the completion of the influence of the reverse dative bond on the energies of the intermolecular interaction at the fourth

Fig. 2.4 Dependence of
vaporization enthalpy on
the number of carbon atoms
in trialkylsilanes: *1–3*
trioctylsilane to tris(decyl)
silane series

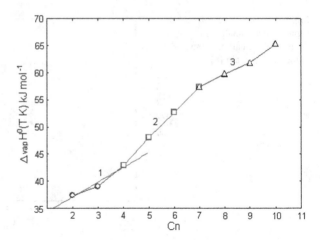

methylene group. We used the dependence presented in Fig. 2.4 for the estimation of the vaporization enthalpy of the unstudied trihexylsilane, tris(undecyl)silane, tris (dodocyl)silane, and tris(tridocyl)silane.

The tetrahedral molecular structure of 2,2-dimethylpropane and tetramethylsilane of one homologous series indicates the significant differences between the types and specificities of the formed intermolecular interactions and those in the molecules of tetraethylsilane and tetrapropylsilane in the condensed condition. The values of the vaporization enthalpies are sharply reduced in the series of compounds, for example:

$$\text{Tetramethylsilane}(26.0) < \text{Tetraethylsilane}(39.0)$$

$$< \text{Tetrapropylsilan}\left(42.2 \ \text{kJ mol}^{-1}\right),$$

Since the carbon and silicon atoms are not able to be present in the eight-coordinated condition, then the presence of four methyl groups in the molecules of 2,2-dimethylpropane and tetramethylsilane provide eight specific interactions $H_3C \rightarrow H\text{–}CH_2$, where the carbon atoms of these groups are in the pentacoordinated condition [8], and the carbon atom of the tetrahedral structure of 2,2-dimethylpropane is the founder of the chain $\text{–}C\text{–}H_2C\text{–}H$ forming the specific interaction.

$$\ce{>C-H2C\cdots{H\cdots H}\cdots CH2-C<}$$

A similar type of specific interaction is formed by the tetramethylsilane molecule, with the silicon atom as the founder of the chain $\text{–}Si\text{–}H_2C\text{–}H$.

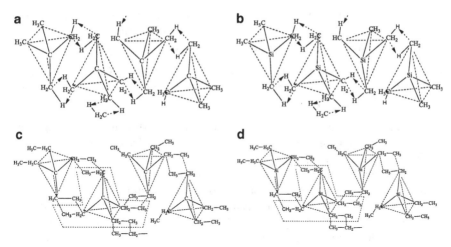

Fig. 2.5 Schematic of the volume-chain network of specific interactions of liquid and crystalline 2,2-dimethylpropane (**a**), tetramethylsilane (**b**), 3,3-dimethylpentane (**c**), and tetraethylsilane (**d**)

Formed by the tetrahedral molecules of 2,2-dimethylpropane and tetramethylsilane in the liquid condition is the volume-chain network of specific interactions shown in Fig. 2.5a, b. The energies of the specific interactions are given in Table 2.1, as calculated using Eq. (2.1a):

$$D - C - H_3C \rightarrow H - CH_2 - C- = D - Si - H_3C \rightarrow H - CH_2 - C$$

$$= \Delta_{vap}H^0(T\,K)/8 \qquad\qquad (2.1a)$$

The replacement of the methyl group in the tetrasilane molecule with the ethyl molecule in tetraethylsilane and 2,2-dimethylpropane to 3,3-dimethylpentane, and, further, with the propyl molecule in tetrapropylsilane, is accompanied by an increase in the difference of the charges at the carbon atoms of the end methylene and methyl groups (Fig. 2.5c, d). This provides its ability to form the specific interactions $DSi-CH_2-CH_3 \rightarrow CH_2-CH_3$ and $DSi-CH_2-CH_2-CH_3 \rightarrow CH_2-CH_2-CH_3$ with increasing stability of the molecular structure and the liquid condition. The interactions are similar to those of tetramethylsilane, whose energies are determined using Eq. (2.1a). The results of the implemented calculations (Table 2.1) lead to the conclusions that, first, they are of quite low stability, and second, the formed specific interaction of tetramethylsilane has an increased stability:

$D-C-H_3C \rightarrow H-CH_2-C-$, 2,2-Dimethylpropane $(2.73) < D-Si-H_3C \rightarrow H-CH_2-C-$, Tetramethylsilane $(3.25 \text{ kJ mol}^{-1})$

A third conclusion is that the energies of the specific interactions of the homologues series of tetraalkylsilanes are described by the natural sequence of its stabilization with an increasing number of carbon atoms in the alkyl ligand:

D–Si–H_3C → H–CH_2–C–, Tetramethylsilane (3.25) < DSi–CH_2–CH_3 → CH_2–CH_3, Tetraethylsilane (4.9) < DSi–CH_2–CH_2–CH_3 → CH_2–CH_2–CH_3, Tetrapropylsilane (5.27 kJ mol^{-1})

Molecules of methane and silane with a tetrahedral structure containing one carbon atom and one silicon atom, respectively, should have differences in the positive charges of the hydrogen atoms. The differences in the charges of the hydrogen atoms of the methyl groups are not an exclusive property of methyl alcohol [9]; it is also known for a number of organic compounds. However, the unique feature of methane and silane is the presence of one carbon and one silicon atom with a negative charge, providing the difference between their charges and those of the hydrogen atoms and the latter with each other. This allows the formation of a specific interaction with the hydrogen atom of the molecules of the near environment in the liquid and solid conditions. Thus, hydrogen atoms with differing positive charges form interactions of reduced stability between each other [8]. As a result, the molecules of methane and silane in the near environment form four specific interactions of the same type C–H → H–C and Si–H → H–Si, respectively, with reduced stability (Fig. 2.6). We note the special calculations indicated for the location of the 2s level of the carbon atom in comparison with the 1s level of hydrogen [10]. As a result, this $2s^2$(c) electron pair of the carbon atom remains essentially undivided or, in other words, both MOs of a_1 symmetry remain significantly unbinding in CH_4. In this regard, the bonds in CH_4 are carried out as the result of the interactions $2p_C$-$1s_H$, caused by the filling of the binding MO of a_1 symmetry. Therefore, the C–H bonds remain rather orbital-scarce hypervalent bonds instead of the usual covalent bonds because the tetrahedral configuration of the ligand is the most favorable for interaction of σ-orbitals of equivalent ligands with the three Cartesian p-orbitals of the central atom. Then, the geometric structure

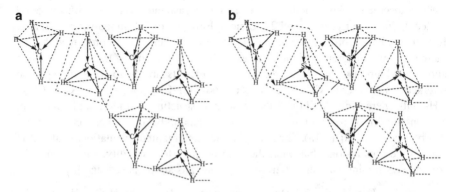

Fig. 2.6 Schematic of the volume-chain network of specific interactions of liquid and crystalline methane (**a**) and silane (**b**)

of the molecule of methane CH_4 is not connected with sp^3 hybridization [10]. A similar situation occurs with the silane SiH_4, with the $3s^2$ electrons being an essentially undivided pair and not connected with the sp^3 hybridization. Therefore, in the silane molecule (like the methane molecule) there is a shifting of electron density to the silicon atom from the hydrogen atoms, resulting in a difference in the positive charges that depends on the location in the molecule and causes the formation of the four hydrogen bonds Si–H•••H–Si. The energies of the interactions (Table 2.2) can be calculated from Eqs. (2.1b) and (2.1c):

$$D - Si - H \rightarrow H - Si- = \Delta_{vap}H^0(TK)sn/4 \qquad (2.1b)$$

$$D - C - H \rightarrow H - C- = \Delta_{vap}H^0(TK)mt/4 \qquad (2.1c)$$

The energies point to the increased stability of the interactions $D–C–H \rightarrow H–C– <$ $D–Si–H \rightarrow H–Si–$ of liquid silane, which is caused by the increased shifting of electron density from the hydrogen atoms in the SiH_4 molecule located at the more distant energy level. The received energies of the specific interactions on the basis of experimental values of the vaporization enthalpies of the compounds with a tetrahedral molecular structure are described by the natural sequence of stabilization:

$D–C–H \rightarrow H–C$, Methane (2.15) $< D–C–H_3C \rightarrow H–CH_2–C–$, 2,2-Dimethylpropane (2.75) $< D–Si–H \rightarrow H–Si–$,

Silane (3.11) $< D–Si–H_3C \rightarrow H–CH_2–$, Tetramethylsilane (3.25) $< DSi–CH_2–$ $CH_3 \rightarrow CH_2–CH_3$, Tetraethylsilane (4.9) $< DSi–CH_2–CH_2–CH_3 \rightarrow CH_2–CH_2–$ CH_3, Tetrapropylsilane (5.27 kJ mol^{-1})

The low energy values of the bonds are caused by the small shifts in electron density from the four-coordinated silicon atom and are in good agreement with the energies of interactions of the alkylamines: $N(CH_3)_3$ (4.25) $< N(C_2H_5)_3$ (5.85) $> N$ $(C_3H_7)_3$ (8.1 kJ mol^{-1}) [8]. With the partial substitution of the hydrogen atoms in the silane molecule by methyl, ethyl, or propyl groups, or by differing alkyl groups in dependence on the location in the molecule, the energies of the specific interactions should be changed by the mutual influence of the ligands on the shifting of electron density.

2.1.2   Energies of Specific Interactions of Liquid Trialkylsilanes, Ethyldialkylsilanes, and Methyldialkylsilanes

According to research using the method of photoelectron spectroscopy [13–17] of the compounds of C_{3v} symmetry, the $HSiR_3$, H_3SiR, R_1SiR_3 and Si–C binding levels α_1 and e are split into 1.2 eV, suggesting a small difference in the energies of intermolecular interactions of Si–C and C–C bonds and the formed intermolecular

Table 2.2 Energies (kJ mol^{-1}) of specific interactions of methane and silane [11, 12]

Compound	Formula	$\Delta H^\circ(T)$ [11, 12]		$\Delta H^\circ(T)$ [11, 12]		$D-C-H \rightarrow H-C$	$D-C-H \rightarrow H-C$
		Subl.	T (K)	Vapor	T (K)	Subl.	Vapor
Methane	CH_4	10.0	84	8.6	175	2.5	2.15
		9.62	77			–	
Silane	SiH_4	–	–	12.43 ± 0.21	161.25		$D-Si-H \rightarrow H-Si = 3.11$

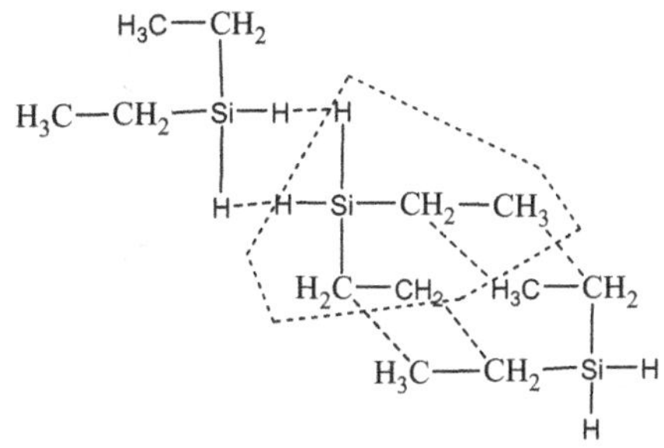

Fig. 2.7 Schematic of the network of specific interactions of liquid and crystalline diethylsilane

interactions. This is confirmed by the obtained energy values of D–C–H → H–C and D–Si–H → H–Si (Table 2.2).

Taking the energy value of the formed hydrogen bond of liquid silane and of compounds with the same type of interaction as the homologous series trialkylsilanes, triethyldialkylsilanes, and methyldialkylsilanes, and using the experimental data on the vaporization enthalpies of the compounds, it is possible to obtain the correct energy values of the specific interactions and trace the influence of the hydrogen atoms on the energies of these bonds. The hydrogen atoms and ethyl groups of triethylsilane and diethylsilane form a network of specific interactions in the liquid and crystal conditions, which include one hydrogen bond and six specific interactions (Fig. 2.7). The energy value of the specific interaction formed by the ethyl ligands can be calculated using Eqs. (2.2a) and (2.2b):

$$DSi - CH_2 - CH_3 \rightarrow CH_2 - CH_3 = \left(\Delta_{vap} H^0(TK) - D - Si - H \right.$$
$$\left. \rightarrow H - Si - \right)/6 \qquad (2.2a)$$

$$DSi - CH_2 - CH_3 \rightarrow CH_2 - CH_3 = \left(\Delta_{vap} H^0(TK) - 2D - Si - H \right.$$
$$\left. \rightarrow H - Si - \right)/4 \qquad (2.2b)$$

Comparison of the calculated values (shown in Table 2.3) indicates that the replacement of two hydrogen atoms in the silane molecule by the two ethyl ligands is accompanied by an insignificant stabilization of the formed specific interaction $DSi–CH_2–CH_3 \rightarrow CH_2–CH_3$ at the two remaining hydrogen bonds:

$DSi–H \rightarrow H–Si = 3.11 < DSi–CH_2–CH_3 \rightarrow CH_2–CH_3$ (5.7),
   Triethylsilane $< DSi–CH_2–CH_3 \rightarrow CH_2–CH_3$ (5.97), Diethylsilane

In turn, one unsubstituted hydrogen atom in the molecule of tripropylsilane (or one hydrogen atom substituted by an ethyl group in the molecule of tetraethylsilane, located in the *trans* condition to the average ethyl group) expresses an even greater stabilizing influence per group on the formed specific interaction at the average value of 5.7 kJ mol^{-1} for the three bonds $DSi–CH_2–CH_3 \rightarrow CH_2–CH_3$. This stabilizing effect of the *trans* influence is expressed by the energy of the interaction formed by the butyl fragment and other alkyl groups of the given series of compounds (Table 2.3) and is described by the correlative dependence of the enthalpy characteristic on the number of carbon atoms of the alkyl chain (Fig. 2.4).

Table 2.3 Energies (kJ mol^{-1}) of specific interactions of liquid trialkylsilanes at 298 K

Compounds	Formula	Structure	$\Delta_{vap}H^0$ (298 K) [6]	DSi–CH$_2$–CH$_3$ → CH$_2$–CH$_3$	DSi–(CH$_2$)$_2$–CH$_3$ → CH$_2$–CH$_2$–CH$_3$
DSi–H → H–Si = 3.11					
Diethylsilane	C$_4$H$_{10}$Si		30.0 ± 0.4	5.97	–
Triethylsilane	C$_6$H$_{16}$Si		37.4 ± 0.6	5.7	–
Tripropylsilane	C$_9$H$_{22}$Si		39.1 ± 0.7	–	6.0
Tributylsilane	C$_{12}$H$_{28}$Si		42.9 ± 0.7	–	DSi–CH$_2$–(CH$_2$)$_2$–CH$_3$ → CH$_2$–(CH$_3$)$_2$–CH$_3$– 6.63
Triisopentylsilane	C$_{15}$H$_{34}$Si		43.8 ± 0.7	DCH$_3$ × 3 = 0.9	6.63
Tripentylsilane	C$_{15}$H$_{34}$Si		48.1 ± 0.8	∑DCH$_2$ × 3 = 5.0	6.63
Trihexylsilane	C$_{18}$H$_{40}$Si		52.7[a]	× 6 = 9.8	6.63
Triheptylsilane	C$_{21}$H$_{46}$Si		57.4 ± 0.8	∑DCH$_2$ × 9 = 14.5	–
Trioctylsilane	C$_{24}$H$_{52}$Si		59.8 ± 0.8	∑DCH$_2$ × 12 = 16.9	6.63
Trinonylsilane	C$_{27}$H$_{58}$Si		61.8 ± 0.8	∑DCH$_2$ × 15 = 19.0	6.63

Tris(decyl)silane	$C_{30}H_{64}Si$	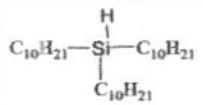 $\begin{array}{c} H \\ C_{10}H_{21}\!-\!Si\!-\!C_{10}H_{21} \\ C_{10}H_{21} \end{array}$	65.3 ± 0.8	$\sum DCH_2 \times 18 = 22.4$	6.63
Tris(undecyl)silane	$C_{33}H_{70}Si$	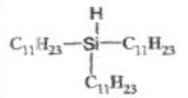 $\begin{array}{c} H \\ C_{11}H_{23}\!-\!Si\!-\!C_{11}H_{23} \\ C_{11}H_{23} \end{array}$	67.7^a	$\sum DCH_2 \times 21 = 26.8$	6.63
Tris(dodecyl)silane	$C_{36}H_{76}Si$	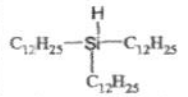 $\begin{array}{c} H \\ C_{12}H_{25}\!-\!Si\!-\!C_{12}H_{25} \\ C_{12}H_{25} \end{array}$	70.4^a	$\sum DCH_2 \times 24 = 27.5$	6.63
Tris((tridocyl)silane	$C_{39}H_{82}Si$	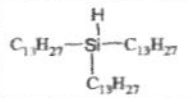 $\begin{array}{c} H \\ C_{13}H_{27}\!-\!Si\!-\!C_{13}H_{27} \\ C_{13}H_{27} \end{array}$	73.0^a	$\sum DCH_2 \times 27 = 30.1$	6.63

[a]Estimated on the basis of the correlation $\Delta_{vap}H^0(T) = f(Cn)$

Of course, the *trans* influence is expressed at the hydrogen atom and the alkyl group. However, one should keep in mind the low stability of the hydrogen bond and its below average change limits and, correspondently, the reduced manifestation of the effect. Also, there are difficulties in the determination of the effect at each fragment. According to the dependence $\Delta_{vap}H^0(TK) = f(Cn)$ (Fig. 2.4), the influence of the reverse dative bond on the shifting of electron density is completed at the third carbon atom of the chain, and the energy of the specific interaction in this series of compounds is reached at tributylsilane. Further increase in the vaporization enthalpy of the compounds is caused by the energy contribution of the increasing number of methylene groups. As is evident from the data, the formed specific interactions are stabilized by an increasing number of carbon atoms in the ligand alkyl chain:

DSi–CH$_2$–CH$_3$ → CH$_2$–CH$_3$, (5.7), Triethylsilane < DSi–(CH$_2$)$_2$–CH$_3$ → CH$_2$–CH$_2$–CH$_3$, (6.0),

Tripropylsilane < DSi–CH$_2$–(CH$_2$)$_2$–CH$_3$ → CH$_2$–(CH$_3$)$_2$–CH$_3$, (6.63 kJ mol^{-1}), Tributylsilane.

Taking into account the low energy contribution of the propyl and butyl fragments to the vaporization enthalpy, there is no doubt that the low contribution of the isostructural methyl group of triisopentylsilane to the vaporization enthalpy can be obtained from the difference between the enthalpy characteristics of triisopentylsilane and tributylsilane:

$$DCH_3 = \left(\Delta_{vap}H^0(TK)\text{itrps} - \Delta_{vap}H^0(TK)\text{trbs}\right)/3 \qquad (2.2)$$

The low energy value of the formed specific interaction DH$_3$C → H–CH$_2$ is equal to 0.15 kJ mol^{-1}, which lies within the error of the experimentally determined vaporization enthalpy of these compounds.

Presented in Fig. 2.8 are the vaporization enthalpies of ethyldialkylsilanes and methyldialkylsilanes. The dependence on the number of carbon atoms of the alkyl

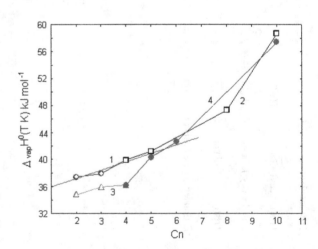

Fig. 2.8 Dependence of vaporization enthalpy on the number of carbon atoms in ethyldialkylsilanes (*1, 2*) and methyldialkylsilanes (*3, 4*)

chains is described by two straight segments, the first of which is caused by the weakening of the influence of the intramolecular reverse dative bond, ending on the butyl fragment, by the corresponding stabilization of the formed specific interactions. The maximum achievable stability is obtained at ethyldibutylsilane and methyldibutylsilane.

The increase in the vaporization enthalpy at the second segment (Fig. 2.8, lines 2 and 4) is caused by the energy contribution of the increasing number of methylene groups of the alkyl chain. The correlation $\Delta_{vap}H^0(T) = f(Cn)$ objectively reflects the nature of the intramolecular and intermolecular interactions of the series of considered compounds, which allows the vaporization enthalpy of the experimentally unstudied compounds to be derived, as given in Tables 2.4 and 2.5. Taking the energy contribution of already obtained energy values of the specific interactions of DSi–H → H–Si (3.11 kJ mol^{-1}) and DSi–CH$_2$–CH$_3$ → CH$_2$–CH$_3$ (5.7 kJ mol^{-1}), the energies of the interactions formed by the alkyl fragments of the ethylalkylsilane series of compounds can be calculated using Eqs. (2.3), (2.3a), and (2.3b)

Triethylsilane :
$$DSi - CH_2 - CH_3 \rightarrow CH_2 - CH_3 = \left(\Delta_{vap}H^0(TK) - D - Si - H \rightarrow H - Si-\right)/6 \tag{2.3}$$

Ethyldipropylsilane :
$$DSi - CH_2 - CH_2 - CH_3 \rightarrow CH_2 - CH_2 - CH_3 =$$
$$\left(\Delta_{vap}H^0(TK) - 2DSi - CH_2 - CH_3 \rightarrow CH_2 - CH_3 - D - Si - H \rightarrow H - Si -\right)/4 \tag{2.3a}$$

Ethyldibutylsilane :
$$DSi - CH_2 - (CH_2)_2 - CH_3 \rightarrow CH_2 - (CH_2)_2 - CH_3 =$$
$$\left(\Delta_{vap}H^0(TK) - 2DSi - CH_2 - CH_3 \rightarrow CH_2 - CH_3 - D - Si - H \rightarrow H - Si-\right)/4 \tag{2.3b}$$

As shown in Table 2.4, the results of the implemented energy calculations of the specific interactions are significantly stabilized with the increasing length of the alkyl chains of the ligand. At the same time, the absolute values are reduced by the replacement of one propyl ligand in tripropylsilane by the ethyl ligand in ethyldipropylsilane and ethyldibutylsilane. An important observation is the increased vaporization enthalpies of the isostructural ethyldiisopropylsilane, ethyldiisobutylsilane, and ethyldiisopentylsilane in comparison to their analogs with a normal structure (Table 2.4). It follows that the contribution of the isostructural methyl group to the change in distribution of electron density to the alkyl fragments is accompanied by an additional increased energy contribution to the enthalpy characteristic of the ethylalkylsilanes, caused by the more significant

Table 2.4 Energies (kJ mol^{-1}) of specific interactions of liquid ethylalkylsilane at 298 K

DSi–H → H–Si = 3.11, DSi–CH$_2$–CH$_3$ → CH$_2$–CH$_3$ = 5.7

Compounds	Formula	Structure	$\Delta_{vap}H^0$ (298 K) [6]	DSi–CH$_2$–CH$_3$ → CH$_2$– CH$_3$	DSi–CH$_2$–CH$_2$– CH$_3$ → CH$_2$–CH$_2$–CH$_3$	DSi–(CH$_2$)$_3$– CH$_3$ → (CH$_2$)$_3$–CH$_3$
Triethylsilane	C$_6$H$_{16}$Si		37.4 ± 0.6	5.7	–	–
Ethyldiisopropylsilane	C$_8$H$_{20}$Si		38.1 ± 0.7	5.7 / DCH$_3$ × 2 = 0.7	–	–
Ethyldipropylsilane	C$_8$H$_{20}$Si		37.9 ± 0.6	–	5.85	–
Ethyldiisobutylsilane	C$_{10}$H$_{24}$Si		39.8 ± 0.7	DCH$_3$ × 2 = 1.9	6.32	–
Ethyldibutylsilane	C$_{10}$H$_{24}$Si		39.9 ± 0.7	–	–	6.35
Ethyldiisopentylsilane	C$_{12}$H$_{28}$Si		42.6 ± 0.7	DCH$_3$ × 2 = 2.7	–	6.35
Ethyldipentylsilane	C$_{12}$H$_{28}$Si		41.2 ± 0.7	$\sum$DCH$_2$ × 2 = 2.2	–	6.35
Ethyldihexylsilane	C$_{14}$H$_{32}$Si		43.6^a	$\sum$DCH$_2$ × 4 = 3.7	–	6.35
Ethyldiheptylsilane	C$_{16}$H$_{36}$Si		45.3^a	$\sum$DCH$_2$ × 8 = 5.4	–	6.35
Ethyldioctylsilane	C$_{18}$H$_{40}$Si		47.3 ± 0.7	$\sum$DCH$_2$ × 10 = 7.4	–	6.35

| Ethyldinonylsilane | $C_{20}H_{44}Si$ | 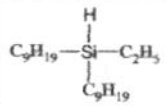 | 53.0^a | $\sum DCH_2 \times 12 = 13.1$ | – | 6.35 |
| Ethyldidecylsilane | $C_{22}H_{48}Si$ | 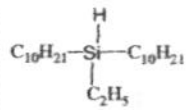 | 58.7 ± 0.8 | $\sum DCH_2 \times 12 = 18.7$ | – | 6.35 |

[a]Estimated on the basis of the correlation $\Delta_{vap}H^0(T) = f(Cn)$

Table 2.5 Energies (kJ mol^{-1}) of specific interactions of liquid methylalkylsilanes at 298 K

Compounds	Formula	Structure	$\Delta_{vap}H^0$ (298 K) [6]	DSi–CH$_2$– CH$_3$ → CH$_2$–CH$_3$	DSi–CH$_2$–CH$_2$– CH$_3$ → CH$_2$–CH$_2$–CH$_3$	DSi–(CH$_2$)$_3$– CH$_3$ → (CH$_2$)$_3$–CH$_3$
DSi–H → H–Si = 3.11, DSi–H$_3$C → H–CH$_2$–Si = 3.25						
Methyldiethylsilane	C$_5$H$_{14}$Si		34.8 ± 0.7	6.3	–	–
Methyldiisopropylsilane	C$_7$H$_{18}$Si		32.4 ± 0.8 35.4[a]	6.3	DCH$_3$ = 0.6	–
Methyldipropylsilane	C$_7$H$_{18}$Si		35.9 ± 0.7	–	6.5	–
Methyldibutylsilane	C$_9$H$_{22}$Si		36.2 ± 0.7	–	–	6.75
Methyldipentylsilane	C$_{11}$H$_{26}$Si		40.3 ± 0.7	$\sum$DCH$_2$ × 2 = 4.2	–	6.75
Methyldihexylsilane	C$_{13}$H$_{30}$Si		42.6 ± 0.7 45.6[b]	$\sum$DCH$_2$ × 4 = 6.4	–	6.75
Methyldiheptylsilane	C$_{15}$H$_{34}$Si		47.0[b]	$\sum$DCH$_2$ × 6 = 10.8	–	6.75
Methyldioctylsilane	C$_{17}$H$_{38}$Si		50.5[b]	$\sum$DCH$_2$ × 8 = 14.3	–	6.75
Methyldinonylsilane	C$_{19}$H$_{42}$Si		54.0[b]	$\sum$DCH$_2$ × 10 = 17.8	–	6.75
Methyldidecylsilane	C$_{21}$H$_{46}$Si		57.4 ± 0.8	$\sum$DCH$_2$ × 12 = 21.2	–	6.75

[a]Estimated
[b]Estimated on the basis of the correlation $\Delta_{vap}H^0(T) = f(Cn)$

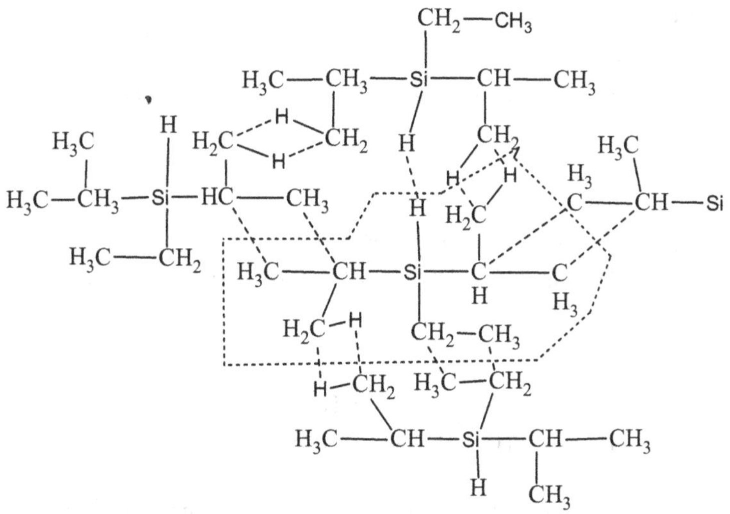

Fig. 2.9 Schematic of the network of specific interactions of liquid and crystalline ethyldiisopropylsilane

participation of these groups in the redistribution of electron density in the molecules. Two isostructural methyl groups of the molecules complicate the network structure of the liquid and crystal conditions of the compounds (Fig. 2.9), forming four interactions of low stability $DH_3C \rightarrow H–CH_2$, whose energies are obtained using Eq. (2.4):

$$DH_3C \rightarrow H - CH_2 = \left(\Delta_{vap}H^0(TK)iso - \Delta_{vap}H^0(TK)nor\right)/4 \qquad (2.4)$$

where $\Delta_{vap}H^0(TK)$ nor represents the vaporization enthalpy of triethylsilane, ethyldipropylsilane, and ethyldibutylsilane The obtained data allow a conclusion to be made on the relation between the stabilization and the presence of isostructural groups with an increasing number of carbon atoms of the ligand alkyl chain:

$DH_3C \rightarrow H–CH_2$: Ethyldiisopropylsilane (0.18) < Ethyldiisobutylsilane (0.5) < Ethyldiisopentylsilane (0.68 kJ mol^{-1})

The compounds of the series of methylalkylsilanes with a combination of the two types of interactions $DSi–H \rightarrow H–Si$ and $DSi–H_3C \rightarrow H–CH_2–Si$ with known energy values allow determination of the energies of the specific interactions formed by the alkylsilanes ligands using the following equations:

Methyldiethylsilane :

$$DSi - CH_2 - CH_3 \rightarrow CH_2 - CH_3 =$$
$$\left(\Delta_{vap}H^0(T\,K) - D - Si - H \rightarrow H - Si - \quad - 2DSi - H_3C \rightarrow H - CH_2 - Si\right)/4$$

$$(2.5)$$

Methyldipropylsilane :

$$DSi - CH_2 - CH_2 - CH_3 \rightarrow CH_2 - CH_2 - CH_3 =$$
$$\left(\Delta_{vap}H^0(T\,K) - D - Si - H \rightarrow H - Si - \quad - 2DSi - H_3C \rightarrow H - CH_2 - Si\right)/4$$

$$(2.5a)$$

Methyldibutylsilane :

$$DSi - CH_2 - (CH_2)_2 - CH_3 \rightarrow CH_2 - (CH_2)_2 - CH_3 =$$
$$\left(\Delta_{vap}H^0(T\,K) - D - Si - H \rightarrow H - Si - \quad - 2DSi - H_3C \rightarrow H - CH_2 - Si\right)/4$$

As shown by the calculation results of the energy parameters, presented in Table 2.5, the specific interactions are stabilized by an increasing number of carbon atoms in the ligand alkyl chain of the series:

$DSi-CH_2-CH_3 \rightarrow CH_2-CH_3$ (6.3),   Methyldiethylsilane $< DSi-CH_2-CH_2-$ $CH_3 \rightarrow CH_2-CH_2-CH_3$ (6.5),   Methyldipropylsilan $< DSi-CH_2-(CH_2)_2-$ $CH_3 \rightarrow CH_2-(CH_2)_2-CH_3$ (6.75 kJ mol^{-1}), Methyldibutylsilane

2.1.3   Energies of Specific Interactions of Liquid Dimethylalkylsilanes, Diethylalkylsilanes, Propylbutylsilanes, and Vinylmethylsilane

The obtained conclusions can be supplemented after analyzing the mutual influence of the different ligand alkylsilanes with the full or part replacement of the hydrogen atoms by propyl and butyl fragments. Note that the set of equations given above for calculation of the energies of the specific interactions take into account the various combinations of alkyl ligands in the molecules and the formed specific interactions; therefore, the calculations are rather simple. The energies of the specific interactions (Table 2.6) of alkylsilanes with methyl and ethyl ligands points to their stabilization by the presence of one hydrogen atom connected to the silicon atom in the molecule of methyldiethylsilane:

$DSi-CH_2-CH_3 \rightarrow CH_2-CH_3$:   Methyltriethylsilane   (5.8) $<$ Methyldiethylsilane (6.4) $<$ Dimethyldiethylsilane (6.5 kJ mol^{-1})

Table 2.6 Energies (kJ mol^{-1}) of specific interactions of liquid methylalkylsilanes at 298 K

Compounds	Formula	Structure	$\Delta_{vap}H^0$ (298 K) [6]	DSi–CH$_2$–CH$_3$–C → CH$_2$– CH$_3$	DSi–CH$_2$–CH$_2$–CH$_3$ → CH$_2$–CH$_2$– CH$_3$
DSi–H → H–Si = 3.11, DSi–H$_3$C → H–CH$_2$–Si = 3.25					
Dimethyldiethylsilane	C$_6$H$_{16}$Si		38.9±0.6	6.5	–
Methyldiethylsilane	C$_5$H$_{14}$Si		34.8±0.7	6.4	–
Methyltriethylsilane	C$_7$H$_{18}$Si		40.5±0.6	5.8	–
Methyldipropylsilane	C$_7$H$_{18}$Si		35.9±0.7	–	6.6
Methyl(2,2-dimethylpropyl) silane	C$_{11}$H$_{26}$Si		38.1±0.1	–$\sum$DCH$_3$ = 2.2:4	6.6
Dimethyldipropylsilane	C$_8$H$_{20}$Si		40.2±0.6	–	6.8
Methyltripropylsilane	C$_{10}$H$_{24}$Si		42.6±0.6	–	6.02

If the molecule of dimethyldiethylsilane contains two methyl groups, the energy of the formed specific interactions remains practically unchanged. This indicates the equivalent influence of the hydrogen atom and methyl group at the location of the ethyl fragment in the *trans* position. The energies of the specific interaction formed by the propyl ligand in alkylsilanes in the presence of one methyl and two propyl ligands and one hydrogen atom in the molecule connected with the silicon atom form a similar sequence of stabilization:

DSi–CH$_2$–CH$_2$–CH$_3$→CH$_2$–CH$_2$–CH$_3$:   Tripropylsilane   (6.0) < Methyltripropylsilane (6.02) < Methyldipropylsilane (6.6) < Methyl (2,2-dimethylpropyl) silane (6.6) < Dimethyldipropylsilane (6.8 kJ mol^{-1})

The maximum stabilization of the specific interaction DSi–CH$_2$–CH$_2$–CH$_3$ $\rightarrow$ CH$_2$–CH$_2$–CH$_3$ is expressed at the two methyl and two propyl groups in the molecule of dimethyldipropylsilane. The compounds with the propyl and butyl fragments in the molecules of dipropyldibutylsilane and propyltributylsilane (Table 2.7) form specific interactions of reduced stability compared with those in the liquid methylpropylsilanes, and the stabilizing effect is within the error range of the experimentally measured vaporization enthalpies of the compounds.

DSi–CH$_2$–(CH$_2$)$_2$–CH$_3$ $\rightarrow$ CH$_2$–(CH$_2$)$_2$–CH$_3$:   Dipropyldibutylsilane   (5.58) < Propyltributylsilane (5.74 kJ mol^{-1})

The specific interactions of the propyl ligand DSi–CH$_2$–CH$_2$–CH$_3$ $\rightarrow$ CH$_2$–CH$_2$–CH$_3$ in the liquid are formed with a reduced energy value:

Ethyltripropylsilane   (5.2) < Propyltriethylsilane   (5.3) < Diethyldipropylsilane (5.54 kJ mol^{-1})

The molecule of tetravinylsilane with four unsaturated ethylene ligands forms eight specific interactions, CH=CH$_2$ $\rightarrow$ CH=CH$_2$, whose energies are determined by the depletion of vaporization enthalpy with the number of bonds. Using the energy value of the formed specific interaction of the methyl group DSi–H$_3$C $\rightarrow$ H–CH$_2$–Si, the energy value of the specific interaction formed by the vinyl fragment of the tetravinylsilane molecule (Table 2.8) can be obtained from Eq. (2.6):

$$DSi - CH = CH_2 \rightarrow CH = CH_2$$
$$= \left(\Delta_{vap}H^0(TK) - 2DSi - H_3C \rightarrow H - CH_2-\right)/6 \qquad (2.6)$$

Its increased stability in comparison with the energy of the same type of specific interaction of liquid tetravinylsilane (Table 2.8) is caused by the manifestation of *trans* influence on the stabilization of the intermolecular interaction. Therefore, the implemented thermodynamic analysis of alkylsilanes revealed the nature of the formed specific interactions and grounded the implementation of the *trans* influence on the stabilization of the interactions.

Table 2.7 Energies (kJ mol^{-1}) of specific interactions of liquid propylbutylsilane and ethyltripropylsilane at 298 K

Compounds	Formula	Structure	$\Delta_{vap}H^0$ (298 K) [6]	DSi–CH$_2$–CH$_2$–CH$_3$ → CH$_2$–CH$_2$–CH$_3$	DSi–CH$_2$–(CH$_2$)$_2$–CH$_3$ → CH$_2$–(CH$_2$)$_2$–CH$_3$
Tetrapropylsilane	$C_{12}H_{28}Si$		42.2 ± 0.7	5.27	–
Dipropyldibutylsilane	$C_{14}H_{32}Si$		44.0 ± 0.8	5.27	5.58
Propyltributylsilane	$C_{15}H_{34}Si$		45.0 ± 0.8	5.27	5.74
				DSi–CH$_2$–CH$_3$ → CH$_2$–CH$_3$	DSi–CH$_2$–CH$_2$–CH$_3$ → CH$_2$–CH$_2$–CH$_3$
Ethyltripropylsilane	$C_{11}H_{26}Si$		41.0 ± 0.7	4.9	5.2
Diethyldipropylsilane	$C_{10}H_{24}Si$		41.5 ± 0.7	4.9	5.54
Propyltriethylsilane	$C_9H_{22}Si$		40.0 ± 0.7	4.9	5.3

Table 2.8 Energies (kJ mol^{-1}) of specific interactions of liquid vinyltrimethylsilane and tetravinylsilane at 298 K

Compounds	Formula	Structure	$\Delta_{vap}H^0$ (298 K) [6]	DSi-H$_3$C → H-CH$_2$-	DSi-CH=CH$_2$ → CH=CH$_2$
Vinyltrimethylsilane	C$_5$H$_{12}$Si	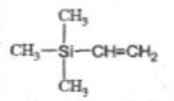	33.1 ± 0.6	3.25	6.8
Tetravinylsilane	C$_8$H$_{12}$Si	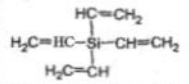	42.7 ± 0.7	–	5.34

2.2 Energies of Specific Interactions of Trisilyls of Nitrogen Main Group Elements and Disilazanes

Research using flame emission spectroscopy (FES) on trisilylamine, phosphine, and arsine is limited by the information that, in the molecule of $N(SiH_3)_3$, the lower effectiveness of the n–$\sigma(SiH)$ interaction contributes to the stabilization of B3MO, expressed in the specific interaction [2]. In this regard, it can be expected that the location of the shifting electron density in the molecule of trisilylphosphine at the more distant energy level of the silicon atom will contribute to its stabilization.

The molecules of trisilylphosphine–trisilylstibine, with the ability of silicon subgroup elements to take the six-coordinated condition, form the volume-chain structure of the liquid condition (Fig. 2.10). On the other hand, the silicon atom with the essentially undivided $3s^2$ electron pair is able to shift and receive electron density (similarly to the carbon atom) and to form the specific intermolecular interactions $DN \rightarrow SiH_3$, $DP \rightarrow SiH_3$, $DAs \rightarrow SiH_3$, and $DSb \rightarrow SiH_3$. Thus, the SiH_3 group expresses the same abilities as CH_3 in trimethylamine, whose energies are obtained by the division of the vaporization enthalpy by the number of bond vacancies, i.e., the number of formed bonds:

$$DE \rightarrow Si - CH_3 = \Delta_{vap}H^0(TK)/6 \qquad (2.7)$$

$DP \rightarrow SiH3\ (6.0) < DAs \rightarrow SiH3\ (6.83\ kJ\ mol^{-1})$
Trimethylamine (4.25) < Trimethylphosphine (4.8) < Trimethylarsine (4.81) < Trimethylstibine (5.2) < Trimethylbismuthine (5.8)

The energies of the specific interactions (Table 2.9) formed at equal temperature intervals reflect its stabilization at the increased values in comparison with the formed interactions in the liquid trimethylamine–trimethylbismuthine series [8].

A molecule of disilazane, with two SiH_3 groups acting as the methyl group in the methylamines and alkyl compounds of the elements of the fifth and sixth main group elements and one functional hydrogen atom, forms six specific interactions with the six bond vacancies: four $N \rightarrow SiH_3$ interactions and two hydrogen bonds N•••H–N (Fig. 2.10b). The result of detailed thermodynamic analysis of the hydrogen bond of the liquid and crystal ammonia in the work by Baev [18] showed that the energies of the hydrogen bonds vary a little in their derivatives and depend

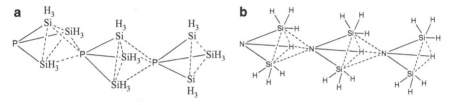

Fig. 2.10 Schematic of the network of specific interactions of liquid and crystalline trisilyl-phosphine (**a**) and disilazane (**b**)

Table 2.9 Energies (kJ mol^{-1}) of specific interactions of liquid trisilylphosphine–trisilylstibine

Compounds	Formula	Structure	$\Delta_{vap}H^0$ (298 K) [6]	T (K)	$DE \rightarrow SiH_3$–E	$DN \bullet\bullet\bullet H$–N
Trisilylphosphine	H_9PSi_3		36.4	243–287	6.0	–
Trisilylarsine	H_9AsSi_3		41.0	258–287	6.83	–
Trisilylstibine	H_9SbSi_3		33.0	–	5.5	–
Disilazane	H_7NSi_2		23.4	213	3.9	3.86
DSi–$H_3C \rightarrow H$–CH_2–Si = 3.25; $DN \bullet\bullet\bullet H$–N = 3.86					$DN \rightarrow Si$–CH_3–N	D–$H_3C \rightarrow H$–CH_2–
Hexamethyldisilazane	$C_6H_{19}NSi_2$		42.2 ± 0.9	298	3.9	0.37

mainly on the coordination of the nitrogen atom. The energies of the hydrogen bond formed by the nitrogen atom of the ammonia molecule with a trigonal pyramid structure (4.23 and 3.86 kJ mol^{-1}) is less stable in comparison with the coordination of the four hydrogen atoms (6.34 and 5.91 kJ mol^{-1}) with a planar molecule structure in the solid and liquid conditions respectively. Therefore, taking the energy value of the formed hydrogen bond of the molecule of disilazane (3.86 kJ mol^{-1}), the energy value of the specific interaction of DN → SiH$_3$ can be estimated using Eq. (2.8):

$$DN \rightarrow SiH_3 = \left(\Delta_{vap}H^0(T\,K) - 2DN \bullet \bullet \bullet H - N\right)/4 \qquad (2.8)$$

The obtained energy value of this type of interaction is presented in Table 2.9. The reduced energy value of the specific interaction DN → SiH$_3$ (3.9 kJ mol^{-1}) compared with the interaction DN → CH$_3$ (4.25 kJ mol^{-1}) is explained by the reduced value of the vertical ionization potential $Iv(n_N)$ at 0.5 eV by tetra(dimethyl)amines rather than $I(n)$ of the molecule of HN(CH$_3$)$_2$ [2]. At the same time, the practical equality of the specific interactions of DN → SiH$_3$ with the energy of the hydrogen bond DN•••H–N (3.86 kJ mol^{-1}) reflects the reduction in the negative charge at the nitrogen atom following the replacement of the hydrogen atom at DN → CH$_3$ [2] and, consequently, at DN → SiH$_3$.

Probably for the first time, the work of Ivanov et al. [9] on the example of the molecule of methyl alcohol demonstrated the special role of the hydrogen atom of the methyl group, which is in the *trans* position with respect to the hydrogen atom of the hydroxyl group forming the interaction with increased stability in comparison with the other two protons formed. A similar special role of the hydrogen atom occurs in the SiH$_3$ group that appears in trialkylsilanes, ethyldialkylsilanes, methyldialkylsilanes, dimethylalkylsilanes, diethylalkylsilanes, propylbutylsilanes, and vinylmethylsilanes, as well as in the isostructural methyl and ethyl groups, noted [8, 18] as the phenomenon of organic and bioorganic chemistry and proteins with the manifestation of the extra stabilizing effect. In this regard, the hexamethyldisilazane is no exception because the methyl groups of each of the two fragments Si(CH$_3$)$_3$, located in the *trans* position to the nitrogen atom on the bond line N–Si(CH$_3$)$_3$, possess exceptional properties in stabilizing the specific interactions of Si–H$_3$C → H–CH$_2$–Si.

These properties differ from the properties of the other two methyl groups of the same fragments of Si(CH$_3$)$_3$. Therefore, there are grounds to consider the other two methyl groups as the isostructural groups participating in the redistribution of electron density in the molecule with the formation of the specific interaction – H$_3$C → H–CH$_2$– of low stability (Fig. 2.11).

The formed network of the specific interactions in liquid hexamethyldisilazane includes 4DN → Si–CH$_3$, 2DN•••H–N, and 4DSi–H$_3$C → H–CH$_2$– with energy

Fig. 2.11 Schematic of the
network of specific
interactions of liquid and
crystalline
hexamethyldisilazane

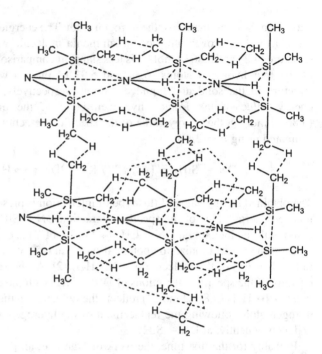

values of 3.9, 3.86, and 3.25 kJ mol^{-1}, respectively, and eight isostructural methyl
groups that form 16 specific interactions $H_3C \rightarrow H–CH_2$ of low stability with an
unknown energy value, which is determined using Eq. (2.8a):

$$DH_3C \rightarrow H - CH_2 = \left(\Delta_{vap}H^0(TK) - 2DN \bullet \bullet \bullet H - N - 4DN \rightarrow Si - CH_3 - \right.$$

$$\left. 4DSi - H_3C \rightarrow H - CH_2 - Si\right)/16 \tag{2.8a}$$

The obtained energy value of the specific interaction formed by the isostructural
methyl group of hexamethyldisilazane is described by the natural sequence of stabilization as follows: ethyldiisopropylsilane (0.17) < methyl(2,2-dimethylpropyl)silane
(0.27) < hexamethyldisilazane (0.37) < ethyldiisobutylsilane (0.48) < ethyldiisopentylsilane (0.68 kJ mol^{-1}), with an increasing number of carbon atoms in the ligand
alkyl chain, which fits the considered compound with eight isostructural methyl
groups.

2.3   Energies of Specific Interactions of Silanamines

According to the data of the X-ray photoelectron spectroscopy of N 1s electrons, the
negative effective charge of the nitrogen atom in the methylamines is reduced
following the replacement of the hydrogen atom by a methyl group, which indicates
the formation of more stable hydrogen bonds in comparison with the methyl group

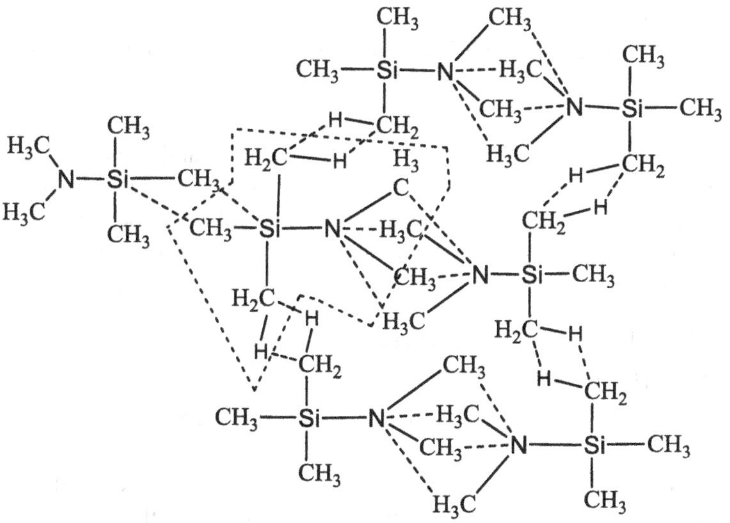

Fig. 2.12 Schematic of the network of specific interactions of liquid and crystalline pentamethylsilanamine

of the specific interaction [2]. This fact has a definite significance in the thermo-dynamic analysis of the linear molecules of alkylsilanamines, where the tetrahedral condition of the silicon atom is not realized and the silicon atom expresses the ability to form the specific interaction and to be in the pentacoordinated condition. At the same time, a rather significant difference in the properties of the carbon and silicon compounds provides the silicon with the reduced negative charge. The similarity in the molecular structures of N,N-dimethyl(methylsilyl)amine with three methyl groups and two functional hydrogen atoms in the first and fifth methyl groups in the last compound the two methyl groups of pentamethylsilanamine act as the isostructural functional groups and form two specific interactions of reduced stability with the pentacoordinated carbon atom. The molecules of these com-pounds form a network of the specific interactions in the liquid condition (Fig. 2.12), including four specific interactions of $DN \rightarrow CH_3-N$, formed by the two methyl groups of the amino groups, and two interactions of the end methyl groups $DSi-CH_3 \rightarrow Si$ with energies of 4.25 [8] and 3.25 kJ mol^{-1}, respectively. The molecule of N,N-dimethyl(methylsilyl)amine forms the two additional hydro-gen bonds $DSi-H \rightarrow H-Si$, and pentamethylsilanamine forms four specific interac-tions $DH_3C \rightarrow H-CH_2$ of the isostructural methyl groups with unknown energies. The energies of these specific interactions are determined from Eq. (2.9) and (2.9a):

N,N-dimethyl(methylsilyl)amine :
$$DSi - H \rightarrow H - Si = \left(\Delta_{vap}H^0(T\,K) - 4DN \rightarrow CH_3 - N - 2DSi - CH_3 \rightarrow Si\right)/2$$

$$(2.9)$$

Pentamethylsilanamine :

$$DH_3C \rightarrow H - CH_2 = \left(\Delta_{vap}H^0(T\,K) - 4DN \rightarrow CH_3 - N - DSi - CH_3 \rightarrow Si\right)/4$$

$$(2.9a)$$

A distinctive feature of N,N-diethyl-1,1,1-trimethylsilanamine is that two ethyl groups of the amine fragment form the specific interaction $DN \rightarrow CH_3$–CH_2–N with an energy of 5.85 kJ mol^{-1} [8]. The energy value of the formed specific interaction of the isostructural methyl group of the silane fragment can be obtained from a similar Eq. (2.9b):

N,N-diethyl-1, 1, 1-trimethylsilanamine :

$$DH_3C \rightarrow H - CH_2 = \qquad\qquad\qquad (2.9b)$$
$$\left(\Delta_{vap}H^0(T\,K) - 4DN \rightarrow CH_3 - CH_2 - N - 2DSi - CH_3 \rightarrow Si\right)/4$$

Given in Table 2.10 are the results of the energy calculations for this type of interaction with the participation of the methyl groups of the silane fragment, which point to its stability. The energy of the interaction is comparable with the energy of the hydrogen bond formed by the hydrogen atoms of the silyl fragment. The energies of the specific interactions with the ethyl group in the silane and amine fragments of pentaethylsilanamine and with a combination of the methyl and ethyl groups of the amine fragment (1,1,1-triethyl-N-(1-methylethyl)silanamine) were calculated using Eqs. (2.10) and (2.10a), where the energy of the specific interaction $DN \rightarrow CH_3$–CH_2–N (5.85 kJ mol^{-1}) is taken from the literature [8] and from the interaction DSi–CH_2–$CH_3 \rightarrow Si$ (5.7 kJ mol^{-1}) (Table 2.4) for the liquid triethylsilane with the location of the ethyl group in the *trans* position:

Pentaethylsilanamine :

$$DisoCH_2 - H_3C \rightarrow CH_2 - CH_3 =$$
$$\left(\Delta_{vap}H^0(T\,K) - 4DN \rightarrow CH_3 - CH_2 - N - 2DSi - CH_2 - CH_3 \rightarrow Si\right)/4$$

$$(2.10)$$

1, 1, 1-triethyl-N-(1-methylethyl)silanamine :

$$DisoCH_2 - H_3C \rightarrow CH_2 - CH_3 = \left(\Delta_{vap}H^0(T\,K) - 2DN \rightarrow CH_3 - CH_2 - N -\right.$$
$$\left.2DN \rightarrow CH_3 - CH_2 - N - 2Dsi - CH_2 - CH_3 \rightarrow Si\right)/4 \qquad (2.10a)$$

The data in Table 2.10 imply that the obtained energy values of the specific interactions formed by the isostructural ethyl group of the silane fragment of these compounds are in good agreement with each other. It allows the use of its average value to estimate the energy of the specific interactions formed by the propyl and butyl ligand and by the hydrogen bond DN•••H–N in different combinations in the silanamine amine fragment. The results of the implemented calculations of the energy parameters with the use of similar equations are given in Table 2.10. The data show that the energies of the formed specific interactions of

Table 2.10 Energies (kJ mol⁻¹) of specific interactions of liquid alkylsilanamines at 298 K

Compounds	Formula	Structure	$\Delta_{vap}H^0$ (298 K) [6]	DSi-CH₃ → Si	DN → CH₃-N	DN → CH₃-CH₂-N	DisoH₃C → H-CH₂
DSi-H → H-Si =2.35; DSi-H₃C → H-CH₂- = 3.25							
N,N-Dimethyl(methylsilyl)amine	C₃H₁₁NSi		28.2	3.25	4.25	–	DSi-H → H-Si = 2.35
Pentamethylsilanamine	C₅H₁₅NSi		33.6 ± 0.8	3.25	4.25	–	2.52
N,N-Diethyl-1,1,1-trimethylsilanamine	C₇H₁₉NSi		37.9 ± 0.8	3.25	–	5.85	2.0
DSi-CH₂-CH₃ → CH₂-CH₃ = 5.7 DSi-H → H-Si =2.35				DSi-CH₂-CH₃ → Si	DN → CH₃-CH₂-N		DisoCH₂- H₃C → CH₂-CH₃
Pentaethylsilanamine	C₁₀H₂₅NSi		42.2 ± 1.0	5.7	5.85 [8]	–	1.85
1,1,1-Triethyl-N-(1-methylethyl)silanamine	C₉H₂₃Si		38.6 ± 0.8	5.7	4.25 5.85	–	1.75
1,1,1-Triethyl-N-propylsilan-amine	C₉H₂₃NSi		41.5 ± 0.8	298 5.7	DN•••H-N =3.86	DN → CH₃-(CH₂)₂-N =7.7	1.8
1,1,1-Triethyl-N-(1-methylbutyl)silanamine	C₁₁H₂₇NSi		46.9 ± 1.0	298 5.7	4.25 9.9	DN → CH₃-(CH₂)₃-N =9.9	1.8
N,N-Dibutyl-1,1,1-triethylsilanamine	C₁₄H₃₃NSi		51.4 ± 0.9	5.7	–	DN → CH₃-(CH₂)₃-N = 6.6?	1.8
1,1,1-Triethyl-N,N-bis(1-methylpro-pyl)silanamine	C₁₄H₃₃NSi		51.4 ± 0.9	5.7	DisoH₃C → H-CH₃=	DN → CH₃-(CH₂)₂-N =7.7	1.8

(continued)

Table 2.10 (continued)

Compounds	Formula	Structure	$\Delta_{vap}H^0$ (298 K) [6]	$DSi\text{-}CH_3 \rightarrow Si$	$DN \rightarrow CH_3\text{-}N$	$DN \rightarrow CH_3\text{-}CH_2\text{-}N$	$DisoH_3C \rightarrow H\text{-}CH_2$
1,1,1-Triethyl-N-octylsilanamine	$C_{14}H_{33}NSi$	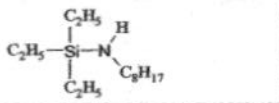	$59.1 \pm 1.$	5.7	$DN\bullet\bullet\bullet H\text{-}N = 3.86$	$DN \rightarrow CH_3\text{-}(CH_2)_3\text{-}N = 18.8 = 9.9 + 4DCH_2$	1.8
N-(1,1-Dimethylethyl)-1,1,1-triethylsilanamine	$C_{10}H_{25}NSi$	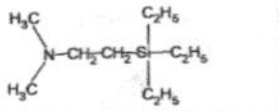	44.3 ± 1.0	5.7	$DN \rightarrow CH_2\text{-}CH_2\text{-}Si = 4.15$	$DN \rightarrow CH_3\text{-}N = 4.25$	1.8

the alkyl ligand of the amine fragment are described by the natural sequence of its stabilization with an increasing number of carbon atoms in the alkyl chain:

$$DN \rightarrow CH_3\text{–}N \ (4.25) < DN \rightarrow CH_3\text{–}CH_2\text{–}N \ (5.85) < DN \rightarrow CH_3\text{–}(CH_2)_2\text{–}N \ (7.7)$$
$$< DN \rightarrow CH_3\text{–}(CH_2)_3\text{–}N \ (9.9 \ kJ \ mol^{-1})$$

The maximum energy value is reached for the four carbon atoms of the butyl ligand. A further increase in the energy parameter is connected with the energy contribution of the methylene groups of the chain. An example is given by 1,1,1-triethyl-N-octylsilanamine with the octyl ligand (Table 2.10).

Molecules of N-(1,1,-dimethylethyl)-1,1,1-triethylsilanamine with similar types and numbers of specific interactions form two additional specific interactions by the methylene groups of $DCH_2 \rightarrow CH_2$, crisscrossing the chains to form networks of the specific interactions.

$$-CH_2\text{—}CH_2-$$
$$-CH_2\text{—}CH_2-$$
$$-CH_2\text{—}CH_2-$$

Taking the energies of $DSi\text{–}CH_2\text{–}CH_3 \rightarrow Si$ (5.7) and $DN \rightarrow CH_3\text{–}N$ (4.25 kJ mol^{-1}) of the specific interactions with reduced stability, formed by the two isostructural methyl groups $DisoCH_2\text{–}H_3C \rightarrow CH_2\text{–}CH_3$ (1.8 kJ mol^{-1}) and their total contribution to the vaporization enthalpy the obtained energy of the formed specific interaction of the methylene group is equal to 4.15 kJ mol^{-1}, as obtained using Eq. (2.10b):

$$D - CH_2 \rightarrow CH_2- = \left(\Delta_{vap}H^0(T\,K) - 2DN \rightarrow CH_3 - CH_2 - N - 2Si - CH_2 - \right.$$
$$\left. CH_3 \rightarrow Si - 4DisoCH_2 - H_3C \rightarrow CH_2 - CH_3\right)/2 \ (2.10b)$$

2.4 Energies of Specific Interactions of Silanetriamine and Disilanylamine

The molecule of N-1,1,1-tetramethyl-[N-(trimethylsilyl)silanamine] (Fig. 2.13a) with two silyl fragments and one N–CH$_3$ group contains four isostructural methyl groups. Being in the *trans* position, the Si–CH$_3$ groups form four specific interactions $DSi\text{–}H_3C \rightarrow H\text{–}CH_2-$ with an energy of 3.25 kJ mol^{-1}, two interactions of the second type $DN \rightarrow CH_3\text{–}N$ with an energy of 4.25 kJ mol^{-1} formed by the amino group, and eight specific interactions $DH_3C \rightarrow H\text{–}CH_2$ with reduced energy formed by the isostructural groups (Table 2.11). The energy contributions to the vaporization enthalpy of the latter interactions are determined from the equality of the energy parameters, as given in Eq. (2.11):

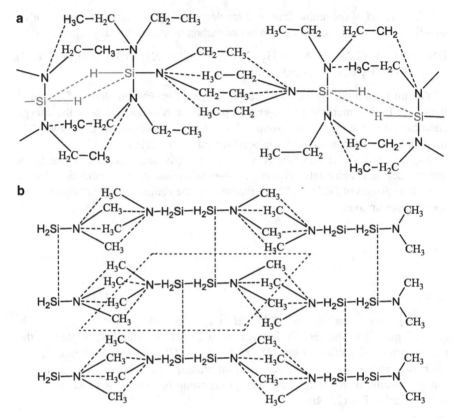

Fig. 2.13 Schematic of the network of specific interactions of liquid and crystalline $N,N,N',N',N'',$ N''-hexaethylsilanetriamine (**a**) and N,N,N',N'-tetramethyldisilanylamine (**b**)

$$DisoH_3C \rightarrow H - CH_2$$

$$= \left(\Delta_{vap}H^0(T K) - 4DSi - H_3C \rightarrow H - CH_2 - Si - 2DN \rightarrow CH_3 - N\right)/8$$

$$(2.11)$$

The molecule of N,N,N',N',N'',N''-hexamethylsilanetriamine with six methyl groups forms 12 specific interactions $DN \rightarrow CH_3-CH_2-N$. The silicon atom of the silane SiH group with three substituted hydrogen atoms expresses the pentacoordinated condition and forms two hydrogen bonds $DSi\cdots H-Si$ (1.56 kJ mol^{-1}) with the same total energy value of $DSi-H\cdots H-Si$ (3.11 kJ mol^{-1}). Taking into account that the vaporization enthalpy is associated with the number of bursting bonds and their energies [8], the following can be written:

Table 2.11 Energies (kJ mol^{-1}) of specific interactions of liquid silanetriamine and disilanylamine at 298 K

Compounds	Formula	Structure	$\Delta_{vap}H^0$ (298 K) [6]	DSi-H$_3$C $\to$ H-CH$_2$-Si	DN $\to$ CH$_3$-N	DisoH$_3$C $\to$ H-CH$_2$
N-1,1,1-Tetramethyl-[N-(trimethylsilyl)silanamine	C$_7$H$_{21}$NSi$_2$	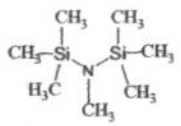	38.1 ± 0.8	3.25	4.25	2.07
N,N,N',N'',N''-Hexaethylsilanetriamine	C$_{12}$H$_{31}$N$_3$Si	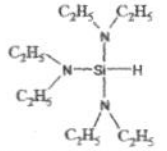	58.4 ± 1.0	DSiH $\to$ H-Si- = 3.11	DN $\to$ CH$_3$-CH$_2$- N = 4.6	–
N,N,N',N'-Tetramethyldisilanylamine	C$_4$H$_{16}$N$_2$Si$_2$	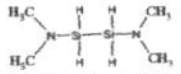	39.3*	–	4.25	DH$_2$Si $\to$ SiH$_2$ = 2.65

$$DN \rightarrow CH_3 - CH_2 - N = \left(\Delta_{vap}H^0(T\,K) - 2DSi - H \bullet \bullet \bullet H - Si\right)/12 \quad (2.12)$$

The obtained energy value of the specific interaction $DN \rightarrow CH_3$–CH_2–N $(4.6$ kJ mol$^{-1})$ has a significantly low value in comparison with the interactions in the liquid pentaethylsilanamine and 1,1,1-triethyl-N-(1-methylethyl)silanamine with one amino group. Thus, the steric effect is shown at the nitrogen atoms interacting with silicon, and one should assume that the silicon atom expresses a significant influence on the reduction of the negative charge of the nitrogen atom, like the molecule of $_3(H_3C)BN(CH_3)_3$ [19].

The molecule of N,N,N',N'-tetramethyldisilanylamine with four methyl groups forms eight specific interactions $DN \rightarrow CH_3$–N with an energy of 4.25 kJ mol^{-1} and, additionally, two SiH_2 groups with properties similar to the methylene groups (Fig. 2.13b). Two interactions of the second type $DH_2Si \rightarrow SiH_2$ form a volume chain structure with the network of specific interactions. The chains are crisscrossed by the interactions of the second type, whose energy is calculated from Eq. (2.13):

$$DH_2Si \rightarrow SiH_2 = \left(\Delta_{vap}H^0(T\,K) - 8DN \rightarrow CH_3 - N\right)/2 \quad (2.13)$$

The calculation results are given in Table 2.11, which shows that the energy of the specific interaction between two silicon methylene groups is significantly reduced because of its location between two strong acceptors of the electron density of the nitrogen atoms. Such a location leads to the increased positive charges of the silicon and hydrogen atoms, with a reduced difference at the hydrogen and silicon atoms [8].

2.5   Energies of Specific Interactions of Alkyldisilanes, Alkyldisilanylamines, and Alkyldisilanamine

For the compounds with Si–Si bonds and with more distant atom energy levels of the fourth main groups elements $M(CH_3)_3$–$M(CH_3)_3$, the three upper levels are the element–element or for the Sn, Ge, and Pb atoms, the metal–metal binding orbitals $4a_{1g}$ and two degenerated M–C–C binding orbitals $4e_g$ and $4e_u$ [20–25]. The results of nonempirical calculations of $Sn_2(CH_3)_6$ [25] point to the fact that the contribution of the orbitals of the tin atoms to the B3MO is equal to more than 60% from the s, p, and d orbitals and 20 and 24% from the two others, respectively. In the same tin molecule, the 5s orbital makes the main contribution to the $2a_{1g}$ molecular orbital. It seems that the important conclusion made by Mochida et al. [21] on the basis of the linear dependence between the vertical ionization potential of Iv M–C I_V M–C at $M(CH_3)_4$ on the weak interaction of M–C and M–M, causes the formation in silanes and alkylsilanes little different from the energy value of the hydrogen bond of D–Si–H $\rightarrow$ H–Si (3.11 kJ mol^{-1}) and specific interaction of D–Si–$H_3C \rightarrow$ H–CH_2–C (3.25 kJ mol^{-1}). The observed similarity in the splitting of the Si–Si level in the linear alkylsilanes $CH_3(Si(CH_3)_2)_nCH_3$ ($n = 2$–4) and Si_nH_{2n+2} [22, 26] indicates that the similarity of the formed specific interactions significantly differ in nature from the formed interactions in the alkylsilanes with a tetrahedral molecular structure. This feature is a part of the participation of the silicone pentacoordinated atom in the formation of hydrogen bonds and specific interactions.

The shifting of electron density in the linear molecule of 1,2-dimethyldisilane provides the carbon atom of the ethyl group with the negative charge and the silicone atom in SiH_2 with the positive charge.

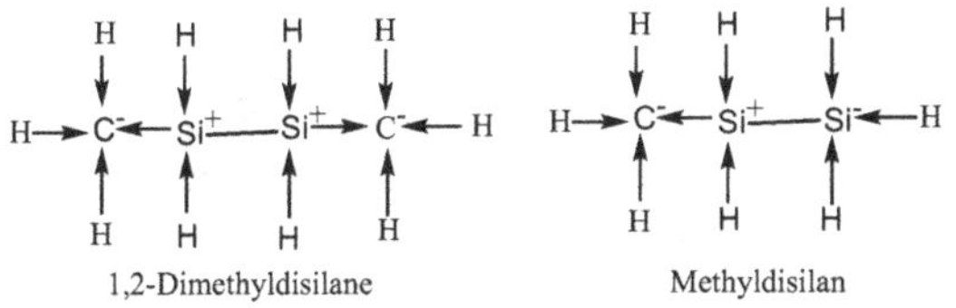

1,2-Dimethyldisilane   Methyldisilan

As a result, the molecule with four bond vacancies of the two silicone atoms SiH_2 and the carbon atoms of the two methyl groups, located in the *trans* position, forms four specific interactions of the same type $DH_3C \rightarrow SiH_2$. The structure of the liquid condition is a network of these bonds (Fig. 2.14a). The energy of the specific interaction is determined from the vaporization enthalpy by dividing it by the four bonds using Eq. (2.14):

$$DH_3C \rightarrow SiH_2 = \Delta_{vap}H^0(T\,K)dmds/4 \qquad (2.14)$$

The energy value of the specific interaction (Table 2.12) was obtained at 258 K and, consequently, is overstated in comparison with the standard conditions. It is about

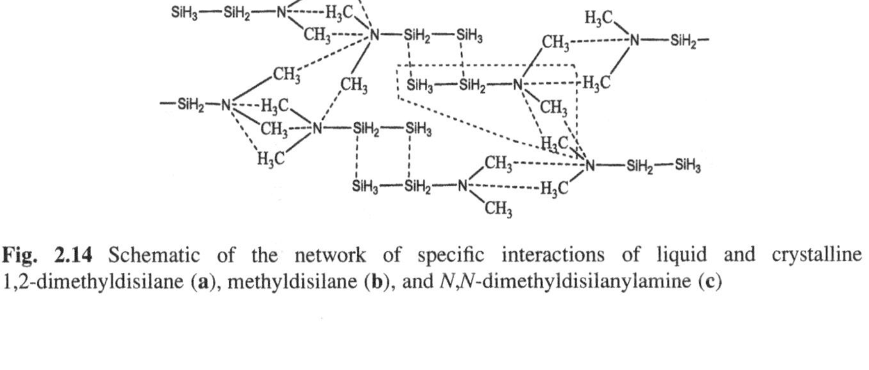

Fig. 2.14 Schematic of the network of specific interactions of liquid and crystalline 1,2-dimethyldisilane (**a**), methyldisilane (**b**), and *N,N*-dimethyldisilanylamine (**c**)

Table 2.12 Energies (kJ mol^{-1}) of specific interactions of liquid disilanes

Compounds	Formula	Structure	$\Delta_{vap}H^0$ (298 K) [6]	T (K)	DSi–CH$_3$ → Si	DSi•••H–Si	DisoCH$_3$ → H–CH$_2$
1,2-Dimethyldisilane	C$_2$H$_{10}$Si$_2$		25.4 / 23.0^a	258	6.35	–	–
Methyldisilane	CH$_8$Si$_2$	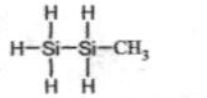	26.8	231	6.35	7.05	–
Hexamethyldisilane	C$_6$H$_{18}$Si$_2$	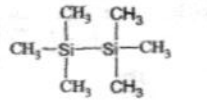	37.4 ± 0.4 / 37.2 / 36.8	– / 288– / 310	6.35	–	1.4
N,N-Dimethyldisilanylamine	C$_2$H$_{11}$NSi$_2$		35.4	240	DN → CH$_3$–N 4.25^a	DSiH$_2$–SiH$_3$ → SiH$_2$–SiH$_3$ 9.2	–
N,1,1,1-Tetramethyldisilaneamine	C$_4$H$_{13}$NSi$_2$	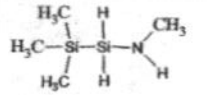	37.4 ± 0.8	298	4.25^a DN•••H–N = 3.86	DH$_2$Si–SiCH$_3$ → SiH$_2$–SiCH$_3$ = 7.8	1.4
Tetrasilane	H$_{10}$Si$_4$	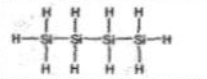	35.6	–	–	8.9	–

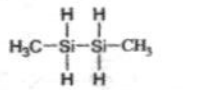

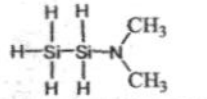

^{a}Correspond to 298 K

twofold greater than the energy of the DSi–H$_3$C $\rightarrow$ H–CH$_2$– (3.25 kJ mol^{-1}) inter-action of liquid tetramethylsilane and trimethylsilane with a tetrahedral molecule structure, which reflects the manifestation of the silicon pentacoordinated atom. It is the coordination of the silicon atom that provides the manifestation of the signif-icantly undivided 3s^2 electron pair in the stabilization of the formed specific interaction that always appears at the reduction of the coordination number of the element atom.

The shifting of the electron density in the molecule of methyldisilane with one methyl group provides the difference in the sign of the methyl group charge and that of the silicon atom of the contacting SiH$_2$ group. The shifting of the electron density from the hydrogen atoms of SiH$_3$ provides the negative charge to the silicon atom, forming a hydrogen bond with the hydrogen atom located in the *trans* position. Thus, the molecule of methyldisilane with four bond vacancies forms two specific interactions DH$_3$C $\rightarrow$ SiH$_2$ and two hydrogen bonds Si•••H–Si, forming a network of the specific interactions in the liquid condition (Fig. 2.14b). The energy of the hydrogen bond can be obtained from the difference between the vaporization enthalpy and the energies of the two specific interactions DH$_3$C $\rightarrow$ SiH$_2$, using the energies of the same bonds of liquid 1,2-dimethyldisilane, with the help of Eq. (2.14a):

$$DSi \bullet \bullet \bullet H - Si = \Delta_{vap}H^0(T\,K) - 2DH_3C \rightarrow SiH_2 \qquad (2.14a)$$

The unaccounted temperature dependence of the energy of the specific interactions of liquid 1,2-dimethyldisilane, obtained at 258 K, exceeding by27°C the measuring temperature of the vaporization enthalpy of liquid methyldisilane, overestimates the obtained energy value of the hydrogen bond DSi•••H–Si of the compound (7.05 kJ mol^{-1}). Nevertheless, one should note that the given energy value of the hydrogen bond is approximately twice as high as the energy of the same type of bond DSi–H $\rightarrow$ H–Si (3.11 kJ mol^{-1}) of liquid trimethylsilane with a tetrahedral molecular structure. It is important to note the influence of the significantly undivided 3s^2 electron pair of the pentacoordinated silicon atom on the shifting of the electron density and on the stabilization of the hydrogen bond.

The molecule of hexamethyldisilane with two methyl groups in the *trans* position forms, like 1,2-dimethyldisilane, four specific interactions DSi–CH$_3$ $\rightarrow$ Si and, additionally, by the four isostructural methyl groups, eight specific interactions DisoCH$_3$ $\rightarrow$ H–CH$_2$ of reduced stability. Taking the energy of the specific interac-tion DH3C $\rightarrow$ SiH$_2$ as being equal to that obtained in the liquid 1,2-dimethyldisilane (6.35 kJ mol^{-1}), the energy of the second type of interaction DisoCH$_3$ $\rightarrow$ H–CH$_2$ with a reduced value can be obtained from Eq. (2.14b):

$$DisoCH_3 \rightarrow H - CH_2 = \left(\Delta_{vap}H^0(T\ K)xmds - 4DH_3C \rightarrow SiH_2\right)/8 \qquad (2.14b)$$

However, the obtained value of 1.4 kJ mol^{-1} differs little from the contribution to the vaporization enthalpy of the isostructural ethyl group (1.8 kJ mol^{-1}) of triethylalkylsilanes.

The available vaporization enthalpies of N,N-dimethyldisilanylamine and $N,1,1,1$-tetramethyldisilaneamine allow the implementation of a thermodynamic analysis and information to be obtained on the interaction specificity in its liquid condition. The molecule of N,N-dimethyldisilanylamine has similar features to ethylmethylamine, where the disilane fragment and ethyl group act as the particle, forming the specific intermolecular interactions $DH_2Si–H_3Si \rightarrow SiH_2–SiH_3$ and $DH_2C–H_3C \rightarrow CH_2–CH_3$, caused by the shifting of electron density.

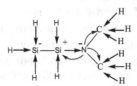

An active enough acceptor of the silicon atom electrons shifts the electron density on the chain from the silicon atom and from the methyl groups providing the hydrogen atoms. As a result, a negative charge is created at the silicon atom of the SiH_3 group and a positive charge is created at SiH_2. The electron density obtained by the silicon atom and located at the $3p_z$ orbital is partly transferred to the contacting silicon atoms and carbon atoms, reducing its negative charge and the positive charge of the silicon atom and, on the other hand, changing the charge of the carbon atoms. Thus, the stability of the formed specific interaction of $DH_2Si–H_3Si \rightarrow SiH_2–SiH_3$ is determined by the difference of charges at the silicon atoms and the ability by the hydrogen atoms and SiH_3 group with $3s^2(Si)$ electron pair.

Therefore, the molecule of N,N-dimethyldisilanylamine with bond vacancies forms two specific interactions $DH_2Si–H_3Si \rightarrow SiH_2–SiH_3$ and four specific interactions of the methyl groups with the nitrogen atom $DN \rightarrow CH_3–N$ with an energy of 4.25 kJ mol^{-1}, forming a network of interactions in the liquid condition of the compound (Fig. 2.14c). Taking into account the correlation of the vaporization enthalpy with the number and energies of the bursting interactions, the energy value of the specific interaction is determined from Eq. (2.14c):

$$DH_2Si - H_3Si \rightarrow SiH_2 - SiH_3$$

$$= \left(\Delta_{vap}H^0(T\,K) - 4DN \rightarrow CH_3 - N\right)/2 \qquad (2.14c)$$

The molecule of $N,1,1,1$-tetramethyldisilaneamine with a methyl group and hydrogen atom of the amino group and three methyl groups at the silicon atom (two of which act as functional isostructural groups) form specific interactions of low stability with the energy not exceeding 1.4 kJ mol^{-1} and shifting the insignificant electron density to acquire a low positive charge.

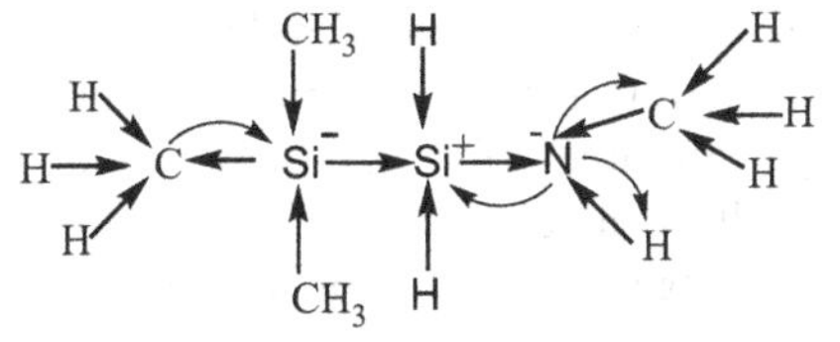

The shifting of electron density on the chain of silicon atoms to the nitrogen atom reduces the negative charge of the carbon atom located in the *trans* position. Partial return of the electron density from the carbon atom as the reverse dative bond to the contacting silicon atom provides the low negative charge and allows the formation of the specific interaction, with reduced stability in comparison with a similar interaction of the liquid N,N-dimethyldisilanylamine. Thus, the molecule of $N,1,1,1$-tetramethyldisilaneamine forms the network of specific interactions in the liquid by two specific interactions $DSiH_2$–CH_3–$Si \rightarrow SiH_2$–Si–CH_3, two $DN \rightarrow CH_3$–N (4.25), two hydrogen bonds $DN \bullet \bullet \bullet H$–$N$ (3.86), and four specific interactions of low stability $DisoH_3C \rightarrow H$–CH_2 (1.4 kJ mol^{-1}) of the two isostructural methyl groups. Taking the obtained interactions of liquid $N,1,1,1$-tetramethyldisilaneamine, the energy value of the specific interaction $DSiH_2$–$CH_3Si \rightarrow SiH_2$–$SiCH_3$ can be calculated using Eq. (2.14d):

$$DSiH_2 - CH_3 - Si \rightarrow SiH_2 - SiCH_3 = \left(\Delta_{vap}H^0(T\,K) - 2DN \rightarrow CH_3 - N \right.$$
$$\left. -2DN \bullet \bullet \bullet H - N - 4DisoH_3C \rightarrow H - CH_2\right)_2 \quad (2.14d)$$

The results of the implemented energy calculations (Table 2.12) of the specific interactions indicate their stabilization in the series of compounds:

$DSiH_2$–$CH_3Si \rightarrow SiH_2$–$SiCH_3$ (7.8) $N,1,1,1$-tetramethyldisilaneamine $< DSiH_2$-$SiH_3 \rightarrow SiH_2$–SiH_3 (9.2, 240 K) N,N-dimethyldisilanylamine $< DSiH_2$–$SiH_3 \rightarrow SiH_2$-SiH_3 (9.2 kJ mol^{-1}, 298 K) tetrasilane,

This implies that the replacement of the hydrogen atoms of the end group by methyl groups leads to destabilization of the specific interaction and that the amino group in N,N-dimethyldisilanylamine has no significant destabilizing effect.

2.6 Specific Interactions of Silacyclobutanes and 1,1-Dimethylsilacyclopentane

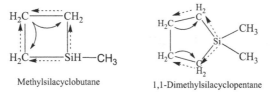

Methylsilacyclobutane   1,1-Dimethylsilacyclopentane

The compound molecules with a heteroatom in the cyclic structure are character-ized by the shifting of electron density with manifestation of the reverse dative bond influence, similar to the linear forms, ending at the sixth methylene group in the cycle [8]. The result of the electron density distribution in the molecule is the increased negative charge of the carbon atoms of the second and fourth methylene groups of the methylsilacyclobutane cycle and similar groups of 1,1-dimethylsilacyclopentane. The planar structure of the cyclic molecule with the heteroatom replaces the properties of the tetrahedral structure of silanes with the properties of the specific interactions formed with the pentacoordinated silicon atom. In turn, this silicon condition causes the different behavior of the methylene groups replacing the hydrogen atoms of the SiH_2 group. The replacement of one hydrogen atom in the methyl group provides it with the functional properties of the isostructural group, with the formation of a specific interaction $-H_2C\diagdown^{H\cdots}_{\diagup H}H_2C-$ of low stability. The replacement of the hydrogen atom located in the *trans* position to the methyl group provides the specific interaction $DSi-H_3C \rightarrow Si$ with the pentacoordinated silicon atom. As a result, the vaporization enthalpy of 1,1-dimethylsilacyclobutane exceeds the value of the given property of methylsi-lacyclobutane by 7.9 kJ mol^{-1}.

Obtained from the difference between the vaporization enthalpies of 1,1-dimethylsilacyclobutane and 1,1,3-trimethylsilacyclobutane, the contribution of the –H–C–H isostructural methyl group is equal to 2.2 kJ mol^{-1}, or 1.1 kJ mol^{-1} for one formed bond. From the equality of the vaporization enthalpies of 1,1-dimethylsilacyclobutane, with two methyl groups at the silicon atom, and 1,2-dimethylsilacyclobutane (Table 2.13), with one group at the silicon atom and the second group at C(1), one can make two conclusions: first, the methyl groups located at the silicon atom form one specific interaction $DSi-H_3C \rightarrow Si$ with equal energy contributions and, second, the second methyl group at the silicon atom in the molecule of 1,1-dimethylsilacyclobutane and at the carbon atom C(1) in the molecule of 1,2-dimethylsilacyclobutane expresses the properties of the isostructural methyl groups and forms specific interactions of low stability that make equal energy contributions (1.5 kJ mol^{-1}) to the vaporization enthalpy. The molecule of methylsilacyclobutane with five bond vacancies forms two specific interactions $DH_3C \rightarrow H–CH_2$ with low energy (1.5 kJ mol^{-1}) and three interactions $DH_2C \rightarrow CH_2$. The molecule of 1,1-dimethylsilacyclobutane additionally forms, by the second methyl group, two specific interactions with the pentacoordinated silicon atom $DSi-H_3C \rightarrow Si$ with a total energy contribution of 7.9 kJ mol^{-1} to the enthalpy characteristics. The energy of the specific interactions $DH_2C \rightarrow CH_2$ of these compounds are determined from Eqs. (2.15) and (2.15a):

Methylsilacyclobutane :

$$DH_2C \rightarrow CH_2 = \left(\Delta_{vap}H^0(TK) - 2DH_3C \rightarrow H - CH_2\right)/3 \qquad (2.15)$$

Table 2.13 Energies of specific interactions of liquid silacyclobutanes and 1,1-dimethylsilacyclopentanes

Compounds	Formula	Structure	$\Delta_{vap}H^0$ (298 K) [6]	$DH_3C \rightarrow H-CH_2$	$DSi-H_3C \rightarrow Si$	$DH_2C \rightarrow CH_2$
1-Methylsilacyclobutane	$C_4H_{10}Si$		25.1	1.5	–	7.4
1,1-Dimethylsilacyclobutane	$C_5H_{12}Si$		33.0 ± 0.8	1.5	3.95	7.4
1,2-Dimethylsilacyclobutane	$C_5H_{12}Si$		33.1	1.5	3.95	7.4
1-Methyl-1-vinylsilacyclobutane	$C_6H_{12}Si$		33.1	$DCH_2 \rightarrow H-C-CH_2 = 1.55$	3.95	7.4
1,1,2-Trimethylsilacyclobutane	$C_6H_{14}Si$		36.0	1.5 C(1)1.5	3.95	7.4
1,1,3-Trimethylsilacyclobutane	$C_6H_{14}Si$		35.3	1.5 C(2)1.1	3.95	7.4
1,1,3,3-Tetramethyl-1,3-disilacyclobutane	$C_6H_{16}Si_2$		36.7 ± 1.1 41.0 ± 2.1	1.5 C(1)1.1; C(3) 0.7	3.95	7.4
1,1-Dimethylsilacyclo pentane	$C_6H_{14}Si$		37.7 ± 2.1	1.5	3.95	5.36

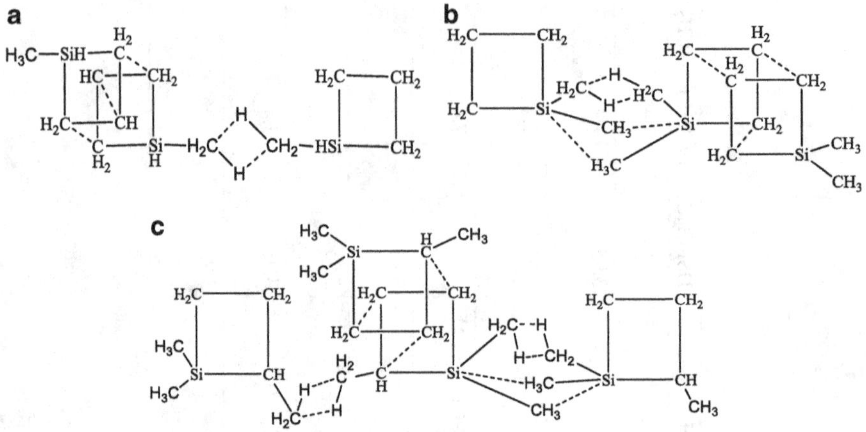

Fig. 2.15 Schematic of the network of specific interactions of liquid and crystalline methylsilacyclobutane (**a**), 1,1-dimethylsilacyclobutane (**b**), and 1,1,2-trimethylsilacyclobutane (**c**)

1, 1-Dimethylsilacyclobutane :

$$DH_2C \rightarrow CH_2 = \left(\Delta_{vap}H^0(T \text{ K}) - 2DSi - H_3C \rightarrow Si - 2DH_3C \rightarrow H - CH_2\right)/3$$

$$(2.15a)$$

The energies of the specific interactions formed by one isostructural methyl group and the second group, forming with the silicon atom the stable interaction of the molecule of 1,2-dimethylsilacyclobutane and 1-methyl-1-vinylsilacyclobutane, are obtained with the help of Eq. (2.15a). The energies of the specific interactions formed by the isostructural methyl groups are determined from the differences in the vaporization enthalpies of the corresponding two differing silacyclobutanes (Table 2.13). An increasing number of isostructural methyl groups in the molecules complicates the network structure of the liquid condition of the compound (Fig. 2.15c). The energies of the formed stable specific interactions of 1,1,2-trimethylsilacyclobutane and 1,1,3-trimethylsilacyclobutane are calculated using Eq. (2.15b) and those of 1,3,3-tetramethyl-1,3-disilacyclobutane using Eq. (2.15c):

$$DH_2C \rightarrow CH_2$$
$$= \left(\Delta_{vap}H^0(T\text{K}) - 2DSi - H_3C \rightarrow Si - 4DH_3C \rightarrow H - CH_2\right)/3 \quad (2.15b)$$
$$DH_2C \rightarrow CH_2$$
$$= \left(\Delta_{vap}H^0(T\text{K}) - 2DSi - H_3C \rightarrow Si - 6DH_3C \rightarrow H - CH_2\right)/3 \quad (2.15c)$$

From the given calculation results (Table 2.13), it is clear that the energies of the specific interactions $DH_3C \rightarrow SiH$ with the pentacoordinated silicon atom and $DH_2C \rightarrow CH_2$ of the methylene groups of the cycle remain unstable for all the considered compounds and the differences are expressed by the interaction energies

of the isostructural methyl groups. The high stability of the formed specific interactions of the methylene groups is caused by the distribution of electron density
under the influence of silicon heteroatom. By taking the energies of the specific
interactions formed by the methyl groups of liquid 1,1-dimethylsilacyclopentane to
be equal to those obtained for liquid 1,1-dimethylsilacyclobutane, equal to 1.5 and
3.95 kJ mol^{-1}, we estimated the energy value of $DH_2C \rightarrow CH_2$ formed by each of
the four methylene groups (Table 2.13).

2.7 Energies of Specific Interactions of Oxygengenated Silanes

2.7.1 Bis(Disilanyl)ether, Trimethylsilanol, and Methoxysilane

According to the photoelectron spectra and quanta chemical calculations [27, 28]
two upper orbitals $2b_1$ of ethers $O(CH_3)_2$ are localized mainly at the oxygen atom.
However, the values of the vertical ionization potential Iv for $n_o(b_1)$ and $n^1{}_o(a_1)$
orbitals are so close that they remain indistinguishable in the photoelectron spectra
of this compound and yield a value for ΔI_{1-2} of 1.9 eV for the dimethyl ether. At the
transition from the dimethyl ether to $O(SiH_3)_2$, the values ΔI_1 and ΔI_2, equal to +1.2
and −0.7 eV, respectively, at the constant energy values of the oxygen bond O 1s
electrons 538.6 eV, correspond to the main influence of the delocalization of the n_o
orbital on the change in the levels' energies. Two pairs of O(SiH)-binding orbitals,
$1a_2$ and $3b_2$ and $3a_1$ and $1b_1$, and o(OSi)-binding $2b_2$ are expressed as being
responsible for the two bands in the photoelectron spectra in the range 12–15 eV.
The photoelectron spectra of the disiloxane derivatives are less reliable [2]. The
assignment of the first two bands at 9.80 and 10.85 eV in the case of $[O(SiCH_3)_3]_2$
and $O(SiH_3)_2$ are adopted similarly [2].

   The molecules of bis(disilanyl)ether with an angled structure (analogous to the
diethyl ether) contact in the liquid condition and are located on two planes at an
angle of 90°. Its four bond vacancies of the two free electron pairs of the oxygen
atom and the two end SiH_3 groups form four specific interactions $DO \rightarrow SiH_3–SiH_2$
(Fig. 2.16a). Taking into account the general interrelation of the vaporization
enthalpy with the number of bursting intermolecular interactions and adding their
energies [29–31], the energy value of the interactions can be obtained directly from
the vaporization enthalpy (Table 2.14). The inequality between the energies of the
specific interactions of bis(disilanyl)ether and diethyl ether $DO \rightarrow SiH_3–SiH_2$
$(9.1) < DO \rightarrow CH_3–CH_2–$ (6.78 kJ mol^{-1}) [8] indicates its increased stability in
the compound with the disilanyl groups. It highlights the increased shifting of the
electron density from the silicon atoms to the oxygen atom in the molecule of bis
(disilanyl)ether, providing the silicon atom with increased positive charge and an
increased difference between the charges of the silicon atom of the end SiH_3 group

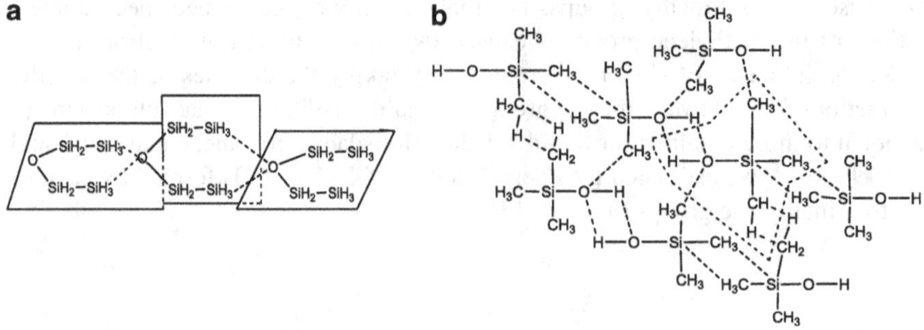

Fig. 2.16 Schematic of the network of specific interactions of liquid and crystalline bis(disilanyl) ether (a) and trimethylsilanol (b)

and the oxygen atom compared with that in the molecule of diethyl ether. Note that the remote energy level of the silicon atom allows it to receive the increased electron density from the oxygen atom, which results in the stabilization of the specific interaction [2].

A distinctive feature of the molecules of trimethylsilanol and methoxysilane as compared with bis(disilanyl)ether is the formation by the silicon atom of a tetrahedral structure where all the methyl groups or oxygen atoms, jointly with the oxygen atom of the hydroxyl group or the corresponding methoxy group, are at the vertices of the tetrahedron. The oxygen atom shifting the electron density from the silicon atom provides it with the positive charge and acceptor properties at the formation of the specific intermolecular interaction DSi–H$_3$C → Si with the carbon atom of the methyl group located in the *trans* position relative to the hydroxyl group (as occurs for methyl alcohol) [9] (Fig. 2.16b). The energy of the formed specific interaction should have an increased stability in comparison with the energy contribution of a similar group in the monosilanes and whose values can be accepted as being equivalent to the energy value of the same type of interaction DSi–H$_3$C → Si (4.35 kJ mol^{-1}), realized in the liquid methyltrimethoxysilane. The remaining two methyl groups implement the functions of the isostructural groups. One group forms two specific interactions DisoH$_3$C → H–CH$_2$ of reduced stability with an energy of 1.5 kJ mol^{-1} (Table 2.13). The second group participates in the formation of the specific interaction DSi–O → CH$_3$–Si of reduced stability with the remaining free electron pair of the oxygen atom of the same hydroxyl group, with an energy value not exceeding the energy of the hydrogen bond (3.1 kJ mol^{-1}) formed by the silane hydrogen atom (Fig. 2.14b). Taking the six formed specific interactions of three types, the energies of the two hydrogen bonds of the same type can be determined from Eq. (2.16):

$$\text{DSi} - \text{O} \bullet \bullet \bullet \text{H} - \text{O} - \text{Si} = (\Delta_{vap}H^0(T\,\text{K})\text{tsol} - 2\text{DSi} - \text{O} \rightarrow \text{CH}_3 - \text{Si} - 2\text{DSi}$$
$$-\text{H}_3\text{C} \rightarrow \text{Si} - - 2\text{DisoH}_3\text{C} \rightarrow \text{H} - \text{CH}_2)/2 \quad (2.16)$$

Table 2.14 Energies (kJ mol^{-1}) of specific interactions of liquid bis(disilanyl)ether, trimethylsilanol, and methoxysilane

Compounds	Formula	Structure	$\Delta_{vap}H^0$ (T K) [6]	T (K)	DO → SiH$_3$–SiH$_2$	DSi–O•••H–O–Si	DSi–O → CH$_3$–O–
DSi•••H–Si = 3.1							
bis(Disilanyl)ether	H$_{10}$OSi$_4$		36.4	273–363	9.1	–	–
Trimethylsilanol	C$_3$H$_{10}$OSi		45.6 ± 7	298	–	13.85	3.1
Methoxysilane	CH$_3$OSi		25.8	201	–	8.2	–

The calculated energy value of the hydrogen bond (Table 2.14), with a large error in the experimentally measured vaporization enthalpy of trimethylsilanol, was comparable with the naturally increased energy value of a similar bond of the liquid methyl alcohol (13.08 kJ mol^{-1}). This value exceed the energy value of hydrogen bond of the liquid water (10.99 kJ mol^{-1}) [8].

In the molecule of methoxysilane, with the oxygen atom possessing high electron acceptor properties, there is shifting of electron density from the silicon atom, providing it with a positive charge in comparison with the silane molecule. This causes the pentacoordinated condition of the silicon atom in the formation of hydrogen bonds, with the hydrogen atom located in the *trans* condition to the methoxy group. As a result, the molecule of methoxysilane forms two hydrogen bonds DSi•••H–Si, whose energies should have increased values in comparison with the energy of bonds formed by each of the two remaining hydrogen atoms. However, it is not possible to determine the energy values. Thus, in order to estimate the energy value of the formed specific interaction DSi–O → CH$_3$–O, the average energy values of hydrogen bonds formed in the liquid silane are used. Therefore, the molecule of methoxysilane forms two hydrogen bonds DSi•••H–Si by the hydrogen atom with the location in the *trans* condition, with energies of 3.1 kJ mol^{-1} and with the same energy value as the other two bonds DSi–H•••H–Si. The energy value of the specific interaction DSi–O → CH$_3$–O is is given in Table 2.14,as calculated using Eq. (2.16a):

$$DSi - O \rightarrow CH_3 - Si$$
$$= \left(\Delta_{vap}H^0(TK)mosi - 2DSi \bullet \bullet \bullet H - Si. - 2DSi - H \bullet \bullet \bullet H - Si \right)/2$$
$$(2.16a)$$

2.7.2 Methoxy-, Ethoxy-, Propoxy-, and Butoxysilanes and Silylacetylene

The compounds with four methoxy, ethoxy, propoxy, or butoxy groups in silanes with a tetrahedral molecular structure, with the equivalent shifting of electron density from the silicon atom, form specific interactions that are not very dependent on the length of the alkyl chain because the shifting of electron density takes place from a single silicon atom. The absence of the other ligands leaves the second free electron pair of the oxygen atom indifferent to the formation of the specific interaction. Therefore, the molecules of these compounds form eight specific interactions DSi–O → CH$_3$–O–Si or DSi–O → R–O–Si, forming a network of specific interactions in the condensed condition (Fig. 2.17). The energy values of these interactions are determined from the vaporization enthalpy by the depletion at the eight specific interactions. From the presented data (Table 2.15), it follows that the energies of the specific interactions are stabilized by an increasing number of carbon atoms in the alkyl chain, as described by the straight dependence DSi–O → R–O–Si = f(Cn):

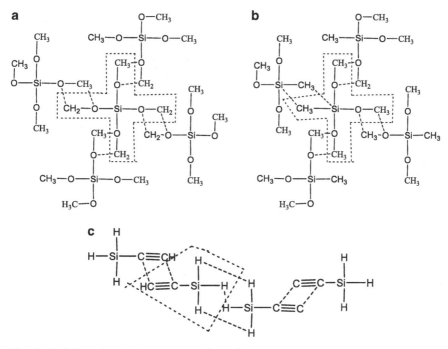

Fig. 2.17 Schematic of the network of specific interactions of liquid and crystalline tetramethoxysilane (**a**), methyltrimethoxysilane (**b**), and silylacetylene (**c**)

DSi–O → R–O–Si: Tetraethoxysilane   (6.06) < Tetrapropoxysilane   (6.22) < Tetrabutoxysilane (6.5 kJ mol^{-1})

This dependence allows the use of the method of extrapolation to clarify the energy of the specific interaction of the liquid tetramethoxysilane (5.82 kJ mol^{-1}) and to obtain a reliable value for the vaporization enthalpy, given in the literature without errors and as equal to the enthalpy characteristic of tetraethoxysilane (Table 2.14).

The molecules of methyltrimetoxysilane and triethoxymethylsilane with one methyl group in the *trans* position suppose the stabilization of the formed specific interaction DSi–H$_3$C → Si with participation of the pentacoordinated silicon and the carbon atom (Fig. 2.15b). The shifting of electron density from the silicon atom to the three oxygen atoms and to the single carbon atom does not practically change the charges of the oxygen atoms and, thus, the energies of the formed specific interactions remain unchanged. This allows the determination of the energies of the specific interactions using Eq. (2.16a), which is converted to the general Eq. (2.16b):

Table 2.15 Energies (kJ mol⁻¹) of specific interactions of methoxy-, ethoxy-, propoxy-, butoxysilanes, and vinylalkyloxysilanes at 298 K

Compounds	Formula	Structure	$\Delta_{vap}H^0$ (298 K) [6]	DSi–O → CH₃–(CH₂)ₙ–O–Si	DSi–H₃C → Si
DSi–H₃C → H–CH₂–Si = 3.25					
Tetramethoxysilane	$C_4H_{12}O_4Si$		48F.4 / 46.5ᵃ	5.82ᵃ n = 0	—
Methyltrimethoxysilane	$C_4H_{12}O_3Si$	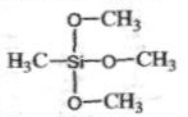	34.3 ± 0.3 / 43.5ᵃ	5.82ᵃ n = 0	4.35
Tetraethoxysilane	$C_8H_{20}O_4Si$		48.5 ± 0.3	6.06 n = 1	—
Triethoxymethylsilane	$C_8H_{18}O_3Si$	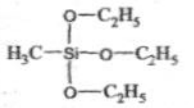	45.1 ± 0.1	6.06 n = 1	4.35
Diethoxydimethylsilane	$C_6H_{16}O_2Si$	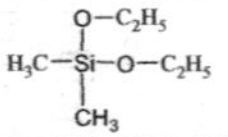	43.1 ± 0.3	7.0 n = 1	4.35
Ethoxytrimethylsilane	$C_5H_{14}OSi$	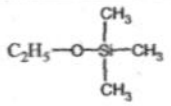	38.4 ± 0.3	8.35 n = 1	4.35
Propoxytrimethylsilane	$C_6H_{16}OSi$	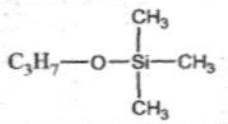	34.3	6.15 n = 2	4.35
Dipropyldiethoxysilane	$C_{10}H_{24}O_2Si$		46.4 ± 0.3	6.22 n = 1	DSi–(CH₂)₂–H₃C → Si 4.55
Tetrapropoxysilane	$C_{12}H_{28}O_4Si$		49.8 ± 0.4	6.22 n = 2	—
Tetraisopropoxysilane	$C_{12}H_{28}O_4Si$	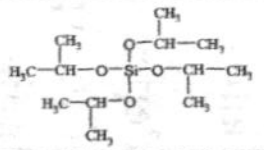	52.7	6.06 n = 1	∑DisoCH₃ = 4.2

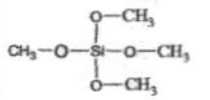

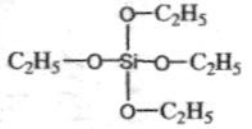

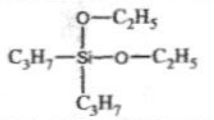

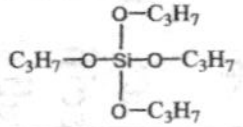

				6.5	—
				$n = 2$	
Tetrabutoxysilane	$C_{16}H_{38}O_4Si$	$C_4H_9-O-Si(O-C_4H_9)(O-C_4H_9)-O-C_4H_9$	52.0 ± 1.0		
Tetravinylsilane	$C_8H_{12}Si$	$H_2C=HC-Si(CH=CH_2)(CH=CH_2)-CH=CH_2$	42.7 ± 0.7	—	5.34
Vinyltriethoxysilane	$C_8H_{18}O_3Si$	$H_2C=HC-Si(O-C_2H_5)(O-C_2H_5)-O-C_2H_5$	50.2 ± 0.4	6.6	5.34
Vinyltripropoxysilane	$C_{11}H_{24}O_3Si$	$C_3H_7-O-Si(O-C_3H_7)(CH=CH_2)-O-C_3H_7$	52.3 ± 0.3	6.93	5.34
Silylacetylene	C_2H_4Si	$H-Si(H)(H)-C\equiv CH$	22.0	$DSi-H\bullet\bullet\bullet H-Si = 3.11$	$D\equiv(H-)C \rightarrow C\equiv(Si-)6.35$

[a]More accurate

$$DSi - H_3C \rightarrow Si = \left(\Delta_{vap}H^0(TK) - 6DSi - O \rightarrow CH_3 - O - Si\right)/2 \quad (2.16b)$$

$$DSi - H_3C \rightarrow Si$$

$$= \left(\Delta_{vap}H^0(TK) - nDSi - O \rightarrow R - O - Si - mDisoH_3C \rightarrow H - CH_2\right)/q$$

$$(2.16c)$$

where n, m, and q are the number of specific interactions DSi–O$\rightarrow$R–O–Si, DisoH$_3$C$\rightarrow$H–CH$_2$, and DSi–H$_3$C$\rightarrow$Si, respectively. The energy of the specific interaction of dipropyldiethoxysilane is determined from Eq. (2.16c):

$$DSi - CH_2 - CH_2 - H_3C \rightarrow Si$$

$$= \left(\Delta_{vap}H^0(TK)dpdet - 4DSi - O \rightarrow CH_3 - CH_2 - O - Si\right)/4$$

$$(2.16d)$$

The results of the implemented calculations are given in Table 2.14. The data show that, for the compounds with a constant energy value of the specific interaction DSi–H$_3$C$\rightarrow$Si (4.35 kJ mol^{-1}), the energy value of the second type of specific interaction starts to increase significantly in the compounds with two and three hydrogen atoms connected tpo the silicon atom:

DSi–O $\rightarrow$ CH$_3$–CH$_2$–O–Si: Tetraethoxysilane   (6.06) = Triethoxymethylsilane (6.06) < Diethoxydimethysilane (7.0) < Ethoxytrimethylsilane (8.35 kJ mol^{-1})

This increase can be attributed to the manifestation of the influence of the *trans* position of the hydrogen atoms on the specific interaction. The increased stabilization of the specific interaction is expressed at the methyl group located in the *trans* position of ethoxytrimethylsilane. The compounds of diethoxydimethysilane (4.35 kJ mol^{-1}) and dipropyldiethoxysilane (4.55 kJ mol^{-1}) illustrate the stabilization tendency of the specific interaction DSi–H$_3$C $\rightarrow$ Si < DSi-(CH$_2$)$_2$–H$_3$C $\rightarrow$ Si with an increasing number of carbon atoms in the alkyl chain.

Using the equality of the energy contribution to the vaporization enthalpy of the ethyl ligands of tetraisopropoxysilane and tetraethoxysilane, the energy contribution of the isostructural methyl groups is obtained from the difference in the enthalpy characteristics of the compounds as being equal to 4.2 kJ mol^{-1} and that of the single specific interaction of low stability as 0.52 kJ mol^{-1}. The compounds with the vinyl ligand of vinyltriethoxysilane illustrate the energy increase of the formed specific interaction DSi–O $\rightarrow$ CH$_2$=CH–O–Si caused by the influence of the unsaturated ligand located in the *trans* position on the shifting of the electron density:

DSi–O $\rightarrow$ CH$_3$–CH$_2$–O–Si: Triethoxymethylsilane (6.06) < DSi–O $\rightarrow$ CH$_2$=CH–O–Si

Vinyltriethoxysilane (6.6 kJ mol^{-1}) < DSi–O $\rightarrow$ CH$_2$=CH–O–Si Vinyltripropoxysilane (6.98 kJ mol^{-1})

Inequality of the energies of the specific interactions of the two compounds with one vinyl ligand reflects the stabilizing influence of one type of specific interaction with increasing of number carbon atom at alkyl chain of propoxy fragment.

Silyl acetylene molecules with the three functional hydrogen atoms of the silyl group and two bond vacancies of the acetylene group –C≡CH form in the liquid condition a network of three hydrogen bonds DSi–H•••H–Si with an energy of 3.11 kJ mol^{-1} and two specific interactions D≡(H–)C → C≡(Si–) (Fig. 2.17c), whose energy value can be determined using Eq. (2.16d), taking into account the energy contributions of the realizing interactions:

$$D \equiv (H-)C \rightarrow C \equiv (Si-)$$
$$= \left(\Delta_{vap}H^0(T\,K)sac - 3\,DSi - H \bullet\bullet\bullet H - Si\right)/2 \qquad (2.16e)$$

The calculated energy value of the specific interaction (Table 2.15) D≡(H–)C → C≡(Si–) (6.35 kJ mol^{-1}) of silyl acetylene differs little from the energy value of the same type of interaction of the liquid 1-propyne D≡(H–)C → C≡ (6.25 kJ mol^{-1}), remaining unchanged for the whole of series of compounds [8].

2.7.3 Phenoxytrimethylsilanes and (Methoxyphenyl) trimethylsilanes

From the data in Table 2.16 for (methoxyphenyl)trimethylsilanes, it is clear that the vaporization enthalpies of these compounds, related to the number and energy of the bursting specific interactions, are characterized by the reduction in the value of the property. The reduction in the enthalpy characteristic in the sequence of the methoxy group location at the carbon atoms of C(2), C(3), and C(4) and according to the permanent number and types of the formed specific interactions indicates the reduction of the energy contribution of the DSi–O → CH$_3$–O–Si interaction. From the schematic presented in Fig. 2.18a of the network structure of the specific interactions of liquid (2-methoxylphenyl)trimethylsilane, it is evident that the trimethylsilane fragment forms two specific interactions DSi–CH$_3$ → SI (3.95 kJ mol^{-1}) by the methyl group located in the *trans* position and four specific interaction of low stability (1.5 kJ mol^{-1}) by the two isostructural methyl groups. The energy contribution of the benzene ring to the vaporization enthalpy of the compounds corresponds to the difference in the vaporization enthalpies (33. 8 kJ mol^{-1}) minus the energy contribution of the two saturated hydrogen atoms (0.6 × 2 kJ mol^{-1}) of the benzene ring [8]. Therefore, the energy of the specific interaction DSi–O → CH$_3$–O–Si can be calculated using Eq. (2.17):

$$DSi - O \rightarrow CH_3 - O - Si = \left(\Delta_{vap}H^0(T\,K)mphsi - \left(\Delta_{vap}H^0(T\,K)bz - 2DH\right)\right.$$
$$\left. - 2DSi - H_3C \rightarrow Si - 2DisoCH_3\right)/2 \qquad (2.17)$$

Table 2.16 Energies (kJ mol^{-1}) of specific interactions of phenoxytrimethylsilanes and (methoxyphenyl)trimethylsilanes at 298 K

Compounds	Formula	Structure	$\Delta_{vap}H^0$ (298 K) [6]	$\sum$DisoCH$_3$	DSi-H$_3$C→Si	DSi-O→CH$_3$-O-Si
($\Delta_{vap}H^0$(T K)bz - 2DH) = 32.6; DHC→CH = 5.63; DHC→C = 5.30						
(2-Methoxylphenyl)trimethylsilane	C$_{10}$H$_{16}$OSi		59.4 ± 0.8	1.5 × 2	3.95	7.65 8.0^a
(3-Methoxylphenyl)trimethylsilane	C$_{10}$H$_{16}$OSi		56.1 ± 0.8	1.5 × 2	3.95	6.0 6.35^a
(4-Methoxylphenyl)trimethylsilane	C$_{10}$H$_{16}$OSi		56.9 ± 0.8	1.5 × 2	3.95	6.4 6.75^a
Triethoxyphenylsilane	C$_{12}$H$_{20}$O$_3$Si		58.3 ± 0.09	–	DSi-H••H-Si =3.1 × 3	2.93
Phenoxytrimethylsilane	C$_9$H$_{14}$OSi		56.9 ± 0.8	1.5 × 4	DSi-H$_3$C→Si =3.25	DSi-O→C- = 8.2
Trimethoxy[(phenylthio)methyl] silane	C$_{10}$H$_{16}$O$_3$Ssi		56.4 ± 0.7	DS→CH$_2$ = 2.0	–	DSi-O→CH$_2$-O = 3.2

^{a}Calculations are fulfilled using Eq. (2.17a)

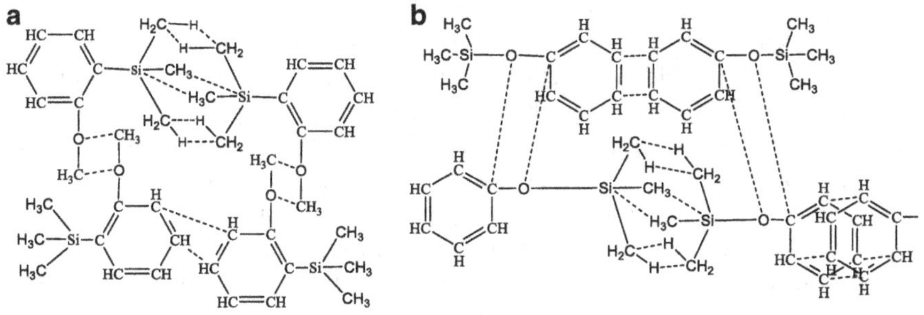

Fig. 2.18 Schematic of the network of specific interactions of liquid and crystalline (2-methoxylphenyl)trimethylsilane (**a**) and phenoxytrimethylsilane (**b**)

The most accurate approach for calculating the energy of specific interactions is based on accounting for the energy of the interactions formed by the CH groups and by the carbon atoms of the benzene ring, between each other DHC → CH, DC → C and DHC → C of the contacting molecules, whose value changes according to the number of saturated hydrogen atoms.

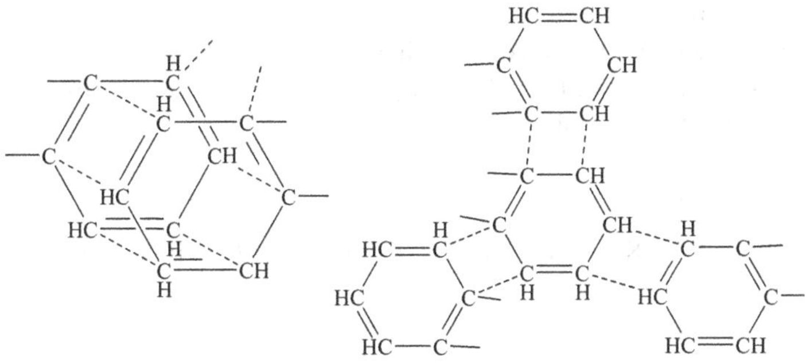

The energy of the first type of specific interaction is determined from the vaporization enthalpy and sublimation enthalpy of benzene and the further interactions, without the energy contribution of the saturated hydrogen atom (0.60 and 1.00 kJ mol^{-1}), correspond to the vaporization process. The energies of these types of specific interactions of liquid benzene are equal to 5.63, 5.00, and 5.30 kJ mol^{-1}, respectively.

The energy calculation of the specific interaction of the formed methoxy group is calculated from Eq. (2.17a), taking into account the number and types of each interaction:

$$DSi - O \rightarrow CH_3 - O - Si = \left(\Delta_{vap}H^0(TK)mphsi - 2DSi - H_3C \rightarrow Si \right.$$
$$\left. -2DisoCH_3 - 2DHC \rightarrow CH - 4DHC \rightarrow C\right)/2 \quad (2.17a)$$

According to the results in Table 2.16, the calculations prove that the stability reduction of this type of specific interaction varies with the location of the methoxy group in the series $C(2) > C(3) \approx C(4)$. This series has the opposite direction in the stabilization of the specific interactions for the other benzene compounds, with the electron acceptor properties of the functional group, in particular, at the C (1) isostructural methyl group, causing stabilization of the hydrogen bond of the hydroxyl group in the series. The silicon atom, shifting part of the electron density to the carbon atoms of the three methyl groups of trimethylsilane, and with the deficit of it, gives the carbon atoms of the benzene ring a very low electron density. As a result, the C(2) carbon atom transfers insignificant electron density to the oxygen atom. It follows that the oxygen receives the increased electron density from the carbon atom of the methyl group of the methoxy fragment, providing the increased difference between the carbon and oxygen atoms and the stability of the formed interaction $DSi-O \rightarrow CH_3-O-Si$. The reduced negative charge of the C (2) carbon atom provides the shifting of the electron density from the C(4) carbon atom of the ring to C(2). Consequently, the location of the trimethylsilane group at C(1) provides the negative charge of the carbon atom, influencing the change in the charges of the carbon atoms of the benzene ring. As a result, the ring carbon atoms have a low ability to shift the electron density, which the oxygen atom receives mainly from the carbon atom of the methyl group of the methoxy fragment. In calculating the energy of specific interactions formed by the ethoxy group of the triethoxyphenylsilane molecule, the energies of the three hydrogen bonds $DSi-H \cdots H-Si$ (3.11 kJ mol^{-1}), formed by the hydrogen atoms of the SiH_3 fragment, are taken into account, as well as the energy contributions of the four saturated hydrogen atoms (0.60 kJ mol^{-1}):

$$DSi - O \rightarrow CH_3 - CH_2 - O - Si$$

$$= \left(\Delta_{vap}H^0(T K)ephsi - \left(\Delta_{vap}H^0(T K)bz - 4DH\right) - 3DSi - H \rightarrow H - Si\right)/2$$

$$(2.17b)$$

The received energy value of the specific interaction, with a more than twofold underestimation of the energy of the $DSi-O \rightarrow CH_3-CH_2-O-Si$ (2.93 kJ mol^{-1}) interaction of the liquid triethoxyphenylsilane, is caused by the strong acceptor properties of the three oxygen atoms with a deficit in the electron density and, thus, having low charge differences with the carbon atoms of the methyl group of the ethyl fragment.

The location of the oxygen atom in the phenoxytrimethylsilane molecule provides the silicon atom with an increased positive charge and the ability to form, with the methyl group, the specific interaction $DSi-CH_3 \rightarrow Si$ (3.95 kJ mol^{-1}) with the essentially undivided $3s^2(Si)$ electron pair , which results in the pentacoordinated condition of the carbon and silicon atoms. Two remaining methyl groups with the properties of the isostructural group form four specific interactions of low stability with an energy of 0.75 or 1.5 kJ mol^{-1} by a single methyl group. The oxygen atom directly connected with the carbon atom of the benzene ring and silicon forms a specific interaction with the carbon atom of the same ring. The remaining five CH groups form five specific interactions of the same type DHC → CH with an energy of 5.63 kJ mol^{-1}. The formed structure, with a network of the specific interactions in the liquid condition (Fig. 2.18b), includes 13 specific interactions of five types, which are used for calculation of the DSi–O → C interaction with unknown energy. The energy value of the specific interaction DSi–O → C is presented in Table 2.16, as obtained using Eq. (2.17c):

$$DSi - O \rightarrow C -$$

$$= \left(\Delta_{vap}H^0(TK)phtmsi - 2DSi - H_3C \rightarrow Si - 2DisoCH_3 - 5DHC \rightarrow CH\right)/2$$

$$(2.17)$$

The location of the methylene group in the trimethoxy[(phenylthio)methyl]silane molecule between the sulfur atom with strong acceptor properties of the electron density and the silicon atom with low positive charge is very important. It was shown in the section titled "Energies of Specific Interactions of Silanamines" that the location of the ethylene group in the molecule of N-(1,1,-dimethylethyl)-1,1,1-triethylsilanamine between dimethylamine (with the strong electron acceptor properties of the nitrogen atoms) and triethylsilane (with low positive charge of the silicon atom) causes the energy of the specific interaction DN→CH$_2$–CH$_2$–Si to reduce to 4.15 kJ mol^{-1} instead of 6.0 kJ mol^{-1} of the DN→CH$_3$–CH$_2$–N interaction. This phenomenon of the destabilizing influence of the oxidant on the specific interaction was revealed for the nitrogenous organic compounds [18]. The location of the methylene group at the oxidant is accompanied by a more than twofold reduction in the energy contribution of the CH$_2$ group to the vaporization enthalpy of the saturated hydrocarbons. According to these data, the energy of the formed specific interaction DS→CH$_2$ of the S–CH$_2$ group is taken to be equal to half of the energy value of DN→CH$_2$–CH$_2$–Si. With the reduced acceptor ability of the nitrogen atom (4.15 kJ mol^{-1}), the energy value of the DS→CH$_2$ of trimethoxy[(phenylthio)methyl]silane can be calculated using Eq. (2.18):

$$DSi - O \rightarrow CH_3 - O - Si$$

$$= \left(\Delta_{vap}H^0(TK)tmphmsi - \left(\Delta_{vap}H^0(TK)bz - DH\right) - 2DS \rightarrow CH_2 - S\right)/6$$

$$(2.18)$$

As presented in Table 2.16, the energy values of the formed specific interaction of the methoxy group of trimethoxy[(phenylthio)methyl]silane points to its low stability, comparable with the energy of the silane hydrogen bond (3.11 kJ mol^{-1}).

2.7.4   Trimethoxy[(methylthio)alkyl)]silanes, [(Ethylthio) alkyl]trimethoxysilanes, and Triethoxy[(ethylthio) alkyl]silane

The revealed phenomenon [18] of the destabilizing influence of the formed alkyl group specific interaction at its location between the element atoms with strong electron acceptor properties underwent its development in the compounds of N-(1,1,-dimethylethyl)-1,1,1-triethylsilanamine   and   trimethoxy[(phenylthio) methyl]silane. Information in this context can be extended to the sulfur-containing silico-organic compounds. From the data in Table 2.17 on the thermodynamic properties of the sulfur-containing silico-organic compounds, it is clear that the difference in the values of the vaporization enthalpies of the first three compounds trimethoxy[(methylthio)methyl]silane, trimethoxy[2-(methylthio)ethyl]silane, and trimethoxy[3-(methylthio)propyl]silane, with similar two first fragments, is determined by the energy contribution of the methyl, ethyl, or propyl ligand. There is no doubt that all the formed specific interactions of the compounds mentioned above are interdependent and that their energies depend on the distribution of electron density inside the molecule. However, the constant value of the low positive charge of the silicon atom has a small influence on the energy value of the formed specific interaction of the methoxy group of these compounds. Thus, its energies should remain practically equal in all the compounds, with a small distinction because of the energy values of the same type of interaction of tetramethoxysilane and methyltrimetoxysilane (5.82 kJ mol^{-1}) (Table 2.15). In turn, the methylene group receives insignificant electron density from the silicon atom because of its shift to the strong electron acceptor oxygen atom of the methoxy group. In this regard, the energy value of the formed interaction of the methoxy group of DSi–O $\rightarrow$ CH$_3$ can be estimated to be equal to the energy value of the same type of interaction of the liquid trimethoxy[(methylthio)methyl]silane (5.82 kJ mol^{-1}). For this reason, the methylene group provides an insignificant shifting of the electron density at the strong acceptor of the sulfur atom and gains a positive charge, the value of which differs insignificantly from the charge of the carbon atom of the methyl group of the S–CH$_3$ fragment. The differences in the energies of the formed specific interactions DSi–O $\rightarrow$ CH$_2$ (2.0 kJ mol^{-1}) and DSi–H$_2$C $\rightarrow$ Si should be insignificant. Therefore, the ten formed specific interactions of three types form a network of these interactions (Fig. 2.19) in the condensed state of trimethoxy[(methylthio) methyl)]silane. The energy of the specific interaction of DS $\rightarrow$ CH$_3$ is determined from Eq. (2.19):

$$DS \rightarrow CH_3 - S$$

$$= \left(\Delta_{vap}H^0(T\,K)\text{mothms} - 6DSi - O \rightarrow CH_3 - 2DS \rightarrow CH_2\right)/2 \quad (2.19)$$

Trimethoxy[2-(methylthio)ethyl]silane   and   trimethoxy[(3-methylthio)propyl] silane form similar structures in the liquid condition, with a network of the specific

Table 2.17 Energies (kJ mol^{-1}) of specific interactions of trimethoxy[(methylthio)alkyl]silanes, [(ethylthio)alkyl]silanes, [(ethylthio)alkyl]trimethoxysilanes, and triethoxy[(ethylthio)alkyl]silane at 298 K

Compounds	Formula	Structure	$\Delta_{vap}H^0$ (298 K) [6]	DS → CH$_2$– Si	DS → CH$_3$– (CH$_2$)$_n$–S	DS → CH$_2$–CH$_2$–Si
DSi–O → CH$_3$ = 5.82; DSi–O → CH$_3$–CH$_3$– = 6.06						
Trimethoxy[(methylthio)methyl]silane	C$_5$H$_{14}$O$_3$SSi	(structure)	40.2 ± 0.6	2.0	$n=0$ 0.9	–
Trimethoxy[2-(methylthio)ethyl]silane	C$_6$H$_{16}$O$_3$SSi	(structure)	45.2 ± 0.7	2.0	$n=1$ 3.15	–
Trimethoxy[(3-methylthio)propyl] silane	C$_7$H$_{18}$O$_3$SSi	(structure)	43.5 ± 0.6	2.0	$n=2$ 2.0	–
[(Ethylthio)methyl]trimethoxysilane	C$_6$H$_{16}$O$_3$SSi	(structure)	41.4 ± 0.6	–	$n=0$ 1.8	1.45
[2-(Ethylthio)ethyl]trimethoxysilane	C$_7$H$_{18}$O$_3$SSi	(structure)	41.4 ± 0.7	–	$n=1$ 1.8	1.45
[3-(Ethylthio)propyl]trimethoxysilane	C$_8$H$_{20}$O$_3$SSi	(structure)	41.8 ± 0.6	–	$n=2$ 2.0	1.45
[(Butylthio)methyl]trimethoxysilane	C$_8$H$_{20}$O$_3$SSi	(structure)	41.6 ± 0.6	–	$n=4$ 1.9	1.45
Trimethoxy[(2-methylpropylthio) methyl]silane	C$_8$H$_{20}$O$_3$SSi	(structure)	38.7 ± 0.6	1.45	$n=2$ 2.0	–
Trimethoxy[(pentylmethylthio)methyl] silane	C$_{11}$H$_{18}$O$_3$SSi	(structure)	56.1 ± 0.7	1.45	$n=0$ 1.8	DCH(CH$_3$)–(CH$_2$)$_2$– CH$_3$=7.3
[2-(Butylthio)propyl]triethoxysilane	C$_{13}$H$_{30}$O$_3$SSi	(structure)	47.1 ± 0.6	–	$n=2$ 3.9	1.45

(continued)

Table 2.17 (continued)

Compounds	Formula	Structure	$\Delta_{vap}H^0$ (298 K) [6]	DS → CH$_2$–Si	DS → CH$_3$–(CH$_2$)$_n$–S	DS → CH$_2$–CH$_2$–Si
Triethoxy[(ethylthio)methyl]silane	C$_9$H$_{22}$O$_3$SSi		42.3 ± 0.6	–	1.5	1.45
Triethoxy[2-(ethyl-thio)ethyl]silane	C$_{10}$H$_{24}$O$_3$SSi		46.9 ± 0.7	–	3.85	1.45

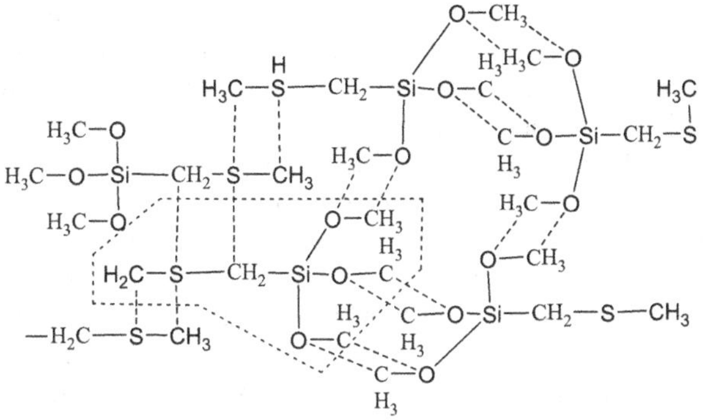

Fig. 2.19 Schematic of the network of specific interactions of liquid and crystalline trimethoxy (methylthio)methyl)silane

interactions. A distinctive feature of these compounds is the formation of the specific interactions DS $\rightarrow$ CH$_3$–CH$_2$–S and DS $\rightarrow$ CH$_3$–CH$_2$–CH$_2$–S, respectively, for which the energies are calculated from Eq. (2.19a), where R is the ethyl or propyl group:

$$\text{DS} \rightarrow \text{R} - \text{S} = \left(\Delta_{\text{vap}}H^0(T\,\text{K}) - 6\text{DSi} - \text{O} \rightarrow \text{CH}_3 - 2\text{DS} \rightarrow \text{CH}_2\right)/2 \quad (2.19a)$$

According to the data in Table 2.17, it is clear that the obtained energy values of the specific interactions of DS $\rightarrow$ R–S are low and that they correspond to the suggested assumption, based on analysis of the electron density shifting in the molecules of these compounds, that the values depend on the location of the methylthio group at the silicon atom. According to the vaporization enthalpies of [(ethylthio)alkyl] trimethoxysilanes and [(butylthio)methyl]trimethoxysilane (Table 2.17), the following conclusions can be made: First, the values of the characteristic of the considered compounds are within the experimental error range; second, the value of the enthalpy characteristic is constant and independent of the ethyl or propyl ligand in the thio group and the methyl, ethyl, or propyl ligand in the (ethylthio)alkyl fragments; and, third, from its place of arrangement at the silicon atom.

This allows the use of the obtained energy values of the DS $\rightarrow$ CH$_3$–S and DS $\rightarrow$ CH$_3$–CH$_2$–CH$_2$–S interactions and calculation of the value of the energy contribution of the formed specific interaction DS $\rightarrow$ CH$_2$–CH$_2$–Si of the ethylene group ethylthio fragment of the compound [3-(ethylthio)propyl]trimethoxysilane, using Eq. (2.19a). The derived energy value of this type of interaction (1.45 kJ mol^{-1}) allows the determination of the energy of the specific interactions DS $\rightarrow$ CH$_3$–(CH$_2$)$_n$–Si, formed by the methyl, ethyl, and propyl fragments of the considered compounds. The energy value of the formed methoxy group of trimethoxy[(2-methylpropyl)thiomethyl]silane of the specific interactions DSi–O $\rightarrow$ CH$_3$ (5.3 kJ mol^{-1}) and (pentylmethyl)thio fragment of the trimethoxy [(pentylmethyl)thiomethyl]silane compounds was defined. The relatively high value of the energy contribution of the DCH(CH$_3$)–(CH$_2$)$_2$–CH$_3$ group is caused by the functional properties of the isostructural group, whose contribution increases with an increase in the length of the carbon chain. The results of the energy calculation of the specific interaction DS → CH_3–CH_2–CH_2–S (3.85 kJ mol^{-1}), formed by the propyl group of the molecule [2-(butylthio)propyl]triethoxysilane reflects the stabilizing influence of the (butylthio)propyl fragment at location 2 at the silicon atom rather than at location 3 (1.9 kJ mol^{-1}) in the molecule of trimethoxy[2-(methylthio)ethyl]silane. A similar manifestation of the stabilizing influence on the specific interaction DS → CH_3–CH_2–S, formed by the ethyl group of the (ethylthio)ethyl fragment at location 2 rather than at location 3 at the silicon atom in the compounds triethoxy[2-(ethylthio)ethyl]silane and triethoxy[(ethylthio)methyl]silane, and trimethoxy[(methylthio)methyl]silane and trimethoxy[2-(methylthio)ethyl]silane, respectively, is caused by the manifestation of the *trans* influence.

2.7.5 1,3,7,5.4-Pentametylcyclopentasiloxane and Decamethylcyclopentasiloxane

The molecule of decamethylcyclopentasiloxane with an equal number of silicon and oxygen atoms in the cycle, and with two methyl groups at the silicon atom and two free electron pairs of the oxygen atom, has a deficit in the ligands able to participate in the formation of the specific interaction. In this situation, the oxygen atom, expressing high acceptor properties, shifts the increased electron density from the silicon atom and provides it with increased acceptor properties at the formation of the specific interaction DSi–H_3C → Si of one methyl group, with participation of the essentially undivided $3s^2$(Si) electron pair, causing its pentacoordinated condition. The energy of the specific interaction has been determined for the liquid dimethylsilacyclobutanes (3.95 kJ mol^{-1}, Table 2.13), the value of which is taken as being equal to the energy of the same type of interaction in liquid decamethylcyclopentasiloxane. The second methyl group of the compound expresses the properties of the isostructural methyl group and interacts with the oxygen atom, forming a specific interaction with reduced stability, DSi–O → CH_3–Si. The molecule of 1,3,7,5.4-pentametylcyclopentasiloxane, with a similar number of silicon and oxygen atoms as the five fragments of $HSiCH_3$ but with different location, expresses similarity in the specificity of the formed interactions. Taking into account the increased stability of the silane specific interaction formed by the methyl group in comparison with the hydrogen bond DSi•••H–Si, it should be supposed the implementation of the condition for $HSiCH_3$ and that is why the hydrogen atom of this group of the molecule of 1,3,7,5.4-pentametylcyclopentasiloxane forms a hydrogen bond with the oxygen atom. The formed specific interaction and the hydrogen bonds, with the planar molecular structure of the compounds, provide the network structure of the liquid compound (Fig. 2.20). The network's crosslinking is implemented by the formation of the specific interactions DSi–H_3C → Si and DSi–O → CH_3–Si in 1,3,7,5.4-pentametylcyclopentasiloxane and decamethylcyclopentasiloxane. The contacting

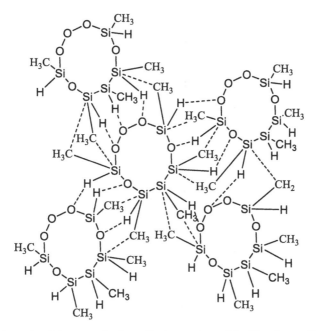

Fig. 2.20 Schematic of the network of specific interactions of liquid and crystalline 1,3,7,5.4-pentametylcyclopentasiloxane

molecules form ten specific interactions of the first type, and the energy of the second type with unknown value is determined with the help of Eqs. (2.20) and (2.20a):

$$DSi - O \rightarrow CH_3 - Si = \left(\Delta_{vap}H^0(TK) - 10DSi - H_3C \rightarrow Si \right)/10 \qquad (2.20)$$

$$DSi - O \bullet \bullet \bullet H - Si = \left(\Delta_{vap}H^0(TK) - 10DSi - H_3C \rightarrow Si \right)/10 \qquad (2.20a)$$

The obtained energy value (Table 2.18) reflects the low stability of the specific interaction DSi–O → CH₃–Si. The relatively high energy value of the hydrogen bond is comparable with the energy of the same bond type realized in aldehydes [8].

2.7.6 Energies of Specific Interactions of Methyldiphenylsilanes

In the section titled "Energies of Specific Interactions of Phenyl Selenides and Diphenyl Mercury" it was shown that, with comparable high acceptor ability, the selene atom of the diphenylselenide molecule forms the specific interaction DSe → CH of low stability (2.5 kJ mol^{-1}) with the carbon atom of the CH group of the benzene ring. In this regard, the methyldiphenylsilanes are of particular

Table 2.18 Energies (kJ mol^{-1}) of specific interactions and hydrogen bonds of liquid cyclic decamethylcyclopentasiloxane and 1,3,7,5,4-pentametylcyclopentasiloxane at 298 K

Compounds	Formula	Structure	$\Delta_{vap}H^0$ (298) [6]	DSi–H$_3$C → Si	DSi–O•••H–Si
Decamethylcyclopentasiloxane	C$_{10}$H$_{30}$O$_3$Si$_5$	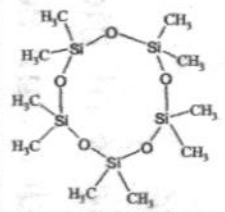	52.3 ± 1.7	3.95	DSi–O → CH$_3$–Si 1.3:2
1,3,7,5,4-Pentametylcyclopentasiloxane	C$_5$H$_{20}$O$_5$Si	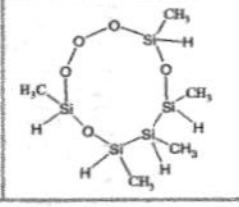	97.0 ± 0.9	3.95	5.75

interest. These compounds are characterized by a reduced value of the vaporization enthalpy, with relatively low experimental error, that is twice the value of the vaporization enthalpy of benzene minus the energy contribution of the two saturated hydrogen atoms (Table 2.18). It follows that the silicon atom shifts electron density to the carbon atom of the CH group distributed in the benzene ring, reducing their charges. Taking the energy values of the formed hydrogen bond DSi–H → H–Si and specific interaction DSi–H$_3$C → H–CH$_2$–Si by the hydrogen and carbon atom of the methyl group of methyldiphenylsilanes as being equal to 3.11 and 3.25 kJ mol^{-1}, the energy contribution of the benzene ring and the energy of the DHC → CH interaction can be obtained. The calculated energy values (Table 2.19) are described by the inequality DHC → CHbz > DHC → CHmdpsi, with a difference of 0.56 kJ mol^{-1}.

The energy contribution of the benzene ring with one saturated hydrogen atom to the enthalpy of sublimation of the molecules of tetraphenylsilane with tetrahedral structure and hexaphenyldisilane with a linear structure can be derived by the depletion of the enthalpy characteristic by the number of phenol fragments in the molecule. Shown in Table 2.20 are the results of the calculations, which reflect the significant reduction in the energies of the specific interactions DHC → CH and DC → C of tetraphenylsilane and hexaphenyldisilane in comparison with those in crystalline benzene.

Attention should be paid to the compounds of the methylphenoxy dimethylsilanes with the methoxy group of the phenol fragment and its strong acceptor properties. Nevertheless, these compounds have reduced vaporization enthalpies in comparison with the methyldiphenylsilanes (Table 2.21). This indicates that the methoxy group with the electron density of the ring carbon atom contributes additional changes to the reduction of the charge differences at the carbon atoms of the CH groups. The contribution of the phenol fragment to the vaporization enthalpy can be derived using Eq. (2.21):

$$\Delta_{vap}H^0(T\,K)\text{phxsi} = \left(\Delta_{vap}H^0(T\,K)2DSi - H_3C \to H - CH_2 - Si\,(3.25 \times 2) \right.$$
$$\left. -2DO \to CH_3 - O\right)/2 \tag{2.21}$$

The mutual influence of the silicon and oxygen atom of the methoxy group complicates the correct estimation of the energy of the specific interaction DO → CH$_3$–O. Therefore, taking it to be equal to the energy of the formed specific interaction of the methoxy group of the methyl phenol ester (5.63 kJ mol^{-1}), the maximum reduced energy contribution of the phenol fragment to the enthalpy characteristic of methylphenoxy dimethylsilanes can be obtained, keeping the tendency of their change in dependence on the location of the methoxy group at the benzene ring.

Table 2.19 Energies (kJ mol^{-1}) of specific interactions of liquid methyldiphenylsilanes at 298 K

Compounds	Formula	Structure	$\Delta_{vap}H^0$ (298 K) [6]	$\Delta_{vap}H^0$ (298 K)bz−DH	$\Delta_{vap}H^0$ (298 K)mdpsi−DH	DHC→CH
DSi−H → H−Si = 3.11, DSi−H$_3$C → H−CH$_2$−Si = 3.25; DHC → CHbz = 5.63						
Methyldiphenylsilane	C$_{13}$H$_{14}$Si		64.8 ± 0.8	33.2 × 2	29.25	4.97
Diphenoxydimethylsilane	C$_{14}$H$_{16}$Si		64.4 ± 0.94	33.2 × 2	29.25	4.97
Ethyldiphenylsilane	C$_{14}$H$_{16}$Si		66.1 ± 0.8	33.2 × 2	29.25	4.97

Table 2.20 Energies (kJ mol^{-1}) of specific interactions of crystalline tetraphenylsilane and hexaphenyldisilane at 298 K

Compounds	Formula	Structure	$\Delta_{sub}H^0$ (298 K) [31]	$\Delta_{sub}H^0$ (298 K)bz − DH	$\Delta_{sub}H^0$ (298 K) − DH	DHC → CH/DC → C
DHC → CHbz = 7.43; DC → Cbz = 6.43						
Tetraphenylsilane	$C_{24}H_{20}Si$		156.9 ± 1.7	43.6−1	39.2	6.7/5.7
Hexaphenyldisilane	$C_{36}H_{30}Si_2$	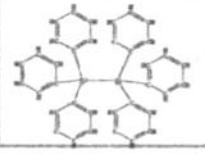	209.9 ± 2.0	43.6 / 30.55	35.0	5.83/4.83

Table 2.21 Energies (kJ mol^{-1}) of specific interactions of liquid methylphenoxy dimethylsilanes at 298 K

Compounds	Formula	Structure	$\Delta_{vap}H^0$ (298 K) [6]	$\Delta_{vap}H^0$ (298 K)bz – DH	$\Delta_{vap}H^0$ (298 K) – DH	DHC → CH
DHC → CHbz = 7.43; DC → Cbz = 6.43						
bis(2-Methylphenoxy)dimethylsilane	C$_{16}$H$_{20}$O$_2$Si		63.6 ± 0.8	33.2 × 2	22.9	4.0
bis(3-Methylphenoxy)dimethylsilane	C$_{16}$H$_{20}$O$_2$Si		61.5 ± 0.8	33.2 × 2	21.85	3.82
bis(4-Methylphenoxy)dimethylsilane	C$_{16}$H$_{20}$O$_2$Si		65,3 ± 0.9	33.2 × 2	23.8	4.25

References

1. Nefedov VI, Vovna VI (1987) Electronic structure of chemical compounds. Nauka, Moscow, 347 p
2. Nefedov VI, Vovna VI (1989) Electronic structure of organic and elementorganic compounds. Nauka, Moscow, 198 p
3. Vovna VI, Gorchakov VV, Cherednichenko AI (1987) Structure of compounds and properties of molecules. University of Vladivostok, Vladivostok, pp 93–123
4. Bancroft GM, Adams I, Creber DK, Eastman DE, Gudat W (1976) High resolution photoelectron studies: electric field gradient splittings of Cd and Sn 4d energy levels in organometallic compounds. Chem Phys Lett 38(1):83–87
5. Bancroft GM, Coatsworth LL, Greber DK, Tse J (1977) High resolution core level photoelectron spectroscopy using He II radiation: electric field gradient splittings in group IIB and group III compounds. Chem Phys Lett 50(2):228–232
6. Chickos JS, Acree WE Jr (2003) Enthalpies of vaporization of organic and organometallic compounds, 1880–2002. J Phys Chem Rev 32(2):519
7. Pauson PL (1967) Organometallic chemistry. Edward Arnold Ltd., London, 238 p
8. Baev AK (2012) Specific intermolecular interactions of organic compounds. Springer, Heidelberg/Dordrecht/London/New York, 434 p
9. Ivanov EV, Puchovskya YuP, Abrosimov VK (2004) Book of abstract of IXth international conference. The problem of solvation and complex formation in solutions. Ivanovo, Russia, p 345
10. Schustorovich EM (1978) Chemical bonds in coordination compounds. Knowledge, Moscow, 110 p
11. Vasiliev IA, Petrov VM (1984) Thermodynamic properties of oxygen organic compounds. Chemistry: Leningrad, Rassia, 238 p
12. Lebedev YA, Miroschnichenko EA (1981) Thermochemistry of the vaporization of organic compounds. The vaporization and sublimation enthalpy and saturated pressure. Nauka, Moscow, p 285
13. Cradock S, Findlay RH, Palmer MH (1974) Bonding in methyl- and silyl-cyclopentadiene compounds: a study by photoelectron spectroscopy and ab initio molecular-orbital calculations. J Chem Soc Dalton Trans 16:1650–1654
14. Drake JE, Glavincevski BM, Gorzelska K (1979) The photoelectron spectra of methylgermane, trifluoro- and trichloro-methylgermane. J Electron Spectrosc 17(2):73–80
15. Drake JE, Glavincevski BM, Gorzelska K (1979) The photoelectron spectra of trimethylgermane, chloro- and fluoro-trimethylgermane. J Electron Spectrosc 16(3):331–337
16. Beltram G, Fehlner TP, Mochida K, Kochi JK (1980) UV photoelectron spectra of group IV alkyl hydrides. J Electron Spectrosc 18(1):153–159
17. Hosomi A, Traylor TG (1975) Studies of interactions of adjacent carbon-methyl bonds by photoelectron spectroscopy. J Am Chem Soc 97(15):3682–3687
18. Baev AK (2014) Specific intermolecular interactions of nitrogenated and bioorganic compounds. Springer, Heidelberg/Dordrecht/London/New York, 579 p
19. Gurianova EN, Goldstein IG, Romm IP (1973) Donor–acceptor bonds. Chemistry, Moscow, 365 p
20. Nefedov VI (1984) X-ray photoelectron spectroscopy of chemical compounds. M. Chemistry, 255 p
21. Mochida K, Worley SD, Kochi JK (1985) UV photoelectron spectra of peralkylated catenates of group 4B elements (silicon, germanium, and tin). Bull Chem Soc Jpn 58(11):3389–3390
22. Bock H, Wittel K, Veith M, Wibory N (1976) Part XXVII. Contributions to the chemistry of silicon and germanium. J Am Chem Soc 98:106
23. Szepes L, Koranyi T, Neray-Szabo G, Modelli A, Distefano G (1981) Ultraviolet photoelectron spectra of group IV hexamethyl derivatives containing a metal–metal bond. J Organomet Chem 217(1):35–41

24. Sacurai H, Ichinose M, Kira M, Traylor TG (1984) Chemistry of organosilicon compounds. 197. Ultraviolet, charge transfer, and photoelectron spectra of phenyl substituted group 4B catenates, phene2E-E^{v}ME3 (E=ST, GE,E^v=C,SI, DE, SN). Chem Lett (8):1383–1384
25. Granozzi G, Bertoncello R, Tondello E, Ajo D (1985) He(I)/He(II) Sn $5p$ photoionization cross sections: definitive evidence from the spectra of Sn2(CH3)6. J Electron Spectrosc 36(2):207–211
26. Bock H, Ensslin W, Feher F, Freund R (1976) Photoelectron spectra and molecular properties. LI. Ionization potentials of silanes SinH2n+2. J Am Chem Soc 98(3):668–671
27. Ermakov AI, Kirichenko EA, Pimkin NI, Chizhov YV, Kleimenov VI (1982) The photoelectron spectra and electronic structure of disiloxanes. Russ J Struct Chem 23(1):62–67
28. Ermakov AI, Kirichenko EA, Pimkin NI, Chizhov YV, Kleimenov VI (1982) Photoelectronic spectra and electronic structure of organotri- and organotetrasiloxanes of linear and cyclic structure. Russ J Struct Chem 23(4):539–545
29. Baev AK (1969) To problems of chemical nature of phase transformation general and applied chemistry, vol 1. Vaisheshika Educating, Minsk, pp 197–206
30. Baev AK (1987) Chemistry of gas-heterogenic systems of elementorganic compounds. Nauka and technique. Science and Technology, Minsk, 175 p
31. Chickos JS, Acree WE Jr (2002) Enthalpies of sublimation of organic and organometallic compounds, 1910–2001. J Phys Chem Rev 31(2):519

Chapter 3
Specific Interactions of Elementorganic Compounds of Germanium Subgroup Elements

3.1 Specific Interactions of Germanium Organic Compounds and Their Energies

3.1.1 Energies of Specific Interactions of Compounds of Germanium with Alkyl Ligands

The electron configuration of $M(CH_3)_4$ compounds with a more symmetrical molecular structure of the germanium subgroup elements, based on data from the literature [1–9], has a quite complex shape: $3t^6{}_2 1t^6{}_1 1e^4 2t^6{}_2 2d^2{}_1 1t^6{}_2 1a^2{}_1$. In the photoelectron spectra, the upper M–C bonding orbital $3t_2$ has a Jahn–Teller splitting value of 0.5 eV. However, for the compound $Pb(CH_3)_4$, the main contribution to the splitting is not oscillating electrons but the spin–orbit interaction [5]. Three C–H bonding molecular orbitals (MOs) ($1t_1$, $1e$, and $2t_2$) in the spectra correspond to a single broad band at 21.2 eV. Note that in a $C(CH_3)_4$ molecule, three MOs determine the bands at 12.6, 14.0, and 15.2 eV and two deep ($1t_2$ and $1a_1$) orbitals are of the CH-binding C 2s-type. Increasing the size of the central atom of the element is accompanied by a reduction in the vertical ionizing potentials [5, 6]. Thus, the silicon atoms and germanium subgroup elements with essentially undivided $3s^2(Si)$–$4s^2(Ge)$ electron pairs express a reduced acceptor ability to join to the electron density in the formation of specific interactions, in comparison with atoms of the zinc and boron subgroup elements.

Compounds of the tetramethylgermane series with a tetrahedral molecular structure illustrate significant changes in vaporization enthalpies (Table 2.1). The relationship of the vaporization enthalpy to the number and energy of the bursting of specific interactions [9–11], structure identity, and the difference in the alkyl ligands of the molecules due to replacement of the methyl ligand with the propyl ligand, points not only to the stabilization of the formed interactions but also to its specificity in comparison with silicon alkyl compounds. At the maximum coordination of the germanium atom in the pentacoordinated condition, the methyl

© Springer International Publishing Switzerland 2015
A.K. Baev, *Specific Intermolecular Interactions of Element-Organic Compounds*,
DOI 10.1007/978-3-319-08563-0_3

ligands interacting with the contacting molecules in the condensed state form a specific interaction of reduced stability $DGe–H_3C \rightarrow H–CH_2$, similar to the silicon atom in organosilicon compounds and methane [12].

With the difference between carbon atoms of the methyl and methylene groups of the chain, ethyl and propyl ligands form specific interactions of increased stability. Therefore, the energies of the eight specific interactions of the same type formed by molecules with a tetrahedral structure are determined from the enthalpy value by the depletion of the number of bonds. Table 3.1 presents the energy values of the specific interactions, which point to a linear dependence on the number of carbon atoms of the alkyl chain. This allows an estimation to be made of the energy value of a similar interaction of the formed butyl ligand of tetrabutylgermane and of its vaporization enthalpy. Taking into account the fact that the influence of the intermolecular interaction on the energy of the specific interaction is completed at the fourth carbon atom of the chain, it is possible to determine the energy contribution of the pentyl ligand during formation of the specific interaction in liquid pentyl(trimethyl)germane for the butyl ligand, $DGe–CH_2–(CH_2)_2–CH_3 \rightarrow CH_2–(CH_2)_2–CH_3$:

$$DCH_2 = \left(\Delta_{vap}H^0(298\,K)pmg - 6DGe - H_3C \rightarrow H - CH_2 - 2DGe - CH_2 \right.$$

$$\left. -(CH_2)_2 - CH_3 \rightarrow CH_2 - (CH_2)_2 - CH_3 \right) \tag{3.1}$$

The contribution of the pentyl group was determined to be 9.7 kJ mol^{-1}, and the remaining energy value of 3.4 kJ mol^{-1} corresponds to the energy contribution of the methylene group to the vaporization enthalpy of the compound. This value corresponds to the energy contribution of each of the four pentyl ligands of the tetrapentylgermane compound, whose vaporization enthalpy is given in Table 3.1.

3.1.2 Energies of Hydrogen Bonds and Specific Interactions of Methylgermanium and Dimethylgermane

Molecules of methylgermanium and dimethylgermane with one and two hydrogen atoms in combination with similar organosilicon compounds allow clarification of the influence of the alkyl ligand location on the stabilization of the formed specific interaction. The obtained energy value of the specific interaction of $D–Ge–H_3C \rightarrow H–CH_2$ of the liquid tetramethylgermane can be taken to be equal to the energy value of the same type of interaction in liquid methylgermanium. Using Eq. (3.2), the energy of hydrogen bond can be obtained considering the energy

Table 3.1 Energies (kJ mol^{-1}) of specific interactions of liquid tetraalkylgermanes

Compounds	Formula	Structure	$\Delta_{vap}H^0$ (298 K) [13]	T (K)	D–Ge– $H_3C \rightarrow H$–CH_2-	DGe–CH_2– $CH_3 \rightarrow CH_2$–CH_3	DGe–CH_2–$(CH_2)_n$–$CH_3 \rightarrow CH_2$–$(CH_2)_n$–CH_3
Tetramethylgermane	GeC$_4$H$_{12}$		28.1 ± 0.1	285	3.50	–	–
Tetraethylgermane	GeC$_8$H$_{20}$		45.7 ± 0.4	298	–	5.7	–
Tetrapropylgermane	GeC$_{12}$H$_{28}$		61.5 ± 4.3	–	–		$n = 1$
			54.7	368			7.7
Tetrabutylgermane	GeC$_{16}$H$_{36}$		77.6^a	298	–	–	$n = 2$
							9.7^a
Pentyl(trimethyl)germane	GeC$_8$H$_{20}$		44.3	318	3.58	DCH$_3$ = 3.4	9.7
Tetrapentylgermane	GeC$_{20}$H$_{44}$		91.2^a	298	–	ΣCH$_2$ = 3.4 × 4	9.7

^{a}Estimated

Table 3.2 Energies (kJ mol^{-1}) of specific interactions of liquid methylgermanium and dimethylgermane

Compounds	Formula	Structure	$\Delta_{vap}H^0$ (298 K) [13]	T (K)	D–Ge– $H_3C \rightarrow H$–CH_2– C–	DGe– H•••H– Ge
Methylgermanium	GeCH$_6$	H H–Ge–H CH$_3$	16.6	196	3.58	3.15
Dimethylgermane	GeC$_2$H$_8$	H H$_3$C–Ge–H CH$_3$	26.5	212	5.05	3.15

contributions of the three functional hydrogen atoms and the methyl group, forming two specific interactions with the molecules of the near environment:

$$DGe\text{–}H \bullet \bullet \bullet H - Ge = \left(\Delta_{vap}H^0(298\,\text{K})mg - 2DGe - H_3C \rightarrow H - CH_2-\right)$$

$$(3.2)$$

As presented in Table 3.2, the energy value of the hydrogen bond of the liquid methylgermanium is equal to the energy value of the same bond in liquid silane:

$$DGe - H \bullet \bullet \bullet H - Ge(3.15; \ T = 196\,\text{K})$$
$$= DSi - H \bullet \bullet \bullet H - Si\left(3.11\,\text{kJ}\,\text{mol}^{-1}; T = 161.25\,\text{K}\right)$$

Low values of these quantities undergo small changes with temperature, corresponding to the experimentally measured vaporization enthalpies of the compounds; thus, the equation can be used in the thermodynamic calculations.

The energy of the specific interaction D–Ge–H$_3$C $\rightarrow$ H–CH$_2$– of liquid dimethylgermane can be determined from Eq. (3.2a), taking into account the energy contributions of all interactions:

$$DGe - H_3C \rightarrow H - CH_2 -$$
$$= \left(\Delta_{vap}H^0(298\,\text{K})dmg - 2DGe - H \bullet \bullet \bullet H - Ge\right)/4 \qquad (3.2a)$$

The obtained energy value of the specific interaction D–Ge–H$_3$C $\rightarrow$ H–CH$_2$– (Table 3.2) of liquid dimethylgermane exceeds approximately half the energy value of the same type of interaction at the location of the methyl group in the *trans* position in the molecule of methylgermanium. It follows that the increased stabilization of the specific interaction is reached at the mutual location of the two functional hydrogen atoms and two methyl groups in the *cis* position in the molecule.

3.1.3 Energies of Hydrogen Bonds and Specific Interactions of Germanium Alkyl Compounds with Metal–Meta Bonds

Chapter 2 has shown that the hydrogen bond Si•••H–SiH$_2$ formed by a molecule of disilane includes the influence of the two hydrogen atoms connected to the silicon atom. A similar influence of two hydrogen atoms connected to a germanium atom should be expressed at the hydrogen bond that is formed by the hydrogen atom in the *trans* position of the GeH$_3$ group in a molecule of digermane. Table 3.3 gives some information on the vaporization enthalpy of digermane derivatives and more precise values on the given compounds can be obtained from the enthalpy characteristics of 1,1-dimethyldigermane, 1,2-dimethyldigermane, and 1,1,2-trimethyldigermane. Replacement of the hydrogen atom in the molecule of 1,1-dimethyldigermane by a methyl group is accompanied by the formation of 1,1,2-trimethyldigermane and an additional energy contribution of 4.2 kJ mol^{-1} to the vaporization enthalpy as a result of the difference in enthalpy characteristics of these compounds . This means that, following the replacement of the hydrogen atom in the *trans* position of 1,1,2-trimethyldigermane by the methyl group, 1,1,2,2-tetramethyldigermane is obtained with a vaporization enthalpy of 37.7 kJ mol^{-1}. Similarly, following the replacement of the methyl group in the molecule of 1,1-dimethyldigermane by a hydrogen atom, 1-methyldigermane with a vaporization enthalpy of 27.6 kJ mol^{-1} is obtained. From the difference in the vaporization enthalpies of 1,1,2-trimethyldigermane and 1,1-dimethyldigermane, it follows that replacement of the chain hydrogen atom by the isostructural methyl group is accompanied by an increase in vaporization enthalpy of 1.7 kJ mol^{-1}. Using this energy value, the vaporization enthalpies of a number of unstudied compounds of the given series can be obtained (Table 3.3).

Molecules of dimethyldigermanes with metal–metal bonds (e.g., disilanes) have a planar structure, with the number of bond vacancies depending on the number and location of the methyl groups. The symmetrical molecule digermane, with four bond vacancies expressed by the two germanium atoms and two functional hydrogen atoms located in the *trans* position to the germanium atoms, forms four hydrogen bonds DGe•••H–GeH$_2$ and a network of these bonds in liquid and crystal conditions (Fig. 3.1a), which causes the pentacoordinated condition of the germanium atom. Based on the relationship between the vaporization enthalpy and the number and energy of specific interactions formed by the compound molecule, the compound's energy is determined directly from the enthalpy value [9–11] through dividing by the number of interactions. The obtained energy value of 6.48 kJ mol^{-1} at 277 K is within the margin of error for the energy value for the same type of interaction of liquid disilane (7.05 kJ mol^{-1} at 258 K). The linear molecule has the planar structure of methyldigermane, with four bond vacancies expressed by the two germanium atoms, a hydrogen atom, and a methyl group. These form two hydrogen bonds DGe•••H–GeH$_2$ and two specific interactions DGe–H$_3$C → Ge,

Table 3.3 Energies (kJ mol⁻¹) of specific interactions and hydrogen bonds of liquid digermane and its methyl derivatives at 298 K

Compounds	Formula	Structure	$\Delta_{vap}H^{0}$ (298 K) [13]	DGe•••H–GeH₂	DGe → CH₃–Ge	DisoGe–H₃C→H–CH₂–G
Digermane	Ge_2H_6	H–GeH₂–GeH₂–H	25.9	6.48	–	–
Methyldigermane	Ge_2CH_8	H–GeH₂–GeH₂–CH₃	30.1	6.48	8.52	–
Dimethyldigermane	$Ge_2C_2H_{10}$	H₃C–GeH₂–GeH₂–CH₃	34.3	–	8.52	1.10 × 2
1-Methyldigermane	Ge_2CH_8	H–GeH–GeH₂–H, CH₃	27.6	6.35	–	1.10 × 2
1,2-Dimethyldigermane	$Ge_2C_2H_{10}$	H–GeH–GeH–H, CH₃ CH₃	29.3ᵃ	6.22	–	1.1 × 4
1,1,2-Trimethyldigermane	$Ge_2C_3H_{12}$	H–Ge–GeH–H, CH₃ CH₃ CH₃	31.0	6.1	–	1.1 × 6
1,1,2,2-Tetramethyldigermane	$Ge_2C_4H_{14}$	CH₃ CH₃ H–Ge–Ge–H, CH₃ CH₃	32.7	6.0	–	1.85 × 8 / 1.1 × 8
1,1-Dimethyldigermane	$Ge_2C_2H_{10}$	H₃C–GeH–GeH₂–H, CH₃	31.8ᵃ	6.3	8.52	1.1 × 2
1,1,2-Trimethyldigermane	$Ge_2C_4H_{14}$	CH₃–GeH–GeH–CH₃, CH₃ CH₃	33.5ᵇ	6.22	8.3	1.85 × 4 / 1.1 × 4
1,1,1,2-Tetramethyldigermane	$Ge_2C_4H_{14}$	CH₃ CH₃–Ge–GeH–H, CH₃ CH₃	35.2	6.1	8.2	1.1 × 6
1,1,2,2-Pentamethyldigermane	$Ge_2C_5H_{16}$	CH₃ CH₃ H₃C–Ge–Ge–H, CH₃ CH₃	36.9	6.0	8.05	1.1 × 8
Germilsilane	GeH_6Si	H H H–Ge–Si–H, H H	25.0***	6.48	–	DGe•••H–Ge = 6.0 / DSi•••H–Si = 5.5

ᵃExperimental data measured at 277 K
ᵇExperimental data measured at 281 K
***Experimental data measured at 220 K

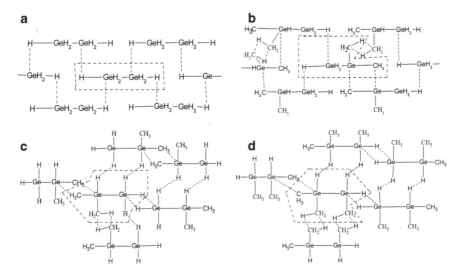

Fig. 3.1 Schematic of the network of specific interactions of liquid and crystalline digermane (**a**), 1,1-dimethyldigermane (**b, c**), and 1,1,2-trimethyldigermane (**d**)

whose energy value can be obtained using Eq. (3.3), taking into account the energy contributions of the hydrogen bonds:

$$DGe - H_3C \rightarrow Ge = \left(\Delta_{vap}H^0(T\,K)mdg - 2DGe \bullet\bullet\bullet H - GeH_2 \right)/2 \quad (3.3)$$

Presented in Table 3.3 are the energy values of the specific interactions, which have increased value caused by shifting of the electron density from the carbon atom of the methyl group to the 4s germanium atom, which is a more distant energy level than that of the 3s silicon atom:

$$DGe - H_3C \rightarrow Ge(8.52) > DG - H_3C \rightarrow Ge(6.35\,kJ\,mol^{-1})$$

Note that the energy contribution of the isostructural methyl group to the vaporization enthalpy of hexamethyldisilane is equal to 1.4 kJ mol^{-1}. In the case of hydrogen atom replacement in the molecule of 1,1,2-trimethyldigermane by the isostructural methyl group, the additional energy expended on vaporization is equal to 1.7 kJ mol^{-1} and that for the substituted hydrogen atom is not more than 0.50 kJ mol^{-1} [12]. Therefore, the energy contribution of the isostructural methyl to the bursting of two bonds D H$_2$C-⟨$^{H}_{H}$⟩-CH$_2$ is 2.20 kJ mol^{-1}, or 1.10 kJ mol^{-1} for one bond. An increasing number of methyl groups in the molecule complicates the liquid and crystal structure and the formed network of specific interactions (Fig. 3.1b). Decreasing the number of hydrogen atoms reduces the total energy value of the hydrogen bond. A detailed picture of the network of specific interactions, with participation of all hydrogen atoms connected with the germanium atom

(Fig. 3.1c, d), illustrates the associations caused by the small differences in their positive charges. Using the value of the energy contribution of the specific interactions of low stability formed by the isostructural methyl group, the energy of the specific interactions and hydrogen bonds for each methyl derivative of digermane was calculated with the help of Eq. (3.4), taking into account the energy contribution to the enthalpy of all marked interaction types:

$$DGe\bullet\bullet\bullet H - GeH_2 = \left(\Delta_{vap}H^0(TK)mdg - DGe - H_3CK \rightarrow Ge - qDisoCH_3\right)/2$$

$$(3.4)$$

If the unknown value is the energy of the specific interaction of $DGe-H_3C \rightarrow Ge$, a different version of this equation is used:

$$DGe - H_3C \rightarrow Ge = \left(\Delta_{vap}H^0(TK)mdg - DGe\bullet\bullet\bullet H - GeH_2 - qDisoCH_3\right)/2$$

$$(3.4a)$$

The calculation results given in Table 3.3 allow a conclusion to be made that the total value of the energy contribution of the saturated hydrogen atoms is reduced in the compounds of the considered series:

$DGe\bullet\bullet\bullet H–GeH_2$: Digermane (6.48) > 1-Methyldigermane (6.30) > 1,2- Dimethyldigermane (6.22) > 1,1,2-Trimethyldigermane (6.1) = 1.1.1.2-Tetramethyldigermane (6.1) > $DGe\bullet\bullet\bullet H–Ge$, 1,1,2,2-Tetramethyldigermane (6.0) = 1,1,1,2.2-Tetramethyldigermane (6.0 kJ mol^{-1})

The real value for the hydrogen bond formed by the hydrogen atom located in the *trans* position is reached following the substitution of four hydrogen atoms in the compounds 1,1,2,2-tetramethyldigermane and 1,1,1,2.2-tetramethyldigermane. A further series of compounds illustrates the reduction in stability of the formed liquid specific interaction with a reduction in the number of hydrogen atoms in the germane group:

$DGe–H_3C{\rightarrow}Ge$: Methyldigermane (8.52) = 1,1-Dimethyldigermane (8.52) > 1,1,2-Trimethyldigermane (6.3) > 1.1.1.2-Tetramethyldigermane (8.1) > 1,1,1,2.2-Pentamethyldigermane (8.05 kJ mol^{-1})

The small differences in the energies of the hydrogen bonds of liquid silane and methylgermanium with a tetrahedral structure and, on the other hand, liquid disilane and digermane with a planar structure allows an analogy to be made with the energies of the same type of interactions formed by the planar molecule germilsilane. Nevertheless, on formation of the intramolecular bond Ge–Si, there is a shift of electron density from the germanium atom to the silicon atom, which

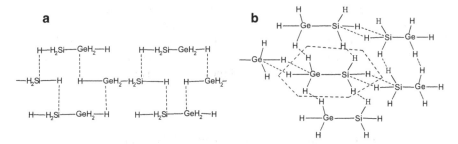

Fig. 3.2 Schematic of the network of specific interactions of liquid (**a**) and crystalline (**b**) germilsilane

increases its negative charge because the charge was generated by the shifting of electron density from hydrogen atoms. Thus, part of the received electron density of the silicon atom returns in the shape of the reverse dative bond, which reduces the difference in the charges at the silicon and hydrogen atoms. By contrast, the electron density shifted from the silicon atom is partly compensated by the transformation from the hydrogen atoms of the germane group, and results in maintenance of the increased difference in the positive charges at the germanium and hydrogen atoms. Therefore, the silicon atom with reduced charge forms the hydrogen bond $DSi\bullet\bullet\bullet H–SiH_2$, which is less stable than the germanium atom. The molecule of germilsilane with two bond vacancies of the silicon and germanium atoms and the two functional hydrogen atoms, located in the *trans* positions to the germanium and silicon atoms, form four hydrogen bonds of two types (with a network of these interactions in the liquid and crystal conditions) (Fig. 3.2a). Taking the energy value of the hydrogen bond of $DGe\bullet\bullet\bullet H–GeH_2$ as being equal to that in the liquid digermane (6.48 kJ mol^{-1}), the energy value of the hydrogen bond (presented in Table 3.3) can be calculated from Eq. (3.5):

$$DSi \bullet \bullet \bullet H - SiH_2 = \left(\Delta_{vap}H^0(T\,K)gsi - 2DGe \bullet \bullet \bullet H - GeH_2\right)/2 \quad (3.5)$$

Note that, due to the lack of information on the specificity of the interparticle interactions of the methyl group and its analogs, it is usually denoted by the associative interactions. In the work of Baev [12], it was shown that the energies of the formed hydrogen bonds of the hydrogen atoms in CH_3, SiH_3, and GeH_3 groups reach their maximum energy value in the *trans* position, the second hydrogen atom has slightly lower stability, and the third hydrogen atom forms with the value of the van der Waals interaction, located outside the experimental error. That is why the accepted energy estimation of 0.5 kJ mol^{-1} for the second hydrogen atom of the GeH_3 group in the hydrogen bond $DGe\bullet\bullet\bullet H–GeH_2$ of methyldigermanes (Table 3.3) takes into account the more detailed and correct representation of the network of specific interactions shown in Fig. 3.2b. Taking into account the energy values of the hydrogen bonds of liquid germilsilane, formed by the hydrogen atom located *trans* to the germanium atom DGe•••H–Ge and silicon atom DSi•••H–Si, the obtained energy values of DGe•••H–GeH$_2$ and DSi•••H–SiH$_2$ minus the energy contribution of the two remaining hydrogen atoms of these groups (0.5 kJ mol^{-1}) are equal to 6.0 and 5.5 kJ mol^{-1}, respectively.

3.1.4   Energies of Specific Interactions of Triethyl (diethylamino)germane

Triethyl(diethylamino)germane is the only compound of this series with a known vaporization enthalpy. The structure and the specificity of this molecule are analogous to those of the alkyl(dialkylamino)silanes. The molecule of triethyl (diethylamino)germane with four bond vacancies of the amino group, with one vacancy of the ethyl group located in the *trans* position to the silicon atom and one group of the atom and four vacancies of the two ethyl groups functioning as isostructural groups, form a liquid with a network of ten specific interactions of three types (Fig. 3.3). Four specific interactions DN → CH$_3$–CH$_2$–N of one type are formed by the amino group with an energy of 5.85 kJ mol^{-1}, equal to the energies of the same bonds of the liquid amines and alkyl(ethylamino)silanes. Small differences in the energies of the hydrogen bonds and the specific interactions of the silicon and germanium of the organic compounds allow the assumption that the energy of the formed specific interaction of the ethyl isostructural group of the triethyl(diethylamino)germane molecule is equal to twice the energy of the isostructural methyl group (1.1 kJ mol^{-1}) (Table 3.3). The obtained value (2.20 kJ mol^{-1}) exceeds the energy value of the same interaction DisoCH$_2$–H$_3$C → CH$_2$–CH$_3$ of the alkylsilanes (1.80 kJ mol^{-1}, Table 2.10). The unknown energy value of the formed specific interaction of the germanium atom with diethyl ligand D–Ge–CH$_2$–H$_3$C → Ge can be determined using Eq. (3.6):

$$D - Ge - CH_2 - H_3C \rightarrow Ge = \left(\Delta_{vap}H^0(TK)edea - 4DisoCH_2 - H_3C \rightarrow CH_2 - CH_3\right)/2$$

$$(3.6)$$

Presented in Table 3.4 are the obtained energy values described by the relation D–Ge–CH$_2$–H$_3$C → Ge (7.3) > DSi–CH$_2$–CH$_3$ → Si (5.7 kJ mol^{-1}), reflecting their correctness.

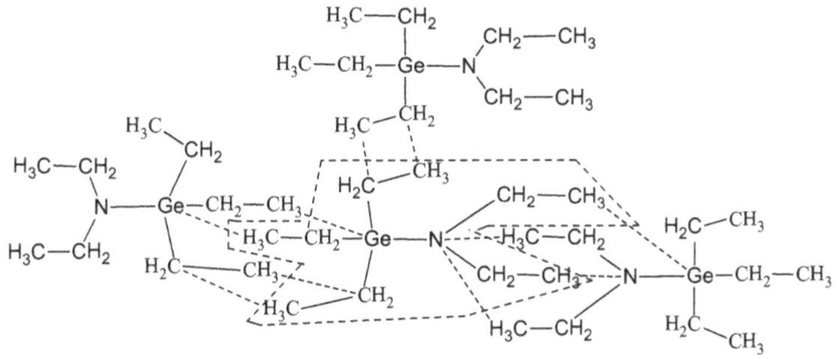

Fig. 3.3 Schematic of the network of specific interactions of liquid and crystalline triethyl (diethylamino)germane

3.1.5 Energies of Specific Interactions of Methoxygermane, Ethoxygermane, Methylthiogermane, and Hexamethyldigermoxane

Molecules of tetramethoxygermane and tetraethoxygermane with tetrahedral structure and location of the oxygen atoms at the vertices of tetrahedron form (like similar silane compounds) eight specific interactions $DGe-O \rightarrow CH_3-O-Ge$ and $DGe-O \rightarrow CH_3-CH_2-O-Ge$. The energies of these interactions can be calculated from the corresponding vaporization enthalpies by the depletion of the number of formed interactions (Table 3.5).

The obtained values illustrate the stabilization of $DGe-O \rightarrow CH_3-O-Ge$ $(5.02) < DGe-O \rightarrow CH_3-CH_2-O-Ge$ $(5.4 \text{ kJ mol}^{-1})$ interactions with an increasing number of carbon atoms in the alkyl ligand.

The trimethylmethoxygermane molecule with one ethyl group at the vertices of the tetrahedron forms six specific interactions by the three methoxy groups, with energies of 5.02 kJ mol^{-1}, and two interactions $D-Ge-H_3C \rightarrow H-CH_2$ of the methyl group, with unknown energy. The energy of this type of specific interaction is calculated using Eq. (3.7):

$$D - Ge - H_3C \rightarrow H - CH_2$$
$$= \left(\Delta_{vap}H^0(T\,K) - 6DGe - O \rightarrow CH_3 - O - Ge\right)/2 \quad (3.7)$$

The calculated energy of this type of specific interaction $(3.73 \text{ kJ mol}^{-1})$ indicates increased stability compared with the $DSi-H_3C \rightarrow H-CH_2-Si$ interaction $(3.25 \text{ kJ mol}^{-1})$ in liquid methyltrimethoxysilane. This draws attention to the inequalities in the energies of the specific interactions:

$DSi-O \rightarrow CH_3-O-Si$ (5.82) Tetramethoxysilane $> DGe-O \rightarrow CH_3-O-Ge$ (5.02) Tetrametoxygermane

Table 3.4 Energies (kJ mol^{-1}) of specific interactions of liquid triethyl(diethylamino)germane at 318 K

Compounds	Formula	Structure	$\Delta_{vap}H^0$ (298 K) [13]	DN → H$_3$C–CH$_2$– N	DisoCH$_2$–H$_3$C → CH$_2$– CH$_3$	D–Ge–CH$_2$– H$_3$C → Ge
Triethyl(diethylamino) germane	GeNC$_{10}$H$_{25}$	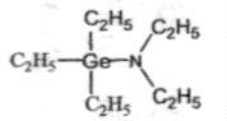	50.9	5.85	2.20	7.3

Table 3.5 Energies (kJ mol^{-1}) of specific interactions of liquid methoxygermane, ethoxygermane, methylthiogermane, and hexamethyldigermoxane

Compounds	Formula	Structure	$\Delta_{vap}H^0$ (298 K) [13]	T (K)	D-Ge-H$_3$C → H-CH$_2$	DGe-O→CH$_3$-O-Ge	DGe-O→CH$_3$-CH$_2$-O-Ge / CH$_2$-O-Ge
Tetramethoxygermane	O$_4$GeC$_4$H$_{12}$		40.2 ± 0.4	–	–	5.02	–
			44.0	344			
Tetraetoxygermane	O$_4$GeC$_8$H$_{20}$		43.1 ± 0.4	–	–	–	5.4
Trimethylmethoxygermane	OGeC$_4$H$_{12}$		32.4	304	3.73	5.02	–
Methylthiogermane	SGeCH$_6$		29.8	257	DGe-H•••H-G = 3.15	DGe-S → CH$_3$-S-Ge 10.2	–
Hexamethyldigermoxane	OGe$_2$C$_6$H$_{18}$		44.1	312	D-Ge-H$_3$C → H-CH$_2$		DGe-O → Ge
					3.25		1.3

DSi–O→CH$_3$–CH$_2$–O–Si (6.06) Tetraethoxysilane > DGe–O→CH$_3$–CH$_2$–O–Ge (5.4 kJ mol^{-1}) Tetraetoxygermane

The inequalities are caused by the increased shifting of electron density, in the shape of the reverse dative bond, from the oxygen atom to the germanium atom. This results in reduced charge at the oxygen atom in comparison with the charge at the oxygen atom in the tetramethoxysilane and tetraethoxysilane .

The molecule of methylthiogermane with three hydrogen atoms of GeH$_3$ forms three hydrogen bonds and two specific interactions DGe–S → CH$_3$–S–Ge. Taking the energy of the hydrogen bonds formed by the hydrogen atoms of GeH$_3$ and SiH$_3$ groups to be equal to 3.15 kJ mol^{-1}, the energy of the specific interaction DGe–S → CH$_3$–S–Ge can be estimated using Eq. (3.7a):

$$DGe - S \rightarrow CH_3 - S - Ge$$

$$= \left(\Delta_{vap}H^0(T\,K)mtg - 6DGe - H \bullet \bullet \bullet H - Ge \right)/2 \qquad (3.7a)$$

The results of these calculations are given in Table 3.5. The energy of the specific interaction DGe–S → CH$_3$–S–Ge is twice the energy of the interaction formed by the methoxy group. This difference is caused by stabilization of the sulfur atom and the 40 K difference in temperatures for the experimentally measured vaporization enthalpies.

The oxygen atom of the hexamethyldigermoxane molecule, with its location at the top of two tetrahedral fragments, contributes significant structural disorder and should be located at an angle of not less than 90°. The hydrogen atoms are located at the other three vertexes, forming hydrogen bonds with an energy of 3.15 kJ mol^{-1}. The oxygen atom, with high acceptor properties, provides the shifting of electron density from the germanium atoms. This shift should be expressed at the positive charges of the hydrogen and germanium atoms and in a reduction of the charge difference of the given atoms. As a result of the electron density distribution, there is a reduction in the energy of the hydrogen bond formed between the germanium atom and hydrogen. At the same time, the oxygen atom with two free electron pairs forms the specific interaction DGe–O → Ge with the essentially undivided 4s^2(Ge) electron pair of the germanium atom. This provides the germanium atom with a pentacoordinated structure in the liquid and crystal conditions (Fig. 3.4). As a result, a network is formed of 12 specific interactions of the six methyl groups and four specific interactions DGe–O → Ge. Taking the energy value of D–Ge–H$_3$C → H–CH$_2$ to be equal to 3.15 kJ mol^{-1}, a reduced estimated energy value of the DGe–O → Ge type of interaction was obtained using Eq. (3.8), the results of which are given in Table 3.5:

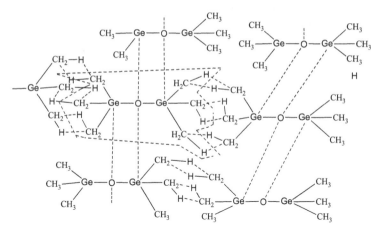

Fig. 3.4 Schematic of the network of specific interactions of liquid and crystalline hexamethyldigermoxane

$$DGe - O \rightarrow Ge$$
$$= \left(\Delta_{vap}H^0(TK)hego - 12DGe - H_3C - H - CH_2 - Ge\right)/4 \quad (3.8)$$

3.2 Energies of Specific Interactions of Tetraalkyltins and Tetramethyllead

Vaporization enthalpies of the elementorganic compounds of tin are presented for the seven alkyl compounds (Table 3.6) for which a thermodynamic analysis can be performed. Molecules with the saturated tetramethyltin, tetraethyltin, and tetrapropyltin and the unsaturated alkyl ligand tetravinyltin with symmetrical tetrahedral structure, form eight specific interactions of the methyl, ethyl, propyl, and vinyl ligand, respectively. This allow the energies of these bonds to be obtained directly from the enthalpy by the division by eight of these interactions. Presented in Table 3.6 are the energy values that allow a conclusion to be made regarding their stabilization by increasing the number of chain carbon atoms:

$DSn–H_3C{\rightarrow}H–CH_2–C–$ (4.1) $<$ $DSn–CH_2–CH_3{\rightarrow}CH_2–CH_3$ (6.33) $<$ $DSn–CH_2–CH_2–CH_3{\rightarrow}CH_2–CH_2–CH_3$ (8.17 kJ mol^{-1})

The energy value of the specific interaction formed by the vinyl $DSn–CH{=}CH_2 \rightarrow CH_2–CH–$ ligand has a reduced value because the temperature used for measuring the vaporization enthalpy was significantly higher than under standard conditions. A further three representatives of the alkyl series of ethyltrimethyltin, trimethylpropyltin, and trimethylvinyltin with three methyl groups form asymmetrical tetrahedral structures and thus allow us to neglect the influence of the methyl ligands on the energies of the specific interactions formed by the

Table 3.6 Energies (kJ mol^{-1}) of specific interactions of liquid tetraalkyltins

Compounds	Formula	Structure	$\Delta_{vap}H^0$ (298 K) [13]	T (K)	D–Sn–H$_3$C → H– CH$_2$–C–	DSn–CH$_2$– CH$_3$ → CH$_2$–CH$_3$	DSn–CH$_2$–CH$_2$–CH$_3$ → CH$_2$– CH$_2$–CH$_3$
Tetramethyltin	SnC$_4$H$_{12}$		31.1 ± 0.1 / 32.8 ± 0.1	298	3.9 / 4.1	–	–
Tetraethyltin	SnC$_8$H$_{20}$		50.6 ± 0.2	298	–	6.33	–
Tetrapropyltin	SnC$_{12}$H$_{28}$		65.4 ± 3.5	298	–	–	8.17
Tetravinyltin	SnC$_8$H$_{12}$		40.5	313–393	–	5.06	–
Ethyltrimethyltin	SnC$_5$H$_{14}$		37.7 ± 1.7	–	4.1	6.55	–
Trimethylpropyltin	SnC$_6$H$_{16}$		41.4	301	4.1	–	8.4
Trimethylvinyltin	SnC$_5$H$_{12}$		37.2 ± 2.1 / 37.7 ± 1.7	–	4.1	6.3 / 6.55	–

ethyl, vinyl, and propyl groups. The energies are obtained from the following equations:

$$DSn - CH_2 - CH_3 \rightarrow CH_2 - CH_3$$
$$= \left(\Delta_{vap}H^0(298\,K)etmt - 6DSn - H_3C \rightarrow H - CH_2 - C-\right)/2$$
$$(3.9a)$$

$$DSn - CH = CH_2 \rightarrow CH_2 - CH -$$
$$= \left(\Delta_{vap}H^0(298\,K)etmt - 6DSn - H_3C \rightarrow H - CH_2 - C - \right)/2$$
$$(3.9b)$$

$$DSn - CH_2 - CH_2 - CH_3 \rightarrow CH_2 - CH_2 - CH_3$$
$$= \left(\Delta_{vap}H^0(298K)etmt - 6DSn - H_3C \rightarrow H - CH_2 - C-\right)/2$$
$$(3.9c)$$

The calculation results given in Table 3.6 allow comparison of the stabilization of specific interactions with similar interactions of liquid symmetrical tetraethyltin, tetravinyltin, and tetrapropyltin.

The data in Table 3.7 on the vaporization enthalpy of alkylleads, obtained at temperatures differing from the standard conditions and without experimental error, exclude their thermodynamic analysis. Using the considered procedure for calculation of the energy of specific interactions, attention should be given to the energy value of the $DPb–H_3C \rightarrow H–CH_2–$ (4.76 kJ mol^{-1}) interaction formed in the liquid tetramethyllead. The energy value continues the natural sequence of stabilization for this type of specific interaction, which reflects the increasing shift of electron density at the energy level located far from the core of the element atom:

$$DSi–H_3C \rightarrow H–CH_2–C– \quad (3.25) < DGe–CH_2–CH_3 \rightarrow CH_2–CH_3 \quad (3.50) < D–Sn–$$
$$H_3C \rightarrow H–CH_2–C– (4.1) < DPb–H_3C \rightarrow H–CH_2–C– (4.76 \text{ kJ mol}^{-1})$$

Table 3.7 Energies (kJ mol^{-1}) of specific interactions of liquid tetramethyllead

Compounds	Formula	Structure	$\Delta_{vap}H^0$ (298 K) [13]	T (K)	DPb– $H_3C \rightarrow H–$ $CH_2–C–$	DPb– H•••H– Pb
Tetramethyllead	PbC$_4$H$_{12}$	CH$_3$ CH$_3$–Pb–CH CH$_3$	38.1 ± 0.4	303	4.76	–
Trimethylplumbane	PbC$_3$H$_{10}$	CH$_3$ CH$_3$–Pb–H CH$_3$	31.1	218	4.76	2.5
Dimethylplumbane	PbC$_2$H$_8$	CH$_3$ CH$_3$–Pb–H H	25.5	198	4.76	3.25

3.3   Energies of Specific Interactions of Tetraphenyl and Hexaphenyl Silicon Subgroup Elements

As revealed by Baev [14], the formation of the specific interaction with reduced stability by the alkyl fragment at its location between two strong acceptors of electron density (Table 2.10) was discussed with the example of N,N,N',N'-tetramethyldisilanylamine compounds. These are strong electron acceptors at the nitrogen atom and the silicon atom of the $-SiH_2-SiH_2-$ fragment, with a low ability to shift the electron density. Thermodynamic analysis showed that the reduced stability of the specific interaction formed by the ethyl group was caused by its location between a strong acceptor and a weak electron donor of the molecule of N-(1,1,-dimethylethyl)-1,1,1-triethylsilanamine (Table 2.11). This problem has received definite development in the example of sulfur-containing silicon organic compounds (see the section in Chap. 2 entitled "Trimethoxy[(methylthio)alkyl] silanes, [(Ethylthio)alkyl]trimethoxysilanes, and Triethoxy[(ethylthio)alkyl] silane"). The information from these studies significantly enriched understanding of the compounds with a benzene ring and element atoms of the silicon subgroup.

The low energy values of the silicon and germanium specific interactions, stabilizing with an increasing number of atoms of this subgroup, are a direct indication of the low coordinating ability of the elements of this group. As a result, the central atom of a compound with an alkyl ligand and symmetrical tetrahedral structure is not able to express increased coordination and can exert an influence only by the shifting of insignificant electron density with essentially undivided $2s^2$, $3s^2$, $4s^2$, $5s^2$, or $6s^2$ electron pairs or by receiving relatively low electron density. Therefore the silicon atom and its analogues are not able to show the high coordination except for the four-coordinated condition with symmetric molecule of tetraphenylsilane (Table 3.8). As a result, all specific interactions in the crystal and liquid conditions are formed by the CH groups and by the carbon atom with saturated hydrogen atom, $DHC \rightarrow CH$ and $DC \rightarrow C$, with different charges that interact with each other by the essentially undivided $2s^2$ electron pair. The silicon atom and its analogs participate in the formation of specific interactions $DSi \rightarrow Si$ with the contacting molecules, forming networks of the specific interactions (Fig. 3.5).

The energy of the specific interaction of the CH group of the benzene ring is determined from the benzene sublimation enthalpy $\Delta_{sub}H^0$(298 K)/6, which is equal to 7.43 kJ mol^{-1}, and the second type of $DC \rightarrow C$ interaction minus the energy contribution of the substituted tetraphenylsilane hydrogen atom (1 kJ mol^{-1}) [12]. The energy of the specific interaction thus has a value of 6.43 kJ mol^{-1}. The experimentally measured value of the sublimation enthalpy of tetraphenylsilane has a value of 156.9 ± 1.7 kJ mol^{-1}, from which the energy contribution of one benzene ring with saturated hydrogen atoms can be obtained as 39.22 kJ mol^{-1}. Without the replacement of hydrogen atoms, the value is 40.22 kJ mol^{-1}. Using these values, the energy values of specific interactions are 6.70 kJ mol^{-1} for $DHC \rightarrow CH$ and 5.70 kJ mol^{-1} for $DC \rightarrow C$. According to these data, it can be concluded that the

Table 3.8 Energies (kJ mol^{-1}) of specific interactions formed by CH groups of the benzene ring of crystalline tetraphenylmetals at 298 K

Compounds	Formula	Structure	$\Delta_{sub}H^0$ (298 K) [15]	$5DHC \rightarrow CH + DC \rightarrow C$	$DHC \rightarrow CH$	$DC \rightarrow C$
$\Delta_{sub}H^0$(298 K) = 44.6, DHC $\rightarrow$ CH = 7.433, DC $\rightarrow$ C = 6.433						
Tetraphenylsilane	$C_{24}H_{20}Si$		156.9 ± 1.7	39.22	6.70	6.20
Tetraphenylgermane	$C_{24}H_{20}Ge$		156.9 ± 4.2	39.22	6.70	6.20
Tetraphenyltin	$C_{24}H_{20}Sn$		161.1 ± 4.7	40.22	6.87	6.37

(continued)

Table 3.8 (continued)

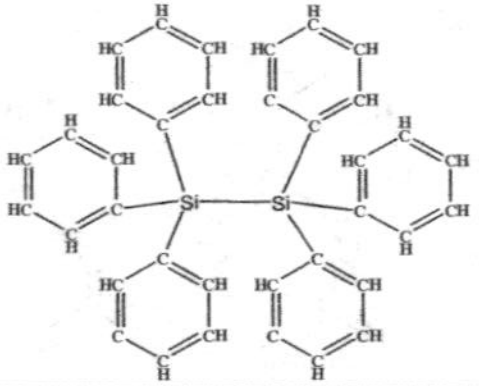

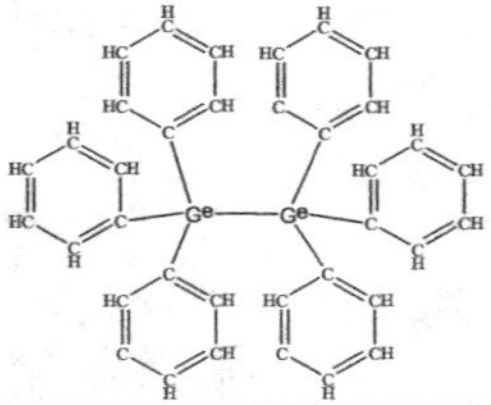

Compounds	Formula	Structure	$\Delta_{sub}H^0$ (298 K) [15]	5DHC→CH+DC→C	DHC→CH	DC→C
Tetraphenyllead	$C_{24}H_{20}Pb$	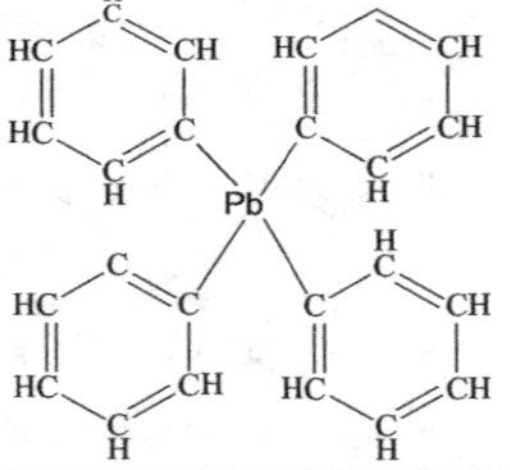	159.1	39.8	6.80	6.30
Hexaphenyldisilane	$C_{36}H_{30}Si_2$		209.9 ± 2.0	39.22 32.5[a]	6.70 5.58[a]	6.20 5.08[a]
Hexaphenyldigermane	$C_{36}H_{30}Ge_2$		209.2 ± 4.2	39.22 32.8[a]	6.70 5.63[a]	6.20 5.13[a]

| Hexaphenylditin | $C_{36}H_{30}Sn_2$ | 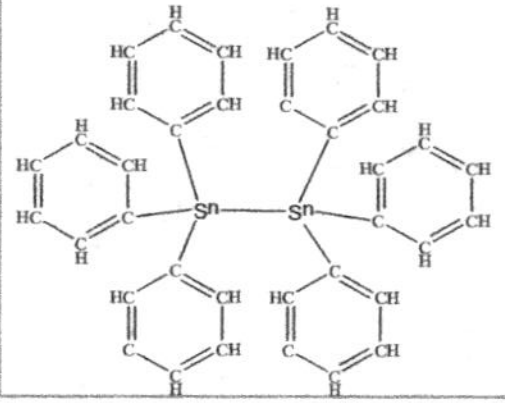 | 188.3 ± 4.2 | 40.22 | | 6.87 | 6.37 |
| | | | | 27.0^a | | 4.67^a | 4.17^a |

[a]Specific interactions formed by the benzene ring at the position of isostructural group

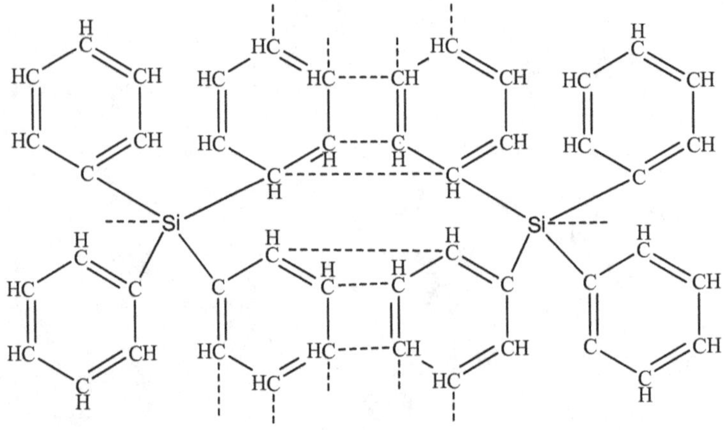

Fig. 3.5 Schematic of the network of specific interactions of liquid and crystalline tetraphenylsilane

shifting of moderate density from the silicon atom to the carbon atom of the ring is accompanied by a significant redistribution of electron density in the benzene ring, reducing the differences in the charges at the carbon atoms and causing a reduction in the energies of the formed specific interactions DHC → CH and DC → C. The calculated energy values of similar specific interactions of the benzene ring of the crystal compounds of tetraphenylgermane, tetraphenyltin, and tetraphenyllead are given in Table 3.8. Based on these data, it can be concluded that the interaction of the carbon atom of a benzene ring with an element atom of low coordinating ability and low donor properties is accompanied by a reduction in the charge difference of the ring carbon atoms and by a reduction in the stability of the formed specific interactions.

Even more significant weakening of the formed specific interactions of DHC → CH and DC → C is expressed at the location of the benzene ring in the position of the isostructural group in the hexaphenyldisilane and hexaphenyld-itin molecules (Table 3.8).

References

1. Potts AW, Price WC (1972) The photoelectron spectra of methane, silane, germane and stannane. Proc R Soc Lond A 326(1565):165–179
2. Caballol R, Catala JA, Problet JM (1986) Jahn–Teller distortions in XH_4^+ radical cations (X = Si, Ge, Sn): an ab initio CI study. Chem Phys Lett 130(4):278–284
3. Ensslin PW, Bergmann H, Elbel S (1975) Photoelectron spectra of polysilanes. Conformational study of tetra- and penta-silane. J Chem Soc Faraday Trans Pt II 71:913–920
4. Bock H, Ensslin W, Feher W, Freund R (1976) Photoelectron spectra and molecular properties. LI. Ionization potentials of silanes Si_nH_{2n+2}. J Am Chem Soc 98(3):668–674

5. Jonas AE, Schweitzer GK, Grimm FA, Carlson TA (1972) The photoelectron spectra of the tetrafluoro and tetramethyl compounds of the group IV elements. J Electron Spectrosc Relat Phenom 1(1):29–66

6. Fernandez J, Lespes G, Dargelos A (1987) Theoretical and experimental study of the vacuum ultraviolet spectrum of tetrasubstituted tin derivatives $SnCl_4$ and $Sn(CH_3)_4$. Chem Phys 111 (1):97–103

7. Novak I, Benson JM, Svensson A, Potts AW (1987) Photoelectron study of tetramethyl stannane using synchrotron radiation. Chem Phys Lett 135(4–5):471–474

8. Evans S, Green JC, Joachim PJ, Orchard AF, Turner DW, Maier JP (1972) Electronic structures of the group IVb tetramethyls by helium-(I) photoelectron spectroscopy. J Chem Soc Faraday Trans Pt II 68(6):905–911

9. Baev AK (1969) Problems of chemical nature of phase transformation, vol 1, General and applied chemistry. Vaisheshika Educating, Minsk, pp 197–206

10. Baev AK (1969) Phase condition and complex formation ability of halogenide metals, vol 1, General and applied chemistry. Vaisheshika Educating, Minsk, pp 207–218

11. Baev AK (1972) Complex formation ability halogenides of second–six groups periodical system, vol 5, General and applied chemistry. Vaisheshika Educating, Minsk, pp 35–51

12. Baev AK (2012) Specific intermolecular interactions of organic compounds. Springer, Heidelberg/Dordrecht/London/New York

13. Chickos JS, Acree WE Jr (2003) Enthalpies of vaporization of organic and organometallic compounds, 1880–2002. J Phys Chem Rev 32(2):519

14. Baev AK (2014) Specific intermolecular interactions of nitrogenated and bioorganic compounds. Springer, Heidelberg/Dordrecht/London/New York

15. Chickos JS, Acree WE Jr (2002) Enthalpies of sublimation of organic and organometallic compounds, 1910–2001. J Phys Chem Rev 31(2):537

Chapter 4
Specific Intermolecular Interactions of Sulfonated Organic Compounds

4.1 Energies of Specific Interactions of Sulfides and Disulfides

4.1.1 Symmetrical Sulfides with Normal Molecular Structure and Isostructural Methyl Groups

Study of the electron structure of dimethyl sulfide using photoelectron spectroscopy [1, 2] and calculation of the electronic structure by the ab initio method [2] have highlighted the sequence matching of the four upper orbitals of dimethyl ether O $(CH_3)_2$ and dimethyl sulfide $S(CH_3)_2$: $1b_1 < 4a_1 < 3b_2 < 1a_2$. Dimethyl sulfides are characterized by increased hyperconjugation, which can be expressed by the stabilization of the formed intermolecular interactions. Studies of the high resolution photoelectron spectroscopy of unsaturated compounds, including vinyl and allyl sulfides [3–7], indicate the existence of a band of the generically less favorable gauche form. The contribution of this form is 6 and 20% at 40 and 200°C, respectively, and can contribute definite errors to the experimentally determined vaporization enthalpy of these compounds.

One piece of interesting initial information can be obtained from the vaporization enthalpies of symmetrical dialkyl sulfides (presented in Fig. 4.1) as a functional dependence $\Delta_{vap}H^0(298\ K) = f(Cn)$, where Cn is the number of carbon atoms in the alkyl chain. Taking into account the correlation of the enthalpy of vaporization with the number and energies of specific interactions in the condensed state [8–10], which burst at the transition to a vapor of monomer molecules (ethers), the two intersecting lines obtained experimentally for the vaporization enthalpy of dipropyl sulfide are caused first by the completion of the intermolecular interaction influence of the reverse dative bond on the energy of the specific interaction $DS \rightarrow CH_3-CH_2-CH_2-S$ formed by the propyl ligand with three carbon atoms in the chain (Fig. 4.1, line 1).

© Springer International Publishing Switzerland 2015
A.K. Baev, *Specific Intermolecular Interactions of Element-Organic Compounds*,
DOI 10.1007/978-3-319-08563-0_4

Fig. 4.1 Dependence of
vaporization enthalpy on
the number of carbon atoms
in the alkyl group of
symmetrical sulfides: (*1*)
weakening of influence of
the reverse dative bond, (2)
influence of the methylene
group

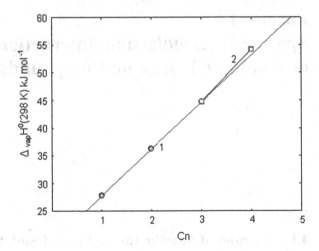

The second line (Fig. 4.1, line 2) is caused by the extra energy contribution of the fourth carbon atom of the methylene group to the vaporization enthalpy of dibutyl sulfide and their further increasing number in the sulfide molecules. A similar dependency to that presented in Fig. 4.1 is expressed by the dialkyl ethers [11]. Dialkyl sulfide molecules, with two bond vacancies of the free electron pairs of the sulfur atom and an essentially undivided $2s^2(c)$ electron pair of the carbon atoms of the two methyl groups of the alkyl ligands, form four specific interactions that produce chains of the ligands located under the "right" angle (Fig. 4.2).

The chains are connected to each other by weak intermolecular interactions, with the energies not exceeding the errors of the measured enthalpy values. The energies of the specific interactions are determined from the enthalpy value by dividing by the number of interactions (Table 4.1). The obtained energy values of the specific interactions are described by the natural sequence of stabilizations, reaching its maximum value at the propyl ligand of dipropyl sulfide:

Dimethyl sulfide (6.92) < Diethyl sulfide (9.05)

< Dipropyl sulfide $(11.17$ kJ mol$^{-1})$

Therefore, the stability of the specific interaction in all further compounds of this series is unchanged, and the increase in the vaporization enthalpy is determined by the energy contribution of the methylene group and its number. A similar range of natural stabilization is described by the specific interactions of the symmetrical simple ethers:

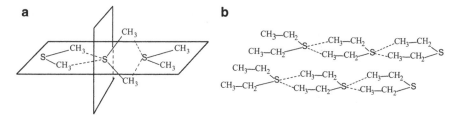

Fig. 4.2 Schematic of the network of specific interactions of liquid symmetrical dimethyl sulfide (**a**) and diethyl sulfide (**b**)

$$\text{Dimethyl ether } (5.63) < \text{Diethyl ether } (6.78)$$

$$< \text{Dipropyl ether } \left(8.92 \text{ kJ mol}^{-1}\right)$$

The energy contribution of the methylene group is equal to the difference in the vaporization enthalpies of dibutyl sulfide and dipropyl sulfide, divided by the two methylene groups of the two butyl ligands (Table 4.1). The increased stability of the specific interactions of sulfides reflects the shifting of the increased electron density from the sulfur atom, located at the more distant third energy level, to the carbon atom.

Presented in Fig. 4.3 is the functional dependence $\Delta_{vap}H^0(298 \text{ K}) = f(\text{Cn})$ of symmetrical dialkyl sulfides and diisoalkyl sulfides, with reduced vaporization enthalpy of diisopentyl sulfide, obtained at a temperature (352 K) that is significantly different from the standard conditions; this illustrates their correlation with each other – the two isostructural methyl groups contribute practically the same energy to the vaporization enthalpy of the compounds. Molecules of symmetrical diisoalkyl sulfides express the same four bond vacancies by the two free electron pairs of the sulfur atoms and two carbon atoms of the methyl groups. Additionally, each isostructural methyl group of the molecule forms two specific interactions $-\text{S}-\text{H}_2\text{C}\overset{\text{H}}{\underset{\text{H}}{\cdots}}\text{CH}_2-\text{S}-$ with a reduced energy value. The compounds of this series are characterized by a reduction in the influence of the reverse dative bond on the stability of the formed specific interaction with an increasing number of carbon atoms in the chain, ending at the propyl ligand (Fig. 4.3). Stable specific interactions crosslink the molecules to the chains and interactions of reduced stability formed by the isostructural methyl groups crosslink the chains to form a network of all interactions in the liquid and crystal conditions (Fig. 4.4). Formed by the alkyl ligands of symmetrical dialkyl sulfides with a normal molecular structure and with an isostructural methyl group, the stable specific interactions remain unchanged for each ligand. That is why the differences in the vaporization enthalpies of these two series of compounds (Fig. 4.3) are the energy contributions of the isostructural methyl groups. However, it is necessary to take into account the energy contribution of the saturated hydrogen atom of the methyl group, which is introduced to the vaporization enthalpy of the compound with a normal molecular structure. The value of this energy contribution is equal to 0.60 kJ mol^{-1} for liquid organic compounds [11].

Table 4.1 Energies (kJ mol^{-1}) of specific interactions of liquid symmetrical dialkyl sulfides at 298 K

Compounds	Formula	Structure	$\Delta_{vap}H^0$ (298 K) [12]	DS → CH$_3$-S	DS → CH$_3$-CH$_2$-S	DS → CH$_3$-(CH$_2$)$_2$-S	DS → CH$_3$-(CH$_2$)$_3$-S
Dimethyl sulfide	C$_2$H$_6$S		27.7	6.92	–	–	–
Diethyl sulfide	C$_4$H$_{10}$S		36.2^a	–	9.05	–	–
Dipropyl sulfide	C$_6$H$_{14}$S		44.7 ± 0.8	–	–	11.17	–
Dibutyl sulfide	C$_8$H$_{18}$S		54.2 ± 0.8	–	–	11.17	11.17 ΣDCH$_2$ = 4.75 × 2
Dipentyl sulfide	C$_{10}$H$_{22}$S		64.1^a	–	–	11.17	4.85 × 4
Dihexyl sulfide	C$_{12}$H$_{26}$S		73.1^a	–	–	11.17	4.73 × 6
Diheptyl sulfide	C$_{14}$H$_{30}$S		82.7^a	–	–	11.17	4.75 × 8

^{a}Estimated on the basis of the correlation $\Delta_{vap}H^0$(298 K)das $= f(\Delta_{vap}H^0$(298 K) ethyl-R sulfides)

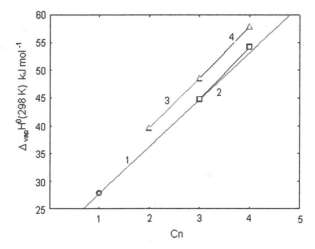

Fig. 4.3 Dependence of the vaporization enthalpy on the number of carbon atoms in the alkyl group of symmetrical sulfides: decreasing influence of the reverse dative bond on the vaporization enthalpy (*1, 2*) and energy contribution of the CH$_2$ groups to the enthalpy property for diisopropyl sulfide, diisobutyl sulfide, and diisopentyl sulfide (*3, 4*)

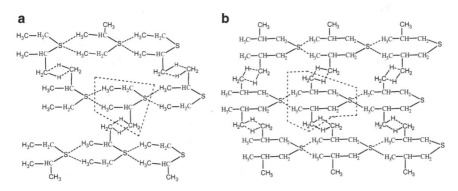

Fig. 4.4 Schematic of the network of specific interactions of liquid symmetrical ethyl isopropyl sulfide (**a**) and diisobutyl sulfide (**b**)

Therefore, the contribution of the single isostructural methyl group of ethyl isopropyl sulfide is equal to the difference in the vaporization enthalpies of this compound and diethyl sulfide (36.2 kJ mol^{-1}), and the energy contribution of the hydrogen atom can be calculated from the enthalpy characteristic:

$$\text{DisoCH}_3 = \Delta_{vap}H^0(298\,\text{K})\text{eips} - \left(\Delta_{vap}H^0(298\,\text{K})\text{des} - \text{DH}\right) \qquad (4.1)$$

Diisopropyl sulfide:

$$\text{DisoCH}_3 = \Delta_{vap}H^0(298\,\text{K}) - \left(\Delta_{vap}H^0(298\,\text{K})\text{des} - 2\text{DH}\right)/2 \qquad (4.1a)$$

The energy contribution of the two isostructural methyl groups to the vaporization enthalpy of diisobutyl sulfide is the difference in the vaporization enthalpies of diisobutyl sulfide and the dipropyl sulfide compound with the same number of carbon atoms in the alkyl chain.

Diisobutyl sulfide:

$$\text{DisoCH}_3 = \Delta_{vap}H^0(298\,\text{K}) - \left(\Delta_{vap}H^0(298\,\text{K})\text{dps} - 2\text{DH}\right)/2 \qquad (4.1b)$$

To calculate the energy contribution of the similar methyl groups of diisopentyl sulfide, the same equation is used but dipropyl sulfide is replaced with dibutyl sulfide. The calculations presented in Table 4.2 indicate the equivalent low energy contribution of the isostructural groups, with their location at the first and third carbon atoms in the molecules of diisopropyl sulfide and diisobutyl sulfide. There is a tendency for an increase in energy contribution at the isostructural methyl group located at the second carbon atom in the diisobutyl sulfide molecule.

4.1.2   Asymmetrical Sulfides of Methyl-R with Normal Structure and Isostructural Methyl Group

It is noteworthy that after completion of the intramolecular reverse dative bond influence following an increase in number of carbon atoms in the chain (up to the propyl ligand), the end methyl group forms a specific interaction of increased stability with the sulfur atom, $\text{DS} \rightarrow \text{CH}_3–\text{CH}_2–\text{CH}_2–\text{S}$. Nevertheless, further increase in the number of methylene groups in the molecule is accompanied by more significant increase in the vaporization enthalpy; a positive deviation from the line reflects the result of the maximum possible stability of this type of interaction (Fig. 4.1). In this regard, it is notable that the negative deviation from a similar line describes the maximizing energy value of the same type of specific interaction of a series of liquid asymmetrical sulfides of methyl-R (Fig. 4.5). A characteristic feature of this series of compounds is the increase of one chain length with an increasing number of carbon atoms with a constant second chain length of one methyl group. Despite the high chain mobility, the specific interactions formed provide it with an increasing energy intensity because of the difference in the length of the two ligands. As a result, each further methylene group of the chain reduces the energy contributed to the vaporization enthalpy. This should be clearly expressed by the structure of the liquid and crystal conditions, as shown in the network of specific interactions of liquid methyl propyl sulfide in Fig. 4.6. The lowest energy structural intensities are expressed by ethyl methylsulfide and methyl propyl sulfide. Therefore, taking the energy of the specific interaction $\text{DS} \rightarrow \text{CH}_3–\text{S}$ of the liquid ethyl methylsulfide to be equal to the value obtained for liquid dimethyl sulfide (6.92 kJ mol^{-1}), the energy of the second type of specific interaction $\text{DS} \rightarrow \text{CH}_3–\text{CH}_2–\text{S}$ can be obtained using Eq. (4.2):

Table 4.2 Energies (kJ mol^{-1}) of specific interactions of liquid symmetrical diisoalkyl sulfides at 298 K

Compounds	Formula	Structure	$\Delta_{vap}H^0$ (298 K) [12]	DS $\rightarrow$ CH$_3$–CH$_2$–S	DS $\rightarrow$ CH$_3$–(CH$_2$)$_2$–S	DisoCH$_3$	DH$_3$C $\rightarrow$ H–CH$_2$
Ethyl isopropyl sulfide	C$_5$H$_{12}$S		37.9 ± 0.8	9.05	–	2.3	1.15
Diisopropyl sulfide	C$_6$H$_{14}$S		39.6 ± 0.8	9.05	–	2.3 × 2	1.15
Diisobutyl sulfide	C$_8$H$_{18}$S		48.5 ± 0.8	–	11.17	2.5 × 2	1.25
Diisopentyl sulfide	C$_{10}$H$_{22}$S		57.9	–	11.17	2.3 × 2	1.15
							$\sum$DCH$_2$
							4.7 × 2

Fig. 4.5 Dependence of the vaporization enthalpy on the number of carbon atoms in the alkyl group of asymmetrical sulfides of methyl-R: (*1*) weakening influence of the reverse dative bond, (*2*) growing influence of the contribution of methylene groups

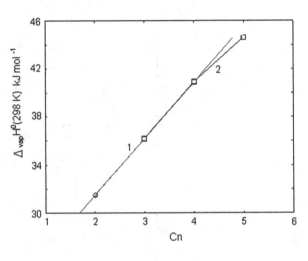

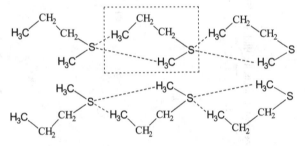

Fig. 4.6 Schematic of the structural strain of liquid methyl propyl sulfide

$$DS \rightarrow CH_3 - CH_2 - S = \left(\Delta_{vap}H^0(298\,K)ems - 2DS \rightarrow CH_3 - S\right)/2 \quad (4.2)$$

This results in an energy of 8.83 kJ mol^{-1}. The given energy value of the specific interaction is less than that of liquid diethyl sulfide (9.05 kJ mol^{-1}) by 0.22 kJ mol^{-1}, which can be attributed to the influence of the steric intensity due to the difference in the two alkyl ligands. The energy value of the specific interaction DS $\rightarrow$ CH$_3$-S should be used in calculating the energy of the specific interaction DS $\rightarrow$ CH$_3$-(CH$_2$)$_2$-S, with the maximum energy value at the methyl propyl sulfide, and for the given series of compounds with large numbers of carbon atoms in the methyl-R chain (Table 4.3):

$$DS \rightarrow CH_3 - (CH_2)_2 - S$$
$$= \left(\Delta_{vap}H^0(298\ K)mps - 2DS \rightarrow CH_3 - S\right)/2 \quad (4.2a)$$

As shown in Table 4.1, the energy value of the specific interaction of the formed propyl ligand is in good agreement with the energy of this type of interaction of dipropyl sulfide and is fully consistent with the distribution of energy values for the

Table 4.3 Energies (kJ mol^{-1}) of specific interactions of liquid asymmetrical alkyl sulfides methyl-R at 298 K

Compounds	Formula	Structure	$\Delta_{vap}H^0$ (298 K) [12]	DS → CH$_3$–S	DS → CH$_3$–CH$_2$–S	DS → CH$_3$–(CH$_2$)$_2$–S	DS → CH$_3$–(CH$_2$)$_3$–S
Ethyl methylsulfide	C$_3$H$_9$S		31.5	6.92	8.83	–	–
Methyl propyl sulfide	C$_4$H$_{10}$S		36.2	6.92	–	11.18	–
Butyl methyl sulfide	C$_5$H$_{12}$S		40.9 ± 0.8	6.92	–	11.18	$\sum$DCH$_2$ = 1.92 × 2
Methyl pentyl sulfide	C$_6$H$_{14}$S		44.6 ± 0.8	6.92	–	11.18	$\sum$DCH$_2$ = 2.15 × 4
Methyl hexyl sulfide	C$_7$H$_{16}$S		49.3	6.92	–	11.18	2.2 × 6
Methyl heptyl sulfide	C$_8$H$_{18}$S		53.9	6.92	–	11.18	2.2 × 8
Methyl octyl sulfide	C$_9$H$_{20}$S		58.4	6.92	–	11.18	2.2 × 10 DisoCH$_3$ /DH$_3$C → H–CH$_2$
Methyl isopropyl sulfide	C$_4$H$_{10}$S		34.1	6.92	8.83	–	3.2/1.60
3-Methyl-2-thiopentane	C$_5$H$_{12}$S		38.3[a]	6.92	–	11.18	2.7/1.35
4-Methyl-2-thiopentane	C$_5$H$_{12}$S		36.9[a]	6.92	–	11.18	1.3/0.65

[a]303 K

specific interactions of the simple ethers of two series, dialkyl ethers and CH_3–O–R [11].

The value of the energy contribution of the methylene group to the vaporization enthalpy at maximum stability of the specific interaction is determined from the difference in the vaporization enthalpies of butyl methyl sulfide and methyl propyl sulfide. The data given in Table 4.3 indicate an approximately twofold reduction (1.92 and 2.15 kJ mol^{-1}) for butyl methyl sulfide and methyl pentyl sulfide, respectively, in comparison with the energy contribution of the methylene group of dibutyl sulfide (4.75 kJ mol^{-1}) in the absence of steric strain in the network structure of the interactions (Fig. 4.2b). Therefore, this is proof of the presence of steric stain in the structure of the liquid and crystal conditions of the compounds of the methyl-R series.

Taking into account the procedure for calculation of the energy contribution of the isostructural methyl group to the vaporization enthalpy, calculations of the energy values for the asymmetrical sulfides were performed. The data presented in Table 4.3 indicate that a relatively increased energy contribution is made by the isostructural methyl group located at the carbon atom of the ethyl ligand connected with the sulfur atom:

Methyl isopropyl sulfide (3.2) > 3-Methyl-2-thiopentane (2.7) < 4-Methyl-2-thiopentane (1.3 kJ mol^{-1})

The increase in the number of carbon atoms in the propyl ligand is accompanied by a reduction in the energy contribution, and the location at the second carbon atom leads to a reduction of the contribution to the enthalpy value. It follows that the participation of the isostructural methyl group in the distribution of electron density is significantly changed from its location in the molecule [11].

4.1.3 Asymmetrical Sulfides of Ethyl-R with Normal Structure and Isostructural Methyl Group

In comparison with ethyl methyl sulfide, a molecule of ethyl propyl sulfide (ethyl-R with an increased number of carbon atoms in the smaller fragment) experiences less structural stress and, so, it appears at least adjacent in this series of compounds. This is clearly manifested in the dependence $\Delta_{vap}H^0(298\ K) = f(Cn)$ (Fig. 4.7), which makes it difficult to break the two initial compounds of the series. Because the influence of the intramolecular reverse dative bond is not expressed at the propyl compound, in the molecule of butyl ethyl sulfide the methylene group contributes to the vaporization enthalpy. The energy contribution of this group is obtained from the difference in the enthalpy characteristics of butyl ethyl sulfide and ethyl propyl sulfide (4.6 kJ mol^{-1}), which differs little from the similar contribution of the methylene group for the symmetrical molecule of dibutyl sulfide (4.75 kJ mol^{-1}), and does not undergo structural defects in the liquid condition following the

Fig. 4.7 Dependence of the vaporization enthalpy $\Delta_{vap}H^0(298\ K)$ of asymmetrical sulfides of ethyl-R (*1, 2*) on the number of carbon atoms in the alkyl chain

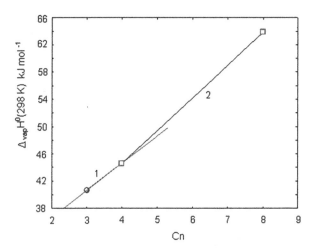

formation of a network of specific interactions. It follows that the molecule of ethyl propyl sulfide with four bond vacancies, forming specific influences, two of them by the ethyl ligand and two by the propyl ligand, should undergo weak structural disorders. Therefore, the energy of the specific interaction of DS $\rightarrow$ CH$_3$–CH$_2$–S, formed by the ethyl ligand of the ethyl propyl sulfide molecule, is taken to be equal to the energy of the same type of interaction (9.05 kJ mol^{-1}) formed by the molecule of diethyl sulfide, which is obtained with the help of Eq. (4.2b):

$$DS \rightarrow CH_3 - CH_2 - CH_2 - S$$
$$= \left(\Delta_{vap}H^0(298\,K)eps - 2DS \rightarrow CH_3 - CH_2 - S\right)/2 \qquad (4.2b)$$

Taking into account the energy contribution of all four interactions of the two types, the energy of the specific interaction DS $\rightarrow$ CH$_3$–CH$_2$–CH$_2$–S is equal to the value of 10.95 kJ mol^{-1} without structural disorder of the DS $\rightarrow$ CH$_3$–CH$_2$–S bond. On the other hand, taking the energy value of DS $\rightarrow$ CH$_3$–CH$_2$–S to be equal to the energy of specific interactions of the same type in liquid ethyl methyl sulfide (8.83 kJ mol^{-1}), a somewhat increased energy value is obtained for the second more stable type of specific interaction, DS $\rightarrow$ CH$_3$–CH$_2$–CH$_2$–S (11.17 kJ mol^{-1}) (Table 4.4).

Presented in Fig. 4.7 is the functional dependence $\Delta_{vap}H^0(298\ K) = f(Cn)$, which is used for estimation of the vaporization enthalpy of unstudied compounds using interpolation and extrapolation. Thus, from the difference in the vaporization enthalpies of ethyl propyl sulfide and the compound in question, the energy contributions of the methylene groups can be determined (Table 4.4).

The reliability of the estimated values of vaporization enthalpy of ethyl-R series compounds presented in Table 4.4 is proved by the straight line in the functional dependence plot for other series of the same class of compounds [11]. Using the dependence of vaporization enthalpies of symmetrical sulfides and asymmetrical

Table 4.4 Energies (kJ mol⁻¹) of specific interactions of liquid asymmetrical ethyl alkyl sulfides at 298 K

Compounds	Formula	Structure	$\Delta_{vap}H^0$ (298 K) [12]	DS → CH₃–CH₂–S	DS → CH₃–(CH₂)₂–S	ΣDCH₂
Ethyl propyl sulfide	$C_5H_{12}S$		40.0	9.05 / 8.83[a]	10.95 / 11.17[a]	–
Butyl ethyl sulfide	$C_6H_{14}S$		44.6 ± 0.8	9.05 / 8.83[a]	10.95 / 11.17[a]	4.6
Pentyl ethyl sulfide	$C_7H_{16}S$	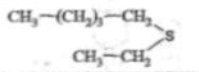	49.2[b]	8.83	11.17	9.2
Hexyl ethyl sulfide	$C_8H_{18}S$	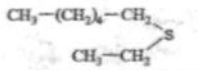	54.0[b]	8.83	11.17	14.0
Heptyl ethyl sulfide	$C_9H_{14}S$	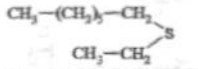	58.9[b]	8.83	11.17	18.9
1-Ethylthiooctane	$C_{10}H_{22}S$	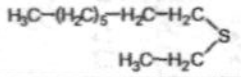	63.9 ± 0.6	8.83	11.17	23.9
1-Ethylthiononyl	$C_{11}H_{24}S$	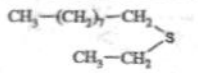	68.6[b]	8.83	11.17	28.6 / DisoCH₃/DH₃C → H–CH₂
Isopropyl propyl sulfide	$C_6H_{14}S$	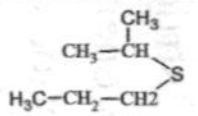	41.8	8.83	11.17	2.40/1.20
Ethyl isobutyl sulfide	$C_6H_{14}S$	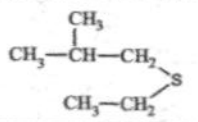	41.3[c]	8.83	11.17	1.90/0.95

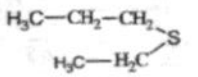

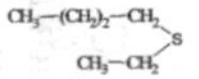

[a]Preferable

[b]Estimated on the base of correlation $\Delta_{vap}H^0$(298 K) = f(Cn)

[c]Corresponding to 320 K

sulfides of the methyl-R series compared with the corresponding vaporization enthalpies of the asymmetrical sulfides of ethyl-R series, the enthalpy values for unstudied symmetrical (Table 4.1) and asymmetrical sulfides of the methyl-R series (Table 4.3) were estimated with high reliability.

For the considered ethyl-R series, there are two compounds with correct values of the vaporization enthalpies (Table 4.4). For calculation of the energy contribution of the isostructural methyl group, it is necessary to use Eq. (4.1) or to obtain the energy contribution of the isopropyl propyl sulfide methyl group from the difference in the vaporization enthalpy of this compound and the ethyl propyl sulfide with the ethyl ligand minus the energy of the substituted hydrogen atom (Eq. 4.3):

$$\text{DisoCH}_3 = \Delta_{vap}H^0(298\,\text{K})\text{ipps} - \left(\Delta_{vap}H^0(298\,\text{K})\text{eps} - \text{DH}\right) \qquad (4.3)$$

For ethyl isobutyl sulfide, the energy contribution is obtained from the difference in the vaporization enthalpies of the same ethyl propyl sulfide minus the energy contribution $(0.60\,\text{kJ mol}^{-1})$ of the substituted hydrogen atom (Eq. 4.3a):

$$\text{DisoCH}_3 = \Delta_{vap}H^0(298\,\text{K})\text{eibs} - \left(\Delta_{vap}H^0(298\,\text{K})\text{eps} - \text{DH}\right) \qquad (4.3a)$$

Given in Table 4.4 are the values of the energy contribution of the isostructural methyl group to the vaporization enthalpy and the energy of the reduced stability of the specific interaction, which are in good agreement with the measured values for the symmetrical and asymmetrical sulfides of the methyl-R series. The energy values presented in Table 4.4 of the same properties for ethyl isobutyl sulfide were derived at 320 K and are therefore underestimated.

4.1.4   Energies of Specific Interactions of tert-Butyl Sulfides

The molecules of the *tert*-butyl sulfides, being analogs of the *tert*-butyl ethers, belong to the compounds with isostructural methyl groups, due to their properties [11]. The molecule of methyl *tert*-butyl sulfide with methyl and ethyl ligands and asymmetrical structure has a structure strain similar to that of ethyl methyl sulfide and has the same four bond vacancies and two isostructural methyl groups. Therefore, it forms, together with the molecules of the near environment, the two specific interactions $DS \rightarrow CH_3-CH_2-S$ by the ethyl fragment with an energy of $8.83\,\text{kJ mol}^{-1}$, two specific interactions of the second type $DS \rightarrow CH_3-S$ by the methyl ligand with an energy of $6.92\,\text{kJ mol}^{-1}$, and, additionally, four specific interactions $DH_3C \rightarrow H-CH_2$ of reduced stability. All three types of interactions form the network structure of the liquid condition, which is schematically presented in Fig. 4.8a.

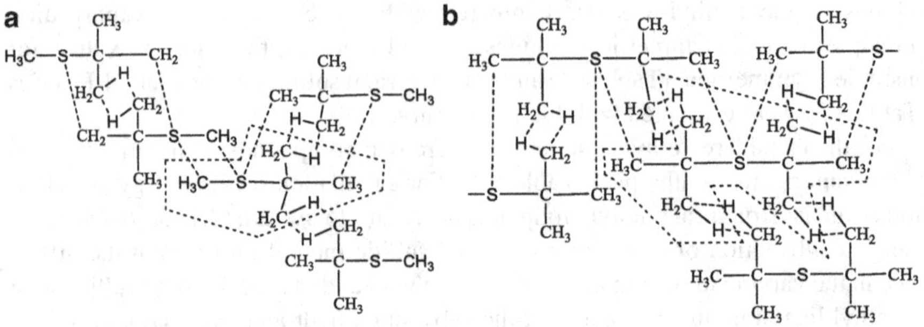

Fig. 4.8 Schematic of the network of specific interactions of liquid methyl *tert*-butyl sulfide (**a**) and D-*tert*-butyl sulfide (**b**)

The value of the energy contribution of the methyl group is calculated using Eq. (4.4) taking into account the energies of all the specific interactions and the energy contributions of the two substituted hydrogen atoms of the methyl groups:

$$
\begin{aligned}
\text{DisoCH}_3 = {} & \Delta_{vap}H^0(298\,\text{K})\text{mtbs} - 2\text{DS} \to \text{CH}_3 - \text{S} - 2\text{DS} \\
& \to \text{CH}_3 - \text{CH}_2 - \text{S} - 2\text{DH})/2
\end{aligned}
\tag{4.4}
$$

The molecule of D-*tert*-butyl sulfide with two butyl fragments contains two ethyl groups and four isostructural methyl groups (Fig. 4.8b). Ethyl ligands form four stable specific interactions DS $\to$ CH$_3$–CH$_2$–S with an energy of 9.05 kJ mol^{-1}, equal to the energy value of the same type of interaction formed by symmetrical diethyl sulfide. Isostructural methyl groups form eight specific interactions of low stability and the network formed in the liquid D-*tert*-butyl sulfide includes 12 specific interactions (Fig. 4.8b). The energies of the specific interactions DH$_3$C $\to$ H–CH$_2$ should be calculated using Eq. (4.4a), which also takes into account the marked interactions and the energy of the four substituted hydrogen atoms of the ethyl ligands:

$$
\text{DisoCH}_3 = \Delta_{vap}H^0(298\,\text{K})\text{mtbs} - 4\text{DS} \to \text{CH}_3 - \text{CH}_2 - \text{S} - 4\text{DH})/2 \tag{4.4a}
$$

The energy calculated for the considered specific interactions are given in Table 4.5. The measured energy values of the specific interactions with low stability DH$_3$C $\to$ H–CH$_2$ of liquid D-*tert*-butyl sulfide (with increased reliability) are in good agreement with the energy value of the same type of interaction (1.15 kJ mol^{-1}) of liquid ethyl isopropyl sulfide and diisopropyl sulfide, obtained with increased errors.

The molecule of *sec*-butyl ethyl sulfide with asymmetrical ligands forms two types of stable specific interactions DS $\to$ CH$_3$–CH$_2$–S with an energy value of 8.83 kJ mol^{-1} and propyl ligand DS $\to$ CH$_3$–CH$_2$–CH$_2$–S. The energy contribution of the isostructural methyl group with a propyl ligand located at the first carbon atom should be taken as being equal to the value contributed by the same group of

Table 4.5 Energies (kJ mol^{-1}) of specific interactions of liquid *tert*-butyl sulfides at 298 K

Compounds	Formula	Structure	$\Delta_{vap}H^0$ (298 K) [12]	$DS \rightarrow CH_3-S$	$DS \rightarrow CH_3-CH_2-S$	$\sum DisoCH_3$	$DH_3C \rightarrow H-CH_2$
Methyl *tert*-butyl sulfide	$C_5H_{12}S$		36.5	6.92	8.83	2.1×2	1.05
D-*tert*-Butyl sulfide	$C_8H_{18}S$		43.8 ± 0.1	–	9.05	2.5×4	1.25
sec-Butyl ethyl sulfide	$C_6H_{14}S$		41.2[a]	10.7	8.83	2.1×2	1.05

[a]319 K

liquid methyl *tert*-butyl sulfide. The energy of the formed specific interaction of the propyl ligand can be determined using Eq. (4.4b):

$$DS \rightarrow CH_3 - CH_2 - CH_2 - S$$
$$= \left(\Delta_{vap}H^0(298\,K)mtbs - 2DS \rightarrow CH_3 - CH_2 - S - DisoCH_3\right)/2 \quad (4.4b)$$

The obtained energies of specific interactions (Table 4.5) are described by the natural sequence of stabilization:

DS $\rightarrow$ CH$_3$–S (6.92) methyl *tert*-butyl sulfide < DS$\rightarrow$CH$_3$–CH$_2$–S (9.05), D-*tert*-butyl sulfide < DS $\rightarrow$ CH$_3$–CH$_2$–CH$_2$–S (10.7 kJ mol^{-1}), *sec*-butyl ethyl sulfide

4.1.5   Sulfides with Unsaturated Ligands

Molecules of the symmetrical divinyl sulfide and diallyl sulfide with the two unsaturated vinyl and allyl groups form, like dialkyl sulfides, chains by the four DS $\rightarrow$ CH$_2$=CH–S and DS $\rightarrow$ CH$_2$=CH–CH$_2$–S specific interactions and the corresponding crosslinked weak van der Waals interactions (Fig. 4.2). This allows the energies of the formed specific interactions to be obtained by the division of the vaporization enthalpies by the number of the given bonds [11]. As presented in Table 4.6, the bond energies point to the increased stability of the formed specific interaction of the vinyl group in comparison with the specific interaction of the ethyl ligand:

DS$\rightarrow$CH$_2$=CH–S (9.57), Divinylsulfide > DS $\rightarrow$ CH$_3$–CH$_2$–S (9.05 kJ mol^{-1}), Diethyl sulfide

Two differing values for the vaporization enthalpy of diallyl sulfide reported in the literature [12] were obtained at differing temperatures without errors, indicating stabilization of the DS $\rightarrow$ CH$_2$=CH–CH$_2$–S interaction by a temperature reduction to 272 K.

Asymmetrical molecules of allyl ethyl sulfide with unsaturated allyl and ethyl ligands undergo a structural stress of the given fragments; therefore, in the energy calculations of the specific interaction DS $\rightarrow$ CH$_2$=CH–CH$_2$–S using Eq. (4.5), one should take the energy value of the specific interaction 2DS $\rightarrow$ CH$_3$–CH$_2$–S to be equal to 8.83 kJ mol^{-1}, obtained for the ethyl propyl sulfide with asymmetrical ligands:

$$DS \rightarrow CH_2 = CH - CH_2 - S$$
$$= \left(\Delta_{vap}H^0(298\,K)aes - 2DS \rightarrow CH_3 - CH_2 - S\right)/2 \quad (4.5)$$

As presented in Table 4.6, the value of 10.6 kJ mol^{-1} is lower than the energy value

Table 4.6 Energies (kJ mol^{-1}) of specific interactions of liquid divinyl sulfide and allyl ethyl sulfides at 298 K

Compounds	Formula	Structure	$\Delta_{vap}H^0$ (298 K) [12]	DisoCH$_3$	DS → CH$_3$-CH$_2$-S	DS → CH$_2$=CH-CH$_2$-S	DH$_3$C → H-CH$_2$
Divinyl sulfide	C$_4$H$_6$S	CH$_2$=CH–S–CH=CH$_2$	38.3 ± 0.7	–	9.57	–	–
Diallyl sulfide	C$_6$H$_{10}$S	H$_2$C=CH–CH$_2$–S–CH$_2$–CH=CH$_2$	46.6^a; 43.2	–	–	11.45^a 10.83	–
Allyl ethyl sulfide	C$_5$H$_{10}$S	H$_2$C=CH–CH$_2$–S–CH$_2$–CH$_3$	38.9	–	8.83	10.6	–
Allyl *tert*-butyl sulfide	C$_7$H$_{14}$S		44.8	3.45 × 2	9.05	10.6	1.70
				2DCH$_2$-CH$_3$ → CH$_2$-CH$_3$		DCH$_2$-CH$_2$-CH=S → CH$_2$-CH$_2$-CH$_2$=S	
Pentamethylene sulfide	C$_5$H$_{10}$S	H$_2$C–H$_2$C–(H$_2$C)–CH=S	42.8	5.6		15.6	

^{a}Obtained at 272 K

of the same type (10.83 kJ mol^{-1}) formed by the symmetrical diallyl sulfide not experiencing the structural stress.

The molecule of allyl *tert*-butyl sulfide with an unsaturated allyl ligand and a *tert*-butyl sulfide fragment with an ethyl and two isostructural groups forms a network of eight specific interactions of the three types (see Fig. 4.9a). The two types of specific interactions formed, DS $\rightarrow$ CH$_2$=CH–CH$_2$–S and DS $\rightarrow$ CH$_3$–CH$_2$–S with energies of 10.6 and 9.05 kJ mol^{-1}, respectively, allow calculation of the contributed energy of the isostructural methyl group to the vaporization enthalpy using Eq. (4.6) and, from this, the energy of the specific interaction of reduced stability (Table 4.6):

$$DisoCH_3 = \left(\Delta_{vap}H^0(298\,K)aes - 2DS \rightarrow CH_3 - CH_2 - S - DS \rightarrow CH_2 \right.$$
$$\left. = CH - CH_2 - S\right)/4 \tag{4.6}$$

The molecule of pentamethylene sulfide with the saturated hydrocarbon chain forms a chain structure, crosslinked by van der Waals interactions to the network of the specific interactions (Fig. 4.9b). The pentamethylene ligand with five chain carbon atoms expresses the ability to shift the electron density to the sulfur atom from the first three chain carbon atoms, which are influenced by the reverse dative bond. This makes it possible for the pentamethylene chain with the ethyl fragment to form two specific interactions: DCH$_2$–CH$_3$ $\rightarrow$ CH$_2$–CH$_3$ with an energy of 5.6 kJ mol^{-1} and the stable specific interaction of the propyl fragment with the sulfur atom of the molecule of the near environment DCH$_2$–CH$_2$–CH–S $\rightarrow$ CH$_2$–CH$_2$–CH$_2$–S [11]. The energy of the specific interaction can be obtained from Eq. (4.6a):

$$DCH_2 - CH_2 - CH = S \rightarrow CH_2 - CH_2 - CH_2 = S$$
$$= \left(\Delta_{vap}H^0(298\,K) - 2DCH_2 - CH_3 \rightarrow CH_2 - CH_3\right)/2 \tag{4.6a}$$

The calculated energy value of the given specific interaction significantly exceeds the energy of the same type of specific interaction of liquid dipropyl sulfide:

Pentamethylene sulfide(15.6) > Dipropyl sulfide$(11.17\,kJ\,mol^{-1})$.

4.1.6 Disulfides with Saturated Ligands

The molecules of the acyclic disulfides have an angle approaching 90° [13] and ΔI (n) for splitting of the n-levels is in the range 0.2–0.3 eV [14–17]. Upon introduction to the molecule, the volume of the *tert*-butyl group increases the angle up to 110° and increases $\Delta I(n)$ to 0.6 eV [14, 17]. This is caused by the replacement of the methylene groups of the butyl ligand by the isostructural groups in the *tert*-butyl fragment, forming specific interactions with reduced energy, defining and complicating the network of interactions of the formed structures of the liquid sulfides,

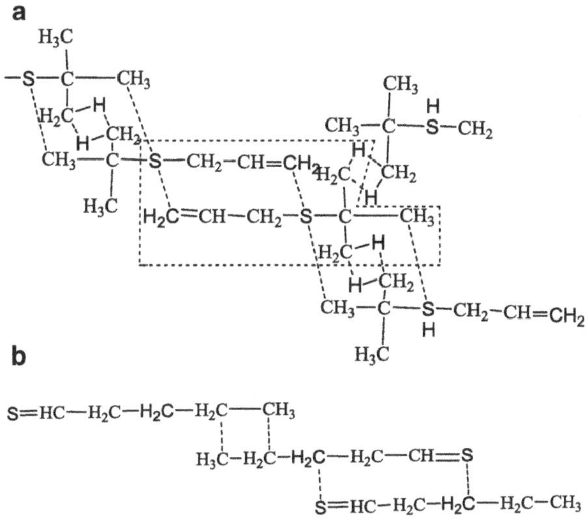

Fig. 4.9 Schematic of the network of specific interactions of liquid allyl *tert*-butyl sulfide (**a**) and pentamethylene sulfide (**b**)

disulfides, and other compounds with the *tert*-butyl fragment. Disulfides of symmetrical molecule structure, with the four bond vacancies of the four free electron pairs of the two sulfur atoms and the two bond vacancies of the essentially undivided $2s^2$ electron pair of the two carbon atoms of the methyl groups, lack these vacancies as compared with the dialkyl sulfide and should, therefore, form more stable specific interactions. As a result, the dependency of the disulfide vaporization enthalpy on the number of carbon atoms in the alkyl chain of the compounds has a clearly expressed deviation from the linear dependency at the propyl ligand of the dipropyl disulfide. This is caused by completion of the influence of the intramolecular reverse dative bond on the energies of the formed specific interaction of the propyl fragment (Fig. 4.10). A positive deviation of the value of the vaporization enthalpy of the dibutyl disulfide reflects the increased energy contribution of the methylene group with the four carbon atoms of this compound chain. Further increase in the vaporization enthalpy of the given series should be subjected to the energy contribution of the increasing number of methylene groups of the given thermodynamic characteristic. Therefore, using the correlation of the enthalpies of disulfides and asymmetrical sulfides of the ethyl-R series, the values for two unstudied disulfides are obtained by the method of vaporization enthalpy extrapolation, as given in Table 4.7.

The formed specific interactions of the two electron pairs of the sulfur atoms with an essentially undivided $2s^2$ electron pair of the carbon atoms of two methyl groups of the alkyl ligands form the chain of interactions, crosslinked by the weak van der Waals interactions (Fig. 4.11).

Transition to a vapor of disulfide monomer molecules is accompanied by breakage of the four specific interactions of one type; therefore, their energies are determined from the enthalpy value divided by the number of these bonds. The

Fig. 4.10 Dependence of the vaporization enthalpy on the number of carbon atoms in the chain of symmetrical dialkyl disulfides (*1, 2*)

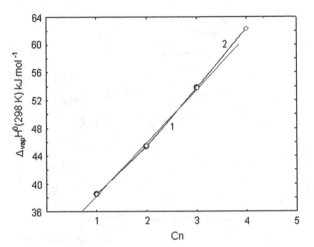

results of the calculations are presented in Table 4.7 and allow the conclusion that the stabilization of the specific interactions with an increasing number of carbon atoms in the alkyl ligands ends at the completion of the intramolecular reverse dative bond influence:

Dimethyldisulfide (9.62) < Diethyl disulfide (11.35) < Dipropyl disulfide (13.45) = Dibutyl disulfide (13.45 kJ mol^{-1})

These data illustrate the increased stability of the specific interactions of disulfides in comparison with each corresponding compound in the series of symmetrical alkyl sulfides, which are characterized by an identical network of specific interactions in its liquid condition with the participation of van der Waals interactions.

The isostructural methyl group contributes to the significant changes in the network structure of disulfide specific interactions , forming specific interactions with the molecules of the near environment that are an order more stable than the van der Waals interactions (Fig. 4.12). The participation of specific interactions with reduced stability provides the crosslinking of the chain. Two representatives of this series of compounds, diisopropyl disulfide and diisobutyl disulfide, form a similar network via eight specific interactions of two types. The energies of the stable specific interactions $DS \rightarrow CH_3-CH_2-S$ and $DS \rightarrow CH_3-(CH_2)_2-S$ corresponding to diisopropyl disulfide and diisobutyl disulfide are given in Table 4.7. The energy values of these compounds' interactions are calculated using Eq. (4.7), where $n = 0.1$:

$$D_3C \rightarrow H - CH_2$$
$$= \left(\Delta_{vap}H^0(298 \text{ K}) - 4DS \rightarrow CH_3 - CH_2 - (CH_2)n - S\right)/4 \qquad (4.7)$$

The reduction in the energy values of interactions of diisopropyl disulfide (1.2 kJ mol^{-1}) and diisobutyl disulfide (1.15 kJ mol^{-1}) reflects the reduction in

Table 4.7 Energies (kJ mol^{-1}) of specific interactions of liquid dialkyl disulfides at 298 K

Compounds	Formula	Structure	$\Delta_{vap}H^0$ (298 K) [12]	DS → CH₃–S	DS → CH₃–CH₂–S	DS → H₃–(CH₂)₂–S	∑DCH₂
Dimethyl disulfide	$C_2H_6S_2$	H_3C-S H_3C-S	38.5 ± 0.6	9.62	–	–	
Diethyl disulfide	$C_4H_{10}S_2$	CH_3-CH_2-S CH_3-CH_2-S	45.4 ± 0.8	–	11.35	–	–
Dipropyl disulfide	$C_6H_{14}S_2$	$CH_3-CH_2-CH_2-S$ $CH_3-CH_2-CH_2-S$	53.8 ± 0.1	–	–	13.45	–
Dibutyl disulfide	$C_8H_{18}S_2$	$CH_3-(CH_2)_2-CH_2-S$ $CH_3-(CH_2)_2-CH_2-S$	62.3 ± 0.2	–	–	13.45	4.25 × 2
Dipentyl disulfide	$C_{10}H_{22}S_2$	$CH_3-(CH_2)_3-CH_2-S$ $CH_3-(CH_2)_3-CH_2-S$	70.2	–	–	13.45	4.1 × 4
Dihexyl disulfide	$C_{12}H_{26}S_2$	$CH_3-(CH_2)_4-CH_2-S$ $CH_3-(CH_2)_4-CH_2-S$	79.8	–	–	13.45	4.33 × 4
Diisopropyl disulfide	$C_6H_{14}S_2$	$H_3C-CH-S$ $H_3C-CH-S$	49	DisoCH₃ = 2.4 × 2	11.35	–	DH₃C → H–CH₂ 1.2
Diisobutyl disulfide	$C_8H_{18}S_2$	$H_3C-CH-CH_2-S$ $H_3C-CH-CH_2-S$	57.2 ± 0.4	DisoCH₃ = 2.3 × 2	–	13.45	1.15

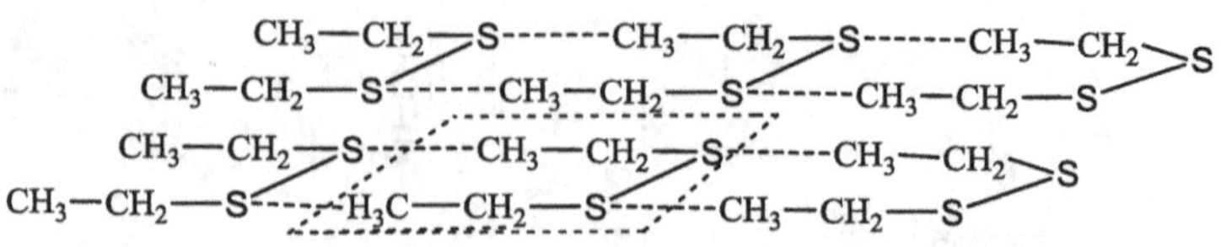

Fig. 4.11 Schematic of the network of specific interactions of liquid diethyl disulfide

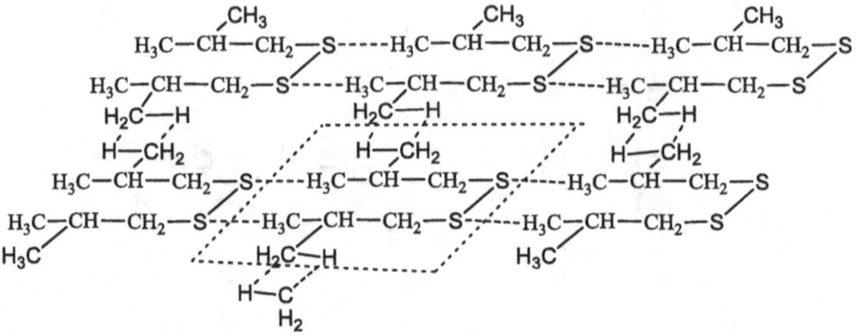

Fig. 4.12 Schematic of the network of specific interactions of liquid diisobutyl disulfide

stability at the removal of the isostructural methyl group from the sulfur atom. A similar reduction in stability of the formed specific interaction of the iso-CH₃ group takes place in isopropyl propyl sulfide and ethyl isobutyl sulfide (Table 4.4).

4.2   Specific Interactions of Cyclopentyl Methylsulfide and Sulfides with Benzoic Cyclic Compounds

4.2.1   Cyclopentyl Methylsulfide

Cyclopentyl methylsulfide is the only representative of compounds with a saturated hydrocarbon ring and for which the vaporization enthalpy is used in thermodynamic analysis for obtaining the specific interaction energy. The sulfur atom, with high acceptor properties, causes an equal shifting of the electron density in the saturated hydrocarbon ring from the carbon atoms in locations C(3) and C(4) through the chain to the carbon atoms in locations C(2) and C(5) and further to the sulfur atom (Fig. 4.13a). Part of the obtained electron density from the ring and the methyl group is transferred as the reverse dative bond to the carbon atoms contacting with the sulfur atom, providing a reduction in the positive charge. The excess electron density at the sulfur atom with the two free electron pairs provides it with the possibility to form two specific interactions, one of which has enhanced stability, together with the methyl group of the contacting molecule of DS → CH₃–S. The energy contribution of the ring of the cyclopentyl methylsulfide molecule is equal to the energy contribution of the five methylene groups of the cyclopentane

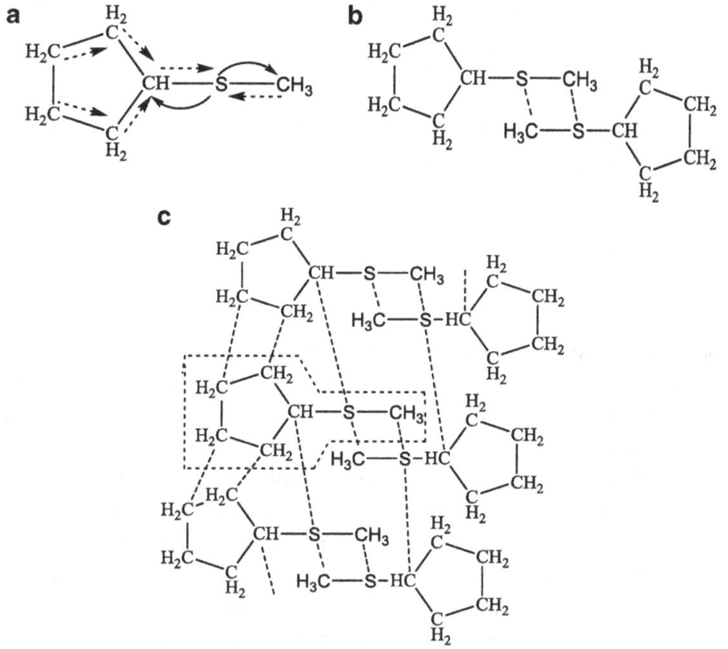

Fig. 4.13 (a) Shift of electron density in cyclopentyl methylsulfide. (b) Specific interactions of cyclopentyl methylsulfide. (c) Schematic of the network of specific interactions of liquid cyclopentyl methylsulfide

minus the energy of the substituted hydrogen atom (27.9 kJ mol^{-1}), including each methylene group (5.7 kJ mol^{-1}) and CH group (5.1 kJ mol^{-}). Thus, the difference between the vaporization enthalpy and the energy contribution of the mentioned groups corresponds to the sum of the energies of the two specific interactions DS $\rightarrow$ CH$_3$–S (Fig. 4.13b). The obtained energy value of the interaction is equal to 8.6 kJ mol^{-1}. This energy value of the specific interaction is in the range for this type of interaction in dimethyl sulfide, with a reduced energy value of the interactions formed by the two free electron pairs of the sulfur atom:

Dimethyl sulfide (6.92) < Cyclopentyl methylsulfide (8.6) < Dimethyldisulfide
     (9.62 kJ mol^{-1})

This is natural, due to the facts that, first, the ring with the five methylene groups shifts more electron density to the sulfur atom in comparison with the methyl group and, second, the sulfur atom of the cyclopentyl methylsulfide molecule coordinates half of the methyl groups compared with liquid dimethyl sulfide. On the other hand, the reduced stability of the specific interaction of the cyclopentyl methylsulfide, compared with liquid dimethyl disulfide, is caused by the increased acceptor ability of each sulfur atom with the S–S bond of CH$_3$–S–S–CH$_3$ molecules.

Table 4.8 Energies (kJ mol^{-1}) of specific interactions of liquid cyclopentane and cyclopentyl methyl sulfide at 298 K

Compounds	Formula	Structure	$\Delta_{vap}H^0$ (298 K) [12]	$\Delta_{vap}H^0$ (298 K)– DH	$DH_2C \rightarrow CH_2$	$DS \rightarrow CH_3$–S
Cyclopentyl methyl sulfide	$C_6H_{12}S$		45.1 ± 0.1	27.9	5.7	8.6
Cyclopentane	C_5H_{10}	–	28.5 ± 0.1	27.9	5.7	–

The second approach to the estimation of the energy value of the specific interactions is based on correctly accounting for all specific interactions. The schematic of the interactions presented in Fig. 4.13c includes the four interactions formed by the methylene groups with an energy of 5.7 kJ mol^{-1}, the same two interactions $DS \rightarrow CH_3$–S, and the two specific interactions of the second electron pair of sulfur atom $DS \rightarrow CH$ with the carbon atom of the CH group.

The difference between the vaporization enthalpy (45.1 kJ mol^{-1}) of the cyclopentyl methylsulfide and the total energy value of the four bonds $DH_2C \rightarrow CH_2$ (22.8 kJ mol^{-1})is the total energy value of $2DS \rightarrow CH$ $+ 2DS \rightarrow CH_3$–S (22.3 kJ mol^{-1}). Assuming that the minimum possible energy value of the $DS \rightarrow CH_3$–S interaction is equal to the energy of the same type of interaction of dimethyl sulfide (6.92 kJ mol^{-1}), the obtained energy of the two $DS \rightarrow CH$ interactions is equal to 8.4 kJ mol^{-1} or one interaction is equal to 4.2 kJ mol^{-1}. The given value indicates its reduced stability compared with the interaction of $DHC \rightarrow CH$ (5.1 kJ mol^{-1}). This points to the fact that the system exceeds the stable condition at 3.0 kJ mol^{-1}. Therefore, the molecule of cyclopentyl methylsulfide forms two specific interactions $DS \rightarrow CH_3$–S with an energy of 8.6 kJ mol^{-1} (Table 4.8) and the second free electron pair of the sulfur atom, like dimethyl disulfide, does not participate in the formation of stable specific interactions. The increased stability of this specific interaction in comparison with the formed dimethyl sulfide is caused by the increased shifting of electron density on the methylene ring to C(1) and further to the sulfur atom, providing the increased difference between the negative charge of the sulfur atom and the carbon atom of the methyl group and the charge of the CH of the dimethyl sulfide group.

4.2.2   Phenyl Sulfides, Benzyl Methyl Sulfide, and (Thio) toluenes

The rigid benzene ring, with differing charges of the carbon atoms located near the ring, is at least subjected to the shifting of electron density in the ring to the contacting sulfur atom. In this situation, the molecules of the methyl phenyl sulfide receive less electron density from the interacting carbon atom of the ring, providing only the reduced electron density of the free electron pairs of sulfur atom. Thus, the main electron density that the sulfur atom receives from the carbon atom of the methyl group provides the increased difference in charge and the stability of the formed specific interactions. As a result, the sulfur atom with two bond vacancies of the free electron pairs forms one specific interaction DS $\rightarrow$ CH$_3$–S with the methyl group and a second with the carbon atom of the benzene ring (Fig. 4.14). Presented in the Fig. 4.14 is a schematic of the formed network of specific interactions of the methyl phenyl sulfide molecule, which includes the five interactions DHC $\rightarrow$ CH with an energy of 5.63 kJ mol^{-1}formed by the benzene ring, two DS $\rightarrow$ C= interactions, and two interactions of the methyl groups with the sulfur atom DS $\rightarrow$ CH$_3$–S, whose energy is taken to be equal to 6.92 kJ mol^{-1} in liquid dimethyl sulfide. The unknown energy value of the specific interaction DS $\rightarrow$ CH$_3$–S was determined with the help of Eq. (4.8):

$$DS \rightarrow CH_3 - S$$
$$= \left(\Delta_{vap}H^0(298\,K)mphs - 5DHC \rightarrow CH - 2D = C \rightarrow C =\right)/2 \qquad (4.8)$$

Using a similar equation, the energy of the specific interaction DO $\rightarrow$ C= for the methyl phenol ester can be obtained. The calculated energy values of the specific interactions of the methyl phenyl sulfide and the methyl phenol ester (Table 4.9) are located in the sequence of stabilization, reflecting the predominant manifestation of the stabilization of the considered compounds by the sulfur atom:

$$DHC \rightarrow CH(5.63) < DS \rightarrow C = (6.17) < DS \rightarrow CH_3 - S\left(6.92\,kJ\,mol^{-1}\right)$$
$$DO \rightarrow C = (3.75) < DHC \rightarrow CH(5.63) = DO \rightarrow CH_3 - O\left(5.63\,kJ\,mol^{-1}\right)$$

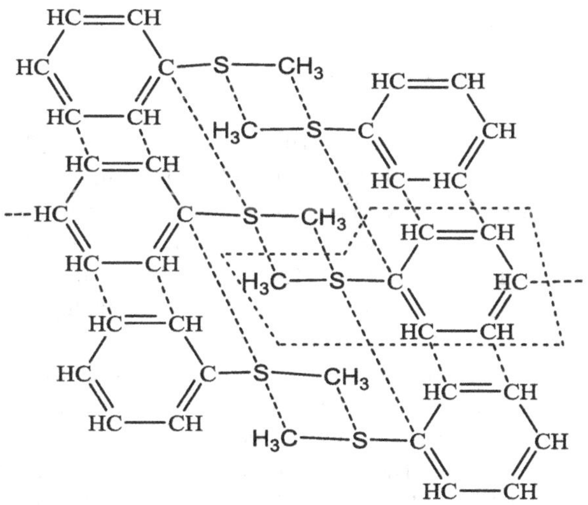

Fig. 4.14 Schematic of the network of specific interactions of liquid methyl phenyl sulfide

Table 4.9 Energies (kJ mol^{-1}) of specific interactions of liquid phenyl sulfides, benzyl methyl sulfide, and (methylthio)toluenes at 298 K

Compounds	Formula	Structure	$\Delta_{vap}H^0$ (298 K) [12]	DS → CH$_3$–S	DS → C=	DS → CH$_3$–CH$_2$–S
DHC → CH = 5.62, DHC → C = 5.33, DC → C = 5.03						–
Benzene	C$_6$H$_6$	–	33.8 ± 0.1	–	–	–
Toluene	C$_7$H$_8$	–	38.0	–	–	–
Methyl phenyl sulfide	C$_7$H$_8$S		54.3 ± 0.1	6.92	6.17	–
Ethyl phenyl sulfide	C$_8$H$_{10}$S		53.6 ± 2.1[a]	–	6.17	6.55[a]
Benzyl methyl sulfide	C$_8$H$_{10}$S		55.2 ± 2.1	6.92	–	DS → CH$_2$–S = 4.08
2-(Methylthio) toluene	C$_8$H$_{10}$S		50.2[b]	6.7	–	DH$_3$C → H–CH$_2$ = 2.2 × 2
4-(Methylthio) toluene	C$_8$H$_{10}$S		50.1[b]	6.65	–	DH$_3$C → H–CH$_2$ = 2.2 × 2

[a]Corresponding to 323–358 K
[b]Temperature is not reported in the literature

The converted Eq. (4.8a) applied to ethyl phenyl sulfide was used for the energy calculation of the specific interaction formed by the ethyl ligand of the compound (see Table 4.9):

$$2DS \rightarrow CH_3 - CH_2 - S$$
$$= \left(\Delta_{vap}H^0(298\,K)eps - 5DHC \rightarrow H - 2DS \rightarrow C =\right)/2 \qquad (4.8a)$$

The obtained energy value corresponds to the temperature range 323–358 K and has a reduced value compared with that of the liquid diethyl sulfide (9.05 kJ mol^{-1}).

A molecule of benzyl methyl sulfide is formed from a molecule of toluene by the replacement of a hydrogen atom by an isostructural methyl group, forming the two specific interactions DH$_3$C → H–CH$_2$. The low stability of this interaction is obtained from the difference between the vaporization enthalpy of this molecule with that of benzene, divided between the two bonds (2.1 kJ mol^{-1}). In this regard, the reduced energy contribution of the ethylene group (4.2 kJ mol^{-1}), located between the benzene ring and the sulfur atom (with high acceptor properties and participating in the formation of the specific interaction DS → CH$_2$) is caused by the rigidity of the benzene ring. The molecule of benzyl methyl sulfide forms two specific interactions DS → CH$_3$–S (6.92 kJ mol^{-1}); four DHC → CH interactions

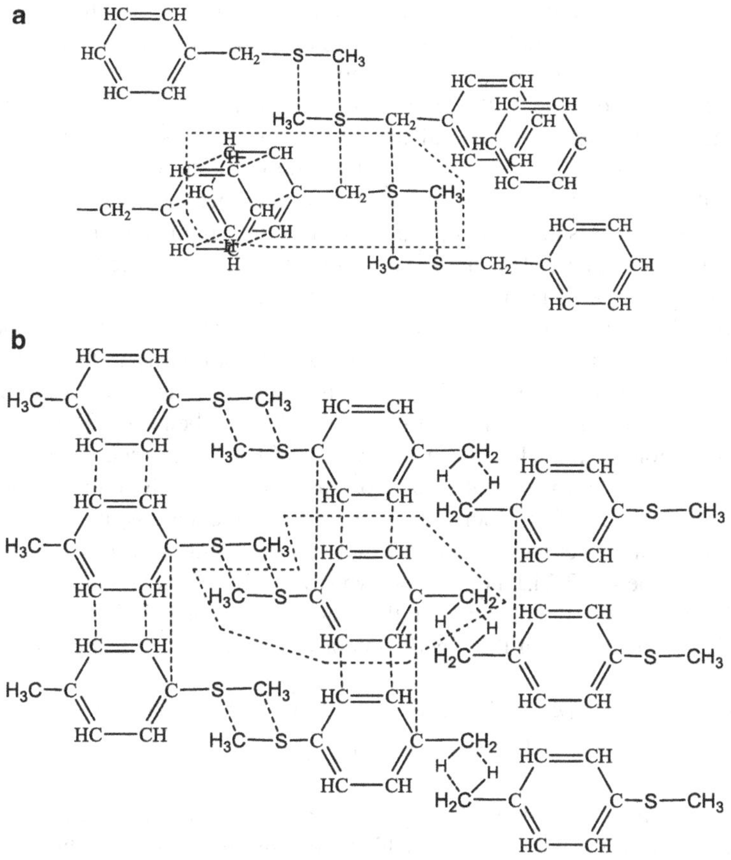

Fig. 4.15 Schematic of the network of specific interactions of liquid benzyl methyl sulfide (**a**) and 4-(methylthio)toluene (**b**)

(5.62 kJ mol^{-1}) and two DHC $\rightarrow$ C interactions (5.32 kJ mol^{-1}) by the benzene ring with a total energy value of 33.2 kJ mol^{-1}; and two interactions originating from the methylene group DS $\rightarrow$ CH$_2$–S to form a network of these interactions (Fig. 4.15a). The energy of the specific interaction DS $\rightarrow$ CH$_2$–S can be determined from Eq. (4.9):

$$DS \rightarrow CH_2 -$$
$$= \left(\Delta_{vap}H^0(298\,K)bms - 4DHC \rightarrow CH - 2DHC \rightarrow C = -2DS \rightarrow CH_3 - S\right)/2$$

$$(4.9)$$

The obtained energy value of the methylene group contribution (4.08 kJ mol^{-1}) to the enthalpy is in good agreement with the energy contribution of the methylene group to the vaporization enthalpy of other compounds [11]. This confirms, first, the destabilizing influence of the specific interaction formed by the functional group at its location between the benzene ring and the sulfur atom with strong acceptor properties and, second, the formation of the specific interaction DS → CH$_3$–S with the two free electron pairs by the sulfur atom.

The toluene molecule with the isostructural methyl group with an energy contribution to the vaporization enthalpy (4.2 kJ mol^{-1}) is significantly different to the other benzene derivatives in the distribution of the electron density and charges at the carbon atoms, as manifested by the specific interactions formed and their energies. It is manifested, in particular, by the significant difference in the vaporization enthalpies (1.0 kJ mol^{-1}) of its two derivatives, 2-(methylthio)toluene and 4-(methylthio)toluene (Table 4.9), and by the lack of influence of the location of functional groups at the carbon atoms of the benzene ring, indicating the identity of the number of formed specific interactions and the similarity in the energies of the specific interactions. On the other hand, the reduced values of the vaporization enthalpies in comparison with the other derivatives of benzene (Table 4.9) are direct indications of the adequate energy values of the formed specific interaction DS → CH$_3$–S. It follows that the molecules of 2-(methylthio)toluene and 4-(methylthio)toluene form four HC → CH specific interactions, two interactions D=C → C= with an energy of 5.03 kJ mol^{-1}, two interactions DH$_3$C → H–CH$_2$ with reduced energy (2.2 kJ mol^{-1}), and two DS → CH$_3$–S interactions forming the network of specific interactions in the liquid condition of these compounds (Fig. 4.15b). The energy of the specific interaction is calculated using Eq. (4.10):

$$DS \rightarrow CH_3 - S = \left(\Delta_{vap}H^0(298\,K)mtt - 4DHC \rightarrow CH - 2D = C \rightarrow C\right.$$
$$= \left. - 2DH_3C \rightarrow H - CH_2\right)/2 \qquad (4.10)$$

The results of the calculations (Table 4.9) allow the conclusion to be made that in the liquid and crystal alkyl phenyl sulfides, the sulfur atoms form two types of specific interactions DS → R–S and DS → C= by both free electron pairs, and that in compounds of the toluene and benzo alkyl sulfide derivatives there is one type of DS → R–S interaction.

The results of thermodynamic analysis of compounds with a single benzene ring allow an extension of the analysis to compounds with two rings. However, the accuracy of the enthalpies of vaporization and sublimation of these compounds given in the literature raises concern about the corresponding values with the relatively high errors (Table 4.10). The known energy values of the specific interactions 10DHC → CH and 2D=C → C= in the liquid and crystal conditions formed by the molecule of diphenyl disulfide (specific interaction forms by CH group; see Fig. 4.16a) allow the energies of the third type of interaction to be calculated using Eq. (4.10a):

$$DS \rightarrow CH = \left(\Delta_{vap}H^0(298\,K) - 10DHC \rightarrow CH - 2D = C \rightarrow C =\right)/4 \quad (4.10a)$$

A similar Eq. (4.10b) is used for the energy calculation of the formed specific interaction DS → CH$_2$ мby the molecule of dicyclohexyl sulfide (specific interaction forms by CH$_2$ group; see Fig. 4.16b):

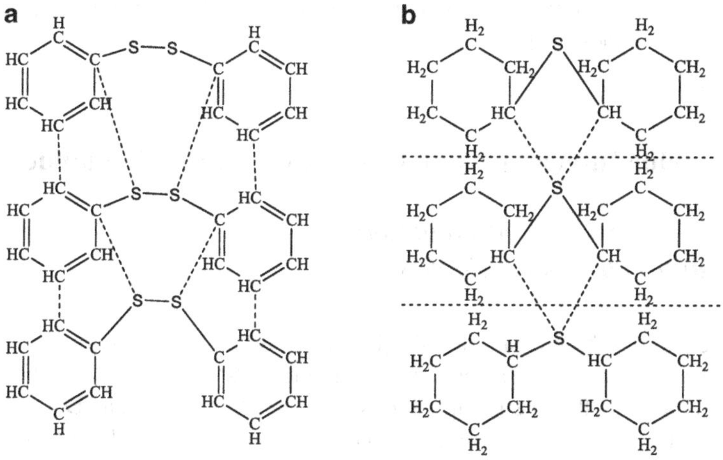

Fig. 4.16 Schematic of the network of specific interactions of liquid and crystalline diphenyl disulfide (**a**) and dicyclohexyl sulfide (**b**)

Table 4.10 Energies (kJ mol^{-1}) of specific interactions of diphenyl disulfide, dibenzyl sulfide, and dicyclohexyl sulfide at 298 K

Compounds	Formula	Structure	$\Delta_{evap}H^0$ (298 K) [12, 18]	DS → CH$_2$–S	DS → CH=
Liquid [12]					
DHC → CH = 5.62, DHC → C = 5.33, DC → C = 5.03					
Diphenyl disulfide	C$_{12}$H$_{10}$S$_2$	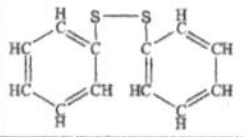	71.1 ± 0.2	–	2.1
			78.7 ± 2.9		4.1
DH$_2$C → CH$_2$– = 5.58; D=CH → CH= = 5.28; D=C → C = 5.03					
Dicyclohexyl sulfide	C$_{12}$H$_{22}$S	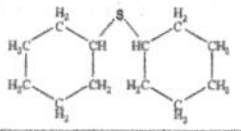	69.0 ± 0.7	0.8	–
Crystalline [18]					
DHC → CH =7.43, DHC → C =6.93, DC → C = 6.43					
Diphenyl disulfide	C$_{12}$H$_{10}$S$_2$	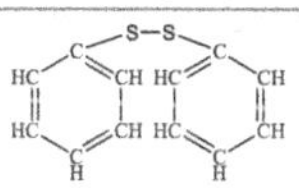	95 ± 3.0	–	2.0

$$DS \to CH_2$$

$$= (\Delta_{vap}H^0(298\,K) - 10DH_2C \to CH_2 - 2D = CH \to CH =)/4 \quad (4.10b)$$

The implemented energy calculations of specific interactions formed by the sulfur atom (Table 4.10) point to the low energy values of these bonds, in which the dissociation occurs at sufficiently low temperatures. In this connection, for the thermodynamic research of compounds, special approaches are required, with the use of static methods [19].

4.3   Specific Intermolecular Interactions of Cyclic Sulfides

4.3.1   Thiiranes, Thiocyclobutanes, and Tetrahydrothiophenes

The molecule of ethylene sulfide (also known as thiirane), with two free electron pairs of the sulfur atom and an essentially undivided $2s^2$ electron pair of the carbon atom of the two methylene groups, forms four specific interactions $DS \rightarrow CH_2$–S, with the participation of the pentacoordinated carbon atom in the liquid and crystal conditions of the chain (Fig. 4.17a). The chains are connected by weak van der Waals intermolecular interactions. The energies of these interactions can be obtained by division of the vaporization enthalpies of the four formed intermolecular interactions:

$$DS \rightarrow CH_2 - S(7.62), \text{Ethylene sulfide} > DS$$
$$\rightarrow CH_3 - S(6.92\,\text{kJ}\,\text{mol}^{-1}), \text{Dimethyl sulfide}$$

As shown in Table 4.11, the energy of this interaction is more stable in comparison with the open structure and methyl group formed by the molecule of dimethyl sulfide. This is caused by the significantly reduced interatomic distances of the C–C and S–C bonds in the molecule with the closed structure, providing the increased shifting of the electron density to the sulfur atom and, consequently, the increased localization of n-electrons to the binding orbitals. This draws attention to the fact that, as in the examples of compounds $E(CH_2)_4$ of the oxygen subgroup elements reported in the literature [20, 21], following replacement of the oxygen atom by further elements of this group, the oscillation circuit of the first band significantly narrows and the degree of *n*-electrons localized at the element atom of this group

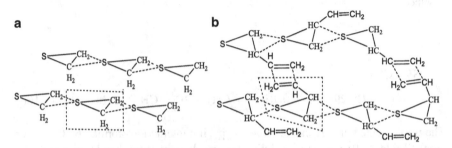

Fig. 4.17 Schematic of the network of specific interactions of liquid ethylene sulfide, thiirane (**a**) and vinylthiirane (**b**)

significantly increases. Based on this and on the thermodynamic analysis, one can make a conclusion regarding the natural character of the observed stabilization of the specific interactions in each series of compounds.

As shown in Table 4.11, the derivatives of thiirane differ in their isostructural groups, but the energy of the stable specific interaction DS → CH$_2$–S remains unchanged. This allows the energy contributions of the isostructural groups to be obtained from the difference in the vaporization enthalpies of the compound and thiirane minus the energy of the substituted hydrogen atom of the isostructural group, or from the vaporization enthalpy of the compound minus the total energy of the four specific interactions. The specific interactions of reduced stability formed by these groups crosslink the formed rings of the chain molecules to form a grid structure (Fig. 4.17b).

The energy contributions of the isostructural group of 2-methylthiirane to the vaporization enthalpy and of the two methyl groups of 2,2-dimethylthiirane are calculated using Eqs. (4.11) and (4.11a), respectively:

$$\text{DisoR} = \left(\Delta_{vap}H^0(298\,K)\right) - 4\text{DS} \rightarrow \text{CH}_2 - \text{S} - \text{DH} \tag{4.11}$$

$$\text{DisoC}_3 = \left(\Delta_{vap}H^0(298\,K)\right) - 4\text{DS} \rightarrow \text{CH}_2 - \text{S} - \text{DH} \tag{4.11a}$$

As presented in Table 4.11, the calculation results indicate, first, that the isostructural methyl groups of the cyclic thiiranes form more stable specific interactions than acyclic sulfides and, second, that the ethyl isostructural group forms specific interactions of greater stability than those of the unsaturated vinyl ligand.

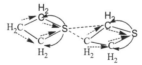

Thiocyclobutane (thiethane)

The electron density in the molecule of thiocyclobutane (thietane) shifts to the sulfur atom and, therefore, directly connects carbon atoms of methylene groups with the sulfur atom, which have a small difference in the positive charges. The sulfur atom with enriched electron density forms, by the two free electron pairs, two intermolecular interactions with the carbon atoms in positions C(2) and C(3) with differing positive charges and unequal energy values of the formed specific interactions DS → CH$_2$–S of one type. Similar bonds are formed by the second coordinating molecule (Fig. 4.18a). The remaining methylene group forms the specific interaction DHC → CH with the third coordinating molecule, whose energy value should be equal to the energy value (6.3 kJ mol^{-1}) of a similar type of interaction, increasing in the liquid cyclobutane. The formed network of specific interactions of the liquid thiocyclobutane includes one DH$_2$C → CH$_2$ and four DS → CH$_2$–S interactions, whose energy value can be obtained using Eq. (4.11b):

Table 4.11 Energies (kJ mol^{-1}) of specific interactions of liquid thioranes, thiocyclobutane, and tetrahydrothiophene at 298 K

Compounds	Formula	Structure	$\Delta_{vap}H^0$ (298 K) [12]	DS→CH$_2$–S	DisoR	DisoH$_3$C→H–CH$_2$	DisoH$_2$C–CH$_3$→CH$_2$–CH$_3$
Ethylene sulfide thiirane	C$_2$H$_4$S	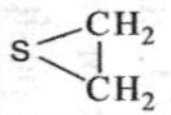	30.5	7.62	–	–	–
2-Methylthiirane	C$_3$H$_6$S	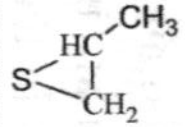	34.8[a]	7.62	4.9	2.45	–
2,2-Dimethylthiirane	C$_4$H$_8$S	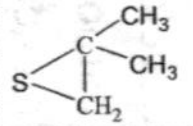	37.8[b]	7.62	4.25x2	2.12	–
2-Ethylthiirane	C$_4$H$_8$S	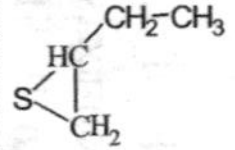	39.7[c]	7.62	9.8	–	4.9
2-Vinylthiirane	C$_4$H$_6$S	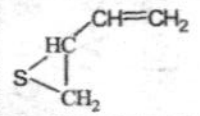	38.7[b]	7.62	8.8	–	D-HC=CH$_2$→CH=CH$_2$=4.4
Thiocyclobutane (thietane)	C$_3$H$_6$S	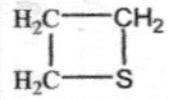	35.8	7.37	–	DH$_2$C→CH$_2$=6.3	–
Tetrahydrothiophene	C$_4$H$_8$S	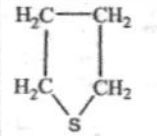	38.8	6.6	–	DH$_2$C→CH$_2$=5.7	–

[a]Corresponding to 287 K
[b]Corresponding to 288 K
[c]Corresponding to 313 K

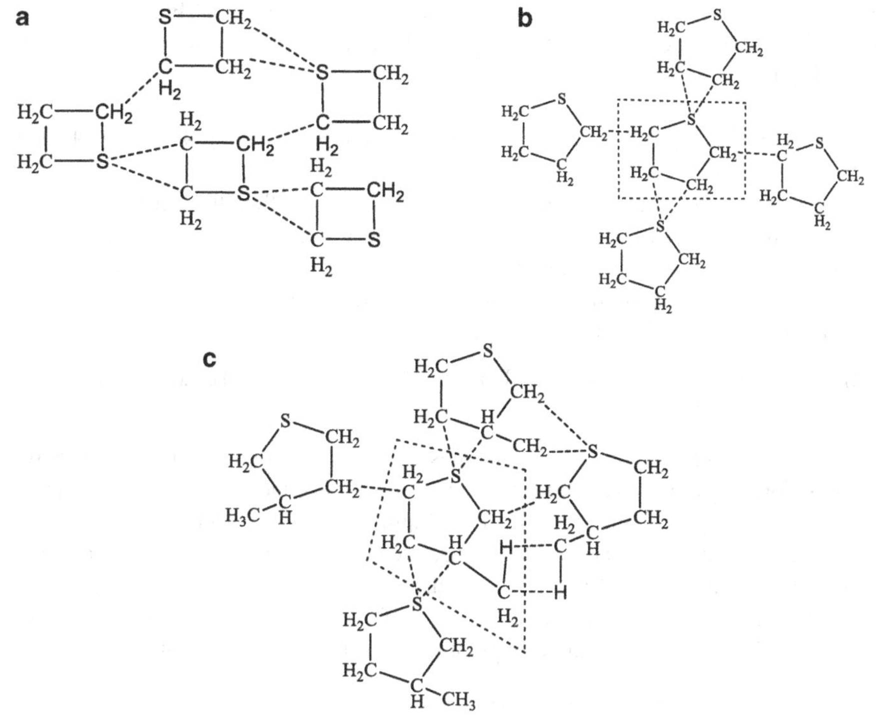

Fig. 4.18 Schematic of the network of specific interactions of liquid thiocyclobutane (thietane) (**a**), tetrahydrothiophene (**b**), and 3-methyltetrahydrothiophene (**c**)

$$DS \rightarrow CH_2 - S = \left(\Delta_{vap}H^0(298\,K) - DH_2C \rightarrow CH_2\right)/4 \qquad (4.11b)$$

The calculated energy value is presented in Table 4.11.

According to the analysis of the vibrational structure of PhE bands and level shifts at the replacements, the sequence of the orbital vertical ionizing potentials of thiophene was established as $1a_2 < 2b_1 < 6a_1 < 1b_1 < 4b_2 < 5a_1 < 3b_2 < 4a_1 < 2b_2 < 3a_1$ [22] and has been proved by the theoretical studies [23]. This indicates the possibility to consider the π orbital of the C_4H_4S molecule in the shape of a linear combination of the two molecular orbitals (MO) of the *cis*-butadiene fragments with the p_z orbitals of the sulfur atom. In this regard, in the molecule of tetrahydrothiophene, the shifting of electron density occurs from the end carbon atoms in locations C (3) and C(4) to the carbon atoms connected with the sulfur atoms and further to the sulfur atom. Part of the received electron density is transferred in the shape of the reverse dative bond to the contacting carbon atoms. Nevertheless, the ring carbon atoms have differing charges. The sulfur atom that received the excess electron density forms, by the two free electron pairs, two specific interactions $DS \rightarrow CH_2-S$ with carbon atoms of the methylene groups in locations C(3) and C(4) with molecules of the near environment. Two other coordinating molecules form the specific interaction $2DH_2C \rightarrow CH_2$ with the remaining methylene groups (Fig. 4.18b). The energy of the specific interaction $DS \rightarrow CH_2$–S tetrahydrothiophene can be obtained from Eq. (4.11c):

$$DS \rightarrow CH_2 - S = \left(\Delta_{vap}H^0(298\,K) - 2DH_2C \rightarrow CH_2\right)/4 \qquad (4.11c)$$

It follows from the obtained data (see Table 4.11) that the formed specific interactions of the sulfur atoms with methylene groups are destabilized with an increasing number of methylene groups in the ring. This is caused by a reduction in the charge difference between the sulfur and the carbon atom of the methylene group with an increasing number of groups:

$DS \rightarrow CH_2$–S: Thiirane (7.62) < Thiocyclobutane (7.37) < Tetrahydrothiophene (6.6 kJ mol^{-1})

The molecules of tetrahydrothiophenes with one and two isostructural methyl groups, forming the above-mentioned interactions of reduced stability, complicate the network of specific interactions of the liquid and crystal condition (Fig. 4.18c). The contribution of the energy of the introduced isostructural methyl group to the vaporization enthalpy of 3-methyltetrahydrothiophene is obtained from the difference between the vaporization enthalpies of 3-methyltetrahydrothiophene and tetrahydrothiophene minus the energy of the substituted hydrogen atom:

$$DisoCH_3 = 2DH_3C \rightarrow H - CH_2$$
$$= \Delta_{vap}H^0(298\ K)mttp - \left(\Delta_{vap}H^0(298\ K)ttp - DH\right) \qquad (4.11d)$$

The energy contribution of the two methyl groups to the enthalpy of *cis*-2,5-dimethyltetrahydrothiophene and *trans*-2,5-dimethyltetrahydrothiophene can also be obtained from the difference in the enthalpy vaporization enthalpies these compounds minus the energy contribution of the two substituted hydrogen atoms (see Table 4.12).

Taking into account that the vaporization enthalpies of the *cis*-2,5-dimethyltetrahydrothiophene and *trans*-2,5-dimethyltetrahydrothiophene were measured at temperatures 28 and 65 K above the standard measurement temperature, the derived energies of the formed specific interactions of the isostructural methyl groups are described by the natural sequence, in which the increased stability is expressed by the methyl group that is more distant from the sulfur atom in the molecule of 3-methyltetrahydrothiophene than in the *cis* and *trans* locations of the 2,5-dimethyltetrahydrothiophene molecule:

3-Methyltetrahydrothiophene   (1.95) > *cis*-2,5-dimethyltetrahydrothiophene (1.02) > *trans*-2,5-dimethyltetrahydrothiophene (0.52 kJ mol^{-1})

Table 4.12 Energies (kJ mol^{-1}) of specific interactions of liquid tetrahydrothiophenes at 298 K

Compounds	Formula	Structure	$\Delta_{vap}H^0$ (298 K) [12]	$DH_2C \to CH_2$	$DS \to CH_2$-S	$DisoCH_3$	$DH_3C \to H$-CH_2
Tetrahydrothiophene	C_4H_8S		38.8	5.70	6.6	–	–
3-Methyltetrahydrothiophene	$C_5H_{10}S$		42.1 ± 0.1	5.70	6.6	3.9	1.95
cis-2,5-Dimethyltetrahydrothiophene	$C_6H_{12}S$		41.7[a]	5.70	6.6	4.1	1.02
trans-2,5-Dimethyltetrahydrothiophene	$C_6H_{12}S$		39.7[b]	5.70	6.6	2.1	0.52

[a]Corresponding to 326 K
[b]Corresponding to 363 K

4.3.2   *Tetrahydro-2H-thiopiranes*

The compounds of the methyltetrahydro-2*H*-thiopiranes series are similar to 3-methyltetrahydrothiophene and methylthiophene and illustrate the tendency for increasing vaporization enthalpy following withdrawal of the location place of the isostructural methyl group from the sulfur atom in the alkyl chain and cycle. This allows the energy contribution of the isostructural methyl group to the vaporization enthalpy of the 4-methyltetrahydro-2*H*-thiopirane to be accepted as being equal to the energy contribution of the same group (3.9 kJ mol^{-1}) to the enthalpy of the 4-methyltetrahydrothiophene at the maximum removal of the CH$_3$ group from the sulfur atom and to obtain the value of the vaporization enthalpy of tetrahydro-2*H*-thiopirane from the difference in the vaporization enthalpies (Table 4.13):

$$\Delta_{vap}H^0(298 \text{ K})thp \ = \ \Delta_{vap}H^0(298 \text{ K})mthp - DisoCH_3.$$

The accepted energy contribution of the CH$_3$ group must be the maximum possible because the measuring temperature of the vaporization enthalpy of the tetrahydro-2*H*-thiopiranes compounds exceeds the standard value by 38°K.

Taking into account that the shifting of the electron density increases the negative charge of the sulfur atom while maintaining the charge difference of the ring carbon atoms, the estimated value of the vaporization enthalpy of tetrahydro-2*H*-thiopirane allows the determination of the energy of the specific interaction DS → CH$_2$–S. The compound's molecule with two enriched electron densities of the two free electron pairs of the sulfur atom of the ring forms four specific interactions DS → CH$_2$–S and the bond vacancy of the essentially undivided 2s^2 electron pair of the remaining three methylene groups forms three specific interactions DH$_2$C → CH$_2$. The energies of these interactions are taken to be equal to the value of 5.70 kJ mol^{-1}, obtained from the vaporization enthalpy of the cyclopentane (Table 4.13). Thus, it is possible to derive Eq. (4.12) whereby the energy of the specific interaction DS → CH$_2$–S for tetrahydro-2*H*-thiopirane and its derivatives can be determined:

$$\text{DS} \rightarrow \text{CH}_2 - \text{S} = \left(\Delta_{vap}H^0(298\,\text{K}) - 3\text{DH}_2\text{C} \rightarrow \text{CH}_2\right)/4 \qquad (4.12)$$

Given in Table 4.13 is the energy value of DS → CH$_2$–S obtained at a temperature that was significantly different from the standard temperature, meaning that it has a reduced value. Nevertheless, it points to a possible insignificantly increased or

Table 4.13 Energies (kJ mol^{-1}) of specific interactions of liquid methyltetrahydro-2H-thiopirane

Compounds	Formula	Structure	$\Delta_{vap}H^0$ (298 K) [12]	T (K)	DS→CH$_2$-S	DisoCH$_3$	DH$_3$C→H–CH$_2$
DH$_2$C→CH$_2$ = 5.70							
2-Methyltetrahydro-2H-thiopirane	C$_6$H$_{12}$S		42.1	333	5.42	3.2	1.6
3-Methyltetrahydro-2H-thiopirane	C$_6$H$_{12}$S		42.5	336	5.42	3.6	1.8
4-Methyltetrahydro-2H-thiopirane	C$_6$H$_{12}$S		42.8	336	5.42	3.9	1.95
Tetrahydro-2H-thiopirane	C$_5$H$_{10}$S		38.9[a]	336	5.42	–	–
						DH$_2$C→CH$_2$	
Cyclopentane	C$_5$H$_{10}$		28.5±0.1	298	–	5.70	–
Spiropentane	C$_5$H$_8$		27.5±0.1	298	–	5.50	–
1,3-Cyclopentadiene	C$_5$H$_6$		28.1±0.1	298	–	5.62	–

(continued)

Table 4.13 (continued)

Compounds	Formula	Structure	$\Delta_{vap}H^0$ (298 K) [12]	T (K)	DS → CH$_2$-S	DisoCH$_3$	DH$_3$C → H-CH$_2$
2,2,5,5-Tetramethyl-3,4-dithiohexene	C$_8$H$_{18}$S$_2$	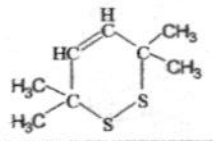	52.5 ± 0.2	298	5.40	5.50	DH$_3$CH-CH$_2$ = 1.05
3,4-Dithiohexene	C$_4$H$_6$S$_2$	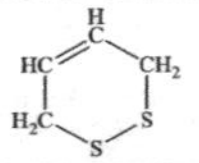	41.7	298	5.40	–	–

^{a}Estimated

approximately equal stability compared with the specific interaction of the second type $DH_2C \rightarrow CH_2$.

Considering the specificity of the interactions of the 2,2,5,5-tetramethyl-3,4-dithiohexene molecule, attention should be paid to cyclopentane and its analogs with one and two double bonds (Table 4.13). Differing from the vaporization enthalpy of cyclopentane by 1.0 and 0.4 kJ mol^{-1}, respectively, the vaporization enthalpies of spiropentane and 1,3-cyclopentadiene reflect the influence of the reverse dative bond on the average energy value of $DH_2C \rightarrow CH_2$ of the formed interaction $DHC \rightarrow CH$. In turn, the difference in the vaporization enthalpies of 2-methyltetrahydro-2H-thiopirane and tetrahydro-2H-thiopirane corresponds to the energy contribution of the isostructural methyl group (3.2 kJ mol^{-1}) to the enthalpy characteristic of the 2-methyltetrahydro-2H-thiopirane compound. It allows the assumption that the four isostructural CH_3 groups make a similar energy contribution to the vaporization enthalpy of 2,2,5,5-tetramethyl-3,4-dithiohexene, with a total contribution of 13.2 kJ mol^{-1}. Using this assumption, the vaporization enthalpy of 3,4-dithiohexene can be estimated using Eq. (4.13), taking into account the energy contributions of the substituted hydrogen atoms:

$$\Delta_{vap}H^0(298 \text{ K})dth = \Delta_{vap}H^0(298 \text{ K})tmdth - 4DisoCH_3 - 4DH \quad (4.13)$$

The molecules of 3,4-dithiohexene and 2,2,5,5-tetramethyl-3,4-dithiohexene with the four bond vacancies of the essentially undivided $2s^2$ electron pairs of the carbon atoms and the four bond vacancies of the free electron pairs of the two sulfur atoms form four specific interactions $DS \rightarrow CH$ and four interactions $DS \rightarrow CH_2$ with an average energy value of 5.40 kJ mol^{-1} (Fig. 4.19). This energy value is within the error range of the energy of $DH_2C \rightarrow CH_2$ of the specific interaction of the experimentally measured vaporization enthalpy (Table 4.13). From this, one can conclude that the given value acts as the limiting value at which the sulfur atom forms the specific interaction with increased stability. In addition to the mentioned interactions, the molecule of 2,2,5,5-tetramethyl-3,4-dithiohexene forms, by the four isostructural methyl groups, eight specific interactions of low stability (1.05 kJ mol^{-1}), which participate in the formation of the network of interactions of the liquid and crystal compounds.

4.3.3 Liquid and Crystalline 2,3-Benzothiophene

The molecule of 2,3-benzothiophene has a rigid benzene ring and the low influence of the thiophene fragment on the change in the structural energy parameters has been proved by the additive energy contribution of the benzoles and benzoic acids to the enthalpies of vaporization and sublimation. In this regard, 2,3-benzothiophene is of special interest for obtaining the energies of the formed specific interaction of the sulfur atom and CH group of the thiophene fragment. Figure 4.20 shows the network of specific interactions of liquid and crystal

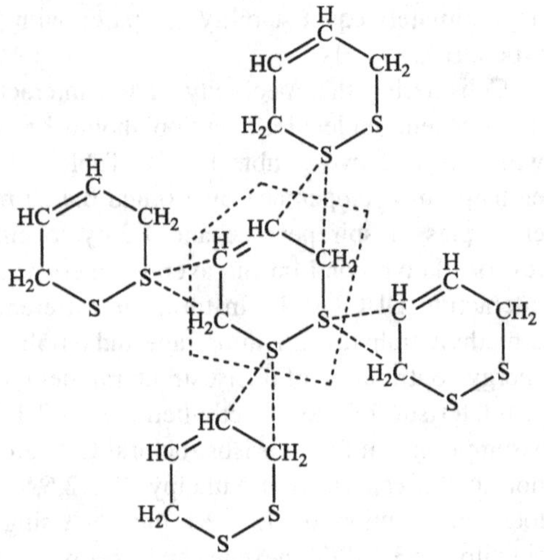

Fig. 4.19 Schematic of the network of specific interactions of liquid and crystalline 3,4-dithiohexene

2,3-benzothiophene, illustrating the four specific interactions DHC → CH formed by the carbon atom of the CH groups, with energy values of 5.63 and 5.03, and one DC → C interaction with energy values of 7.43, and 6.63 kJ mol^{-1}, for the liquid and crystal states respectively.

The two formed specific interactions of DHC → C of the carbon atom of the benzene ring and the CH groups of the thiophene fragment were taken to be equal to 5.33 and 6.92 kJ mol^{-1}, respectively, for a similar type of interaction in liquid and crystal benzene. The thiophene fragment forms two more stable specific interactions DS → CH. The energy of this type of interaction can be determined using Eq. (4.14):

$$DS \rightarrow CH = \Delta_{evap}H^0(298\ K) - 4DHC \rightarrow CH - 2DHC \rightarrow C - DC \rightarrow C \tag{4.14}$$

The calculation results presented in Table 4.14 point to the small difference in the energy values of the specific interactions of the liquid and crystal conditions, which indicates the increased error of the experimentally measured vaporization enthalpy given in the literature without any indication of the errors. At the same time, the small difference between the energies of the specific interactions of DS → CH and the energies of DHC → CH of the formed CH groups of the benzene ring is caused by the low ability of these groups to shift the electron density to the sulfur atom of the thiophene fragment.

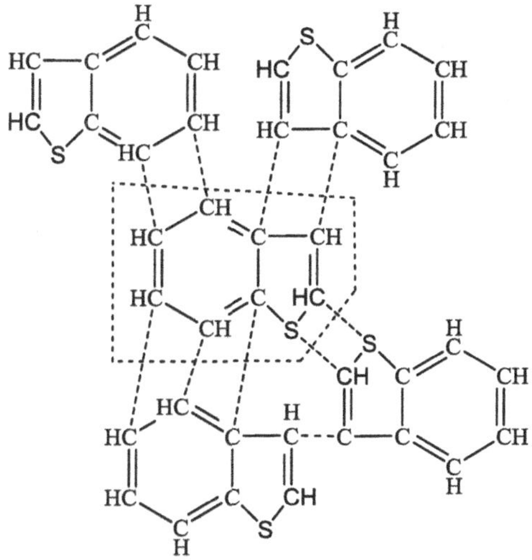

Fig. 4.20 Schematic of the network of specific interactions of liquid and crystalline 2,3-benzothiophene

Table 4.14 Energies (kJ mol^{-1}) of specific interactions of 2,3-benzothiophene

Compounds	Formula	Structure	$\Delta_{evap}H^0$ (298 K)	DS → CH
		Liquid [12]		
DHC → CH = 5.63, DHC → C = 5.33, DC → C = 5.03				
Benzo[b]thiophene	C$_8$H$_6$S	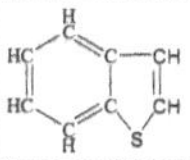	54.3^a	7.75
		Crystalline [18]		
DHC → CH = 7.42, DHC → C = 6,92, DC → C = 6.65				
2,3-Benzothiophene	C$_8$H$_6$S	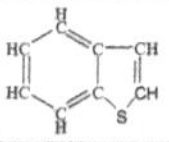	65.7 ± 0.2^b	7.85

^{a}320 K
^{b}298 K

4.3.4 Thiophenes

The molecule of thiophene forms a network of specific interactions in the liquid and crystal conditions (Fig. 4.21), with the three coordinating molecules forming three specific interactions DHC → CH with energies of 5.63 and 7.42 kJ mol^{-1} for the liquid and crystal conditions, respectively, and also two interactions of increased stability, whose energies are determined from Eq. (4.15):

$$DS \rightarrow CH = \Delta_{evap}H^0(298 \text{ K}) - 3DHC \rightarrow CH \qquad (4.15)$$

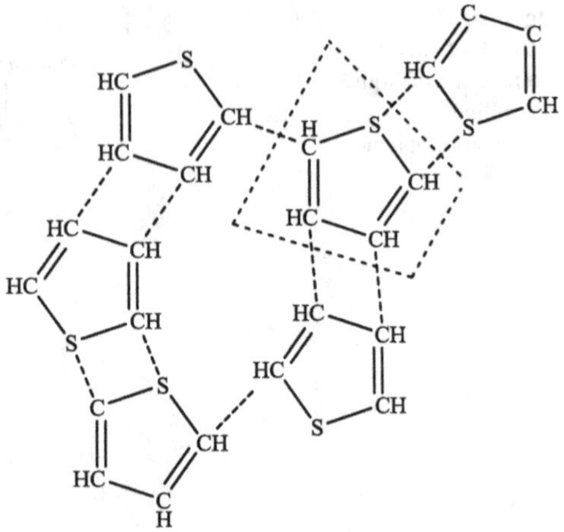

Fig. 4.21 Schematic of the network of specific interactions of liquid thiophene

Due to the unaccounted temperature of the energy of specific interaction $DHC \rightarrow CH$, the obtained energy of the second type of interaction $DS \rightarrow CH$ has an increased value (Table 4.15).

The obtained energy value of the specific interaction $DS \rightarrow CH$ remains unstable for all compound derivatives of the thiophene series with the methyl, ethyl, and other isostructural alkyl functional groups. The energy contribution of the isostructural groups to the vaporization enthalpy of these compounds and the formed specific interactions can be obtained from the difference in the vaporization enthalpies of compounds with the isostructural group and thiophene minus the energy contributed by the substituted hydrogen atom (0.60 kJ mol^{-1}) using Eq. (4.16):

$$DisoCH_3 = 2DH_3CH - CH_2 = DisoR$$

$$= \Delta_{vap}H^0(298\,K)mtph - \left(\Delta_{vap}H^0(298\,K)tph - DH\right) \qquad (4.16)$$

According to the obtained data (Table 4.15), it follows that the maximum energy contribution is given by the isostructural methyl group located at C(3):

$$3\text{-Methylthiophene}(2.45) > 2\text{-Methylthiophene}\left(2.05\,kJ\,mol^{-1}\right)$$

The stability of the same specific interactions at the two isostructural methyl groups in the molecule is described by the natural sequence, with the location at carbon atoms C(2) and C(4), at C(3) and C(4), and, further, at C(2) and C(5) reflecting the shifting of the electron density and the distribution of charges at the carbon atoms:

2,4-Dimethylthiophene $(1.7) \approx 3.4$-Dimethylthiophene $(1.67) > 2,5$-Dimethylthiophene $(1.27$ kJ mol$^{-1})$

Table 4.15 Energies (kJ mol^{-1}) of specific interactions of liquid thiophene and its derivatives

Compounds	Formula	Structure	$\Delta_{vap}H^0$ (298 K) [12]	T(K)	DS → CH	DisoCH$_3$	DH$_3$C → H-CH$_2$
[18] DHC → CH = 7.42,							
Thiophene	C$_4$H$_4$S		46.8[a]	213	12.25	–	–
Liquid DHC → CH = 5.63							
Thiophene	C$_4$H$_4$S		34.6	298	8.85	–	–
2-Methylthiophene	C$_5$H$_6$S		38.7	298	8.85	4.1	2.05
3-Methylthiophene	C$_5$H$_6$S		39.5	298	8.85	4.9	2.45
2,4-Dimethylthiophene	C$_6$H$_8$S		41.4	338	8.85	6.8:2	1.7
2,5-Dimethylthiophene	C$_6$H$_8$S		39.7	348	8.85	5.1:2	1.27
3,4-Dimethylthiophene	C$_6$H$_8$S		41.1	343	8.85	6.7:2	1.67
							D(CH$_2$)$_2$-CH$_3$ → CH$_2$-(CH$_2$)$_n$-CH$_3$
2-Ethylthiophene	C$_6$H$_8$S		33.7[b]	348	8.85	–	–
3-Ethylthiophene	C$_6$H$_8$S		40.7	333	8.85	6.1	$n = 0$ 3.05
2-Propylthiophene	C$_7$H$_{10}$S		46.0	273	8.85	11.2	$n = 1$ 5.6

[a]Crystalline
[b]Underestimated

The energies of the specific interactions formed by the ethyl and propyl ligand are subordinated by the general rule of their stability increasing with an increased number of carbon atoms in the chain, independently of its location at C(3) in the molecule of 3-ethylthiophene and at C(2) in the molecule 2-propylthiophene.

4.4   Hydrogen Bonds and Specific Intermolecular Interactions of Thieles

4.4.1   1-Propanethiol Series

The nature of the interactions in a molecule of water and a molecule of hydrogen sulfide SH_2 are analogous. Existence in this molecules $1b_1$-orbital the "real" unbinding, $2a_1$-H-H connecting with the insignificant contribution to the E–H-bond and $1b_2$-E–H-binding stipulate their low associated possibility. Adam et al. [24] set 11 conditions, caused by the violence of the one-electron ionization. With all probability, it should be assumed that precisely these peculiarities cause the significantly low association of hydrogen sulfide in the liquid condition. Like hydrogen sulfide (and unlike water and the alcohols, which are unassociated in vapors), mercaptanes have low vaporization enthalpies as shown by the boiling points of its first representatives CH_3SH (280.75 K) and C_2H_5SH (327.85 K). In the case of methyl mercaptan, the sequence of the energies of orbitals takes the form: $2a''$ $(n) < 5a'$ $(n') < 4a'$ (HSC) $< 1a''$ (CH) $< 3a'$ (CH) [23]. The electron configurations of CH_3OH and CH_3SH coincide; however, the binding character of $3a'$-MO and $4a'$-MO differ [25], causing the intermolecular interactions of these two classes of organic compounds to differ. The lack of dimeric forms of the mercaptan molecules and the underestimated (by about half) values of the vaporization enthalpies of the first representatives of these series of compounds compared with the values of alcohols reflect the weak intermolecular interactions. There is a similar dependence of the vaporization enthalpies of alcohols and mercaptanes on the number of specific interactions of the carbon atoms of the alkyl chain and their energies, but the compounds differ significantly in the processes that occur during the vaporization (Fig. 4.22). The vaporization process of alcohols consists more than 80% of the transition to vapor of dimeric molecules of methyl alcohol and about 5% of butyl alcohol, and their dissociation process.

$$CH_3-CH_2-O-H \atop H-O-CH_2-CH_3 \rightleftharpoons 2\left(H-O-CH_2-CH_3\right)$$

It is so high because the presence of these molecular forms in the vapors smoothes the influence of the reverse dative bond, with its completion at the third carbon atom of the chain or, more precisely, excludes the manifestation of the intersection of two lines for the vaporization enthalpies of propyl or at least butyl

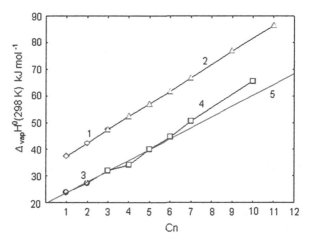

Fig. 4.22 Dependence of the vaporization enthalpy on the number of carbon atoms in the series of 1-alcohols (*1, 2*) and 1-mercaptanes (*3, 4*). See text for description of *line 5*

alcohol. It should be noted that the dissociation energies of the dimeric molecules of isopropyl and isobutyl alcohols are comparable with the value of the vaporization enthalpy [26]. In the absence of the dimeric forms of molecules in the vapors of alcohols with the largest number of carbon atoms in the chain, the vaporization process is accompanied by the breakage of the two hydrogen bonds of high stability.

$$H-O-CH_2-CH_3$$
$$CH_3-CH_2-O-H \rightleftharpoons CH_3-CH_2-O-H$$
$$H-O-CH_2-CH_3$$

There is also breakage of two specific interactions $DO \rightarrow CH_3-(CH_2)_n-O$, formed by the end terminal methyl group of the alkyl ligand and the second free electron pair of the oxygen atom, with the energy value of the given specific interaction of the diethyl alcohol being 6.78 kJ mol^{-1} [11]. The increase in the alcohol vaporization enthalpy after the completion of the reverse dative bond influence at the transition to vapor only in monomeric molecules is caused by the energy contribution of the increasing number of methylene groups of the alkyl chain (Fig. 4.22, line 2).

The low values of the vaporization enthalpies of mercaptanes (Fig. 4.22, lines 3 and 4) reflect the breakage of the low stability specific interactions. The intersection of the two lines showing the dependency of vaporization enthalpy on the number of chain carbon atoms is the result of the completion of the reverse dative bond influence, after which (line 5) the contribution of the methylene groups gradually increase and, starting from 1-hexanethiol, significantly increases (line 4). The lack of association in the vapors of the mercaptanes and the low values of

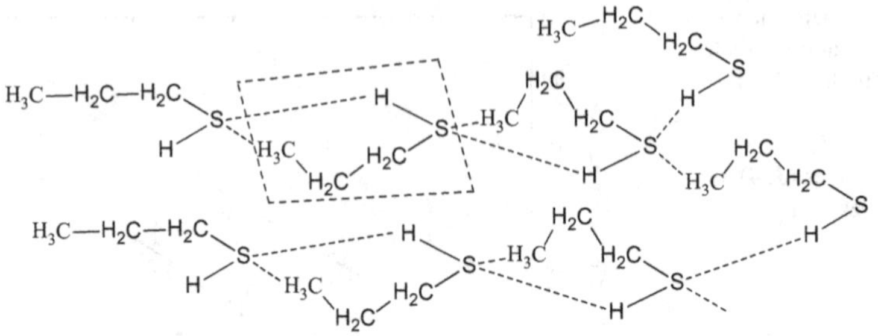

Fig. 4.23 Schematic of the network of specific interactions of liquid 1-propanethiol

the vaporization enthalpies are indications of the low stability of the formed specific interaction compared with liquid alcohols. The presence of two ligands in the mercaptan molecule provides the sulfur atom with the specific interaction of the alkyl ligand DS $\rightarrow$ CH$_3$–S or DS $\rightarrow$ CH$_3$–CH$_2$–S and the hydrogen bond DS•••H–S, forming a network of the specific interactions in the liquid compounds (Fig. 4.23).

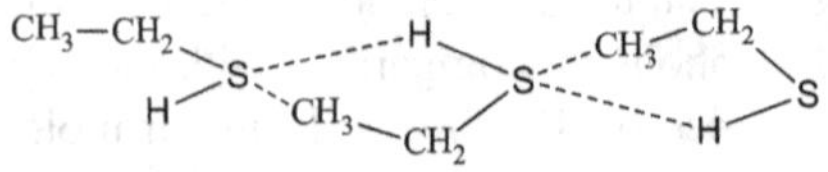

The increase in the number of carbon atoms of the chain is accompanied by the effect of the structural stress of the destabilizing specific interaction DS $\rightarrow$ R–S, causing the location of the vaporization enthalpy of 1-butanethiol and 1-pentanethiol at line 5 in Fig. 4.22. Further increase in the number of methylene groups provides the description of the dependency $\Delta_{vap}H^0(298\ \mathrm{K}) = f(\mathrm{Cn})$ by the straight line (Fig. 4.22). Small differences in the interatomic distances of the methyl group and the hydrogen atom in the molecule of methyl mercaptan permit the energy of the specific interaction of DS $\rightarrow$ CH$_3$–S to be taken as being equal to the energy of the same type of interaction of the liquid dimethyl sulfide, 6.92 kJ mol^{-1}. Such an assumption is based on the similar assumptions for ethers with differing alkyl ligands [8]. The assumption allows estimation of the energy of the hydrogen bond using the vaporization enthalpy of the methyl mercaptan:

$$\mathrm{DS} \bullet \bullet \bullet \mathrm{H} - \mathrm{S} = \left(\Delta_{vap}H^0(298\ \mathrm{K})\mathrm{mc} - 2\mathrm{DS} \rightarrow \mathrm{CH}_3 - \mathrm{S}\right)/2 \qquad (4.17)$$

The obtained energy value of the hydrogen bond is in good agreement with the energy value of the same type for liquid aldehydes [11], remaining unchanged in the compounds of these series. This shows that the energy value of the hydrogen bond of the liquid mercaptanes remains stable at 5.0 kJ mol^{-1} and that the energies of the formed specific interactions of alkylthiols can be obtained using the modified Eq. (4.17a), where n is the number of methylene groups:

$$\mathrm{DS} \bullet \bullet \bullet \mathrm{H} - \mathrm{S} = \left(\Delta_{vap}H^0(298\,\mathrm{K})\mathrm{amc} - 2\mathrm{DS} \rightarrow \mathrm{CH}_3 - (\mathrm{CH}_2)n - \mathrm{S}\right)/2 \quad (4.17a)$$

0The results of the calculations (Table 4.16) lead to the following conclusion on the stabilization of specific interactions, which have reduced values in comparison with the energies of the same sulfide interactions:

DS → CH_3–S, methyl mercaptan $(6.92) <$ DS → CH_3–CH_2–S, ethyl mercaptan $(8.65) <$ DS → CH_3–CH_2–CH_2–S, 1-propanethiol $(10.95$ kJ mol$^{-1})$

Dimethyl sulfide $(6.92) <$ Diethyl sulfide $(9.05) <$ Dipropyl sulfide $(11.17$ kJ mol$^{-1})$

The difference in the energy values of the interactions of these compounds corresponds to the destabilizing effect of the structural stress of the ethyl $(0.40$ kJ mol$^{-1})$ and propyl $(0.22$ kJ mol$^{-1})$ ligands. The energy contribution of the methylene groups $\sum DCH_2$ to the vaporization enthalpy of 1-butanethiol to 1-decanethiol was obtained from the difference between the vaporization enthalpy of the compounds of this series and that of 1-propanethiol (Table 4.16).

4.4.2 Series of 2-Propanethiol and Alkylthiols with Isostructural Methyl Groups

Thioalcohols of the 2-propanethiol series are compounds with an isostructural methyl group located at the first carbon atom of the alkyl chain. Therefore, the compounds of the thioalcohol series form the stable specific interaction not by the propyl group in the case of 2-propanethiol nor by the butyl group of 2-butanethyol, but by the ethyl or, correspondingly, the propyl ligand with energies of 8.65 and 10.95 kJ mol^{-1}, respectively. The specific interactions of the isostructural methyl group participate in the crosslinking of the chains to form a united network of specific interactions of two types and two hydrogen bonds with an energy of 5.0 kJ mol^{-1} (Fig. 4.24). The energy contributed by the formed specific interaction of the isostructural methyl group to the vaporization enthalpy can be determined from Eq. (4.18):

$$DH_3C \rightarrow H - CH_2$$
$$= \left(\Delta_{vap}H^0(298\,K) - 2DS \rightarrow CH_3 - (CH_2)n - S - 2DS \bullet\bullet\bullet H - S\right)/2$$
(4.18)

The energy contribution of the methylene groups to the vaporization enthalpy of 2-pentanethiol and 2-hexanetyiol are determined using Eq. (4.19):

$$\sum DCH_2 = \left(\Delta_{vap}H^0(298\,K) - 2DS \rightarrow CH_3 - (CH_2)n - S \right.$$
$$\left. - 2DS \bullet\bullet\bullet H - S - DH_3C \rightarrow H - CH_2\right)/2$$
(4.19)

The results of the implemented calculations can be seen in Table 4.17.

Table 4.16 Energies (kJ mol^{-1}) of hydrogen bonds and specific interactions of liquid compounds of the 1-propanethiol series at 298 K

Compounds	Formula	Structure	$\Delta_{vap}H^0$ (298 K) [12]	$DS \rightarrow CH_3-(CH_2)_n-S$	$DS \bullet \bullet \bullet H-S$	$\sum DCH_2$
Methyl mercaptan	CH_4S	CH$_3$–S–H	23.8	$n = 0$	5.0	–
				6.92		
Ethyl mercaptan (ethanethiol)	C_2H_6S	CH$_3$–CH$_2$–S–H	27.3	$n = 1$	5.0	–
				8.65		
1-Propanethiol	C_3H_8S	H$_3$C–H$_2$C–H$_2$C–S–H	31.9	$n = 2$	5.0	–
				10.95		
1-Butanethiol	$C_4H_{10}S$	CH$_3$–CH$_2$–(CH$_2$)–S–H	34.1	10.95	5.0	2.2
1-Pentanethiol	$C_5H_{12}S$	H$_3$C–(CH$_2$)$_2$–CH$_2$–CH$_2$–S–H	41.1	10.95	5.0	8.1
			40.0^a			
1-Hexanethiol	$C_6H_{14}S$	CH$_3$–(CH$_2$)$_3$–CH$_2$–CH$_2$–S–H	44.6 ± 0.2	10.95	5.0	12.7
1-Heptanethiol	$C_7H_{16}S$	CH$_3$–(CH$_2$)$_5$–CH$_2$–S–H	50.6 ± 0.2	10.95	5.0	18.7
1-Octanethiol	$C_8H_{18}S$	CH$_3$–(CH$_2$)$_6$–CH$_2$–S–H	55.3^a	10.95	5.0	23.4
1-Nonanethiol	$C_9H_{20}S$	CH$_3$–(CH$_2$)$_7$–CH$_2$–S–H	60.4^a	10.95	5.0	28.5
1-Decanethiol	$C_{10}H_{22}S$	CH$_3$–(CH$_2$)$_8$–CH$_2$–S–H	65.5 ± 0.5	10.95	5.0	33.6

^{a}Elucidated on the base of dependence $\Delta_{vap}H^0$(298 K) $= f(Cn)$

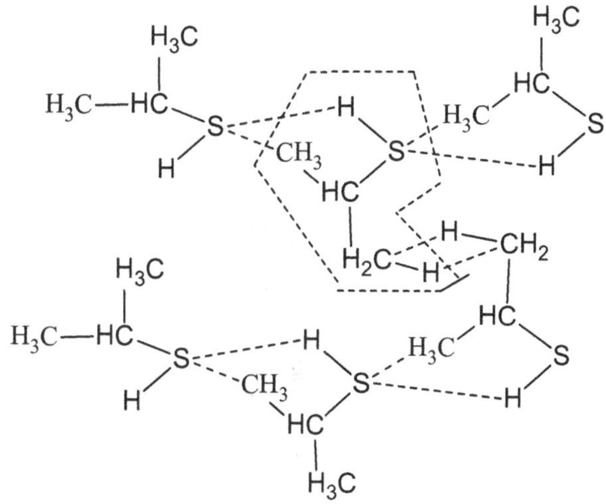

Fig. 4.24 Schematic of the network of specific interactions of liquid 2-propanethiol

Using Eqs. (4.18) and (4.19), the energies of the specific interactions formed by the isostructural methyl group of methyl-1-propanthiols and 3-methyl-1-butanethyol can be determined, as well as the energy contributions of the methylene groups. The obtained data do not reflect the influence of the methyl group location at the second and third carbon atoms of the alkyl chain on the stability change of these interactions. The average energy values of the specific interactions contributed by the two isostructural methyl groups of the alkylthiols are calculated using the following equation:

$$DH_3C \rightarrow H - CH_2$$
$$= \left(\Delta_{vap}H^0(298\,K) - 2DS \rightarrow CH_3 - (CH_2)n - S - 2DS \bullet \bullet \bullet H - S\right)/4$$

(4.20)

The data presented in Table 4.18 allow a conclusion to be made on the significant destabilizing influence of the location of the two isostructural methyl groups at the first carbon atom in the *tert*-butyl mercaptan and at the second carbon atom of the chain, interacting with the sulfur atom of 2-methyl-2-butanethyol and 3-methyl-2-butanethyol, and then at the second 2-methyl-1-propanthiol, 2-methyl-1-butanethyol, and at the third carbon atom of the chain of 3-methyl-1-butanethyol.

The location of the two methyl groups at the second carbon atom of the chain in the molecule of 2,2-dimethyl-1-propanethyol is accompanied by stabilization of the formed specific interactions (from 0.77 to 1.12 kJ mol^{-1}), the two methyl groups located at C(1) and C(2), 2,3-dimethyl-2-butanethiol (1.23 kJ mol^{-1}) at C(3), and the S–H group at C(2) of the molecule of 3-methyl-2-butanethyol (1.40 kJ mol^{-1}). The location of the two methyl groups at the first carbon atom and one at the second carbon atom in the molecule of 2,3-dimethyl-2-butanethiol is accompanied by

Table 4.17 Energies (kJ mol^{-1}) of hydrogen bonds and specific interactions of liquid compounds of series 1-propanethiol

Compounds	Formula	Structure	$\Delta_{vap}H^0$ (298 K) [12]	T (K)	$DS \to CH_3\text{--}(CH_2)_n\text{--}S$	$DS\cdots H\text{--}S$	$DH_3C \to H\text{--}CH_2$
2-Propanethiol	C_3H_9S	$H_3C\text{--}HC\text{--}H_3C$ $\quad$ S--H	29.5; 30.1	298	8.65	5.0	1.55
2-Pentanethiol	$C_5H_{12}S$	$H_3C\text{--}(CH_2)_2\text{--}CH\text{--}CH_3$ $\quad$ S--H	38.4	302	10.95	5.0	1.55 $DCH_2 = 3.4$
2-Hexanetyiol	$C_6H_{14}S$	$CH_3\text{--}(CH_2)_3\text{--}CH\text{--}CH_2$ $\quad$ S--H	42.7	325	10.95	5.0	1.55 $\sum DCH_2$
2-Heptanethiol	$C_7H_{16}S$	$CH_3\text{--}(CH_2)_4\text{--}CH\text{--}CH_2$ $\quad$ S--H	44.1	356	10.95	5.0	–

Table 4.18 Energies (kJ mol^{-1}) of hydrogen bonds and specific interactions of liquid alkylthiols with isostructural methyl groups

Compounds	Formula	Structure	$\Delta_{vap}H^0$ (298 K) [12]	$DS \to CH_3-$ $(CH_2)_n-S$	$DS \cdots H-S$	DCH_3	$DH_3C \to H-CH_2$
3-Pentanethiol	C$_5$H$_{12}$S	H$_3$C–H$_2$C–CH–CH$_2$–CH$_3$ S–H	38.3^a	$n = 2$ 10.95	5.0	6.4	D(CH$_2$)$_2$–CH$_3 \to$ (CH$_2$)$_2$– CH$_3$ = 3.2
2-Methyl-1-propanthiol	C$_4$H$_{10}$S	CH$_3$–CH–CH$_2$–S–H CH$_3$	34.8	$n = 2$ 10.95	5.0	2.9	1.45
2-Methyl-1-butanethyol	C$_5$H$_{12}$S	H$_3$C–CH$_2$–CH–CH$_2$–S–H CH$_3$	39.9 ± 0.1	$n = 2$ 10.95	5.0	2.9	1.45
3-Methyl-1-butanethyol	C$_5$H$_{12}$S	H$_3$C–CH–CH$_2$–CH$_2$–S–H CH$_3$	39.9 ± 0.1	$n = 2$ 10.95	5.0 DCH$_2$ = 5.1	2.9	1.45
tert-Buthylmercaptan	C$_4$H$_{10}$S	CH$_3$ CH$_3$–C–S–H CH$_3$	30.8	$n = 1$ 8.65	5.0 DCH$_2$ = 5.1	3.5;2	0.77
2-Methyl-2-butanethyol	C$_5$H$_{12}$S	CH$_3$ H$_3$C–CH$_2$–C–CH$_3$ S–H	35.0	$n = 2$ 10.95	5.0	3.1;2	0.77
2-Methyl-2-pentanethiol	C$_6$H$_{14}$S	CH$_3$ CH$_3$–CH$_2$–CH$_2$–C–CH$_3$ S–H	40.0 ± 0.1	$n = 2$ 10.95	5.0 DCH$_2$ = 5.0	3.1;2	0.77
3-Methyl-2-butanethyol	C$_5$H$_{12}$S	H$_3$C–CH–CH–CH$_3$ CH$_3$ S–H	37.5 ± 0.1	$n = 2$ 10.95	5.0	5.6;2	1.40
2,2-Dimethyl-1-propanethyol	C$_5$H$_{12}$S	CH$_3$ H$_3$C–C–H$_2$C–S–H CH$_3$	36.4 ± 0.1	$n = 2$ 10.95	5.0	4.5;2	1.12
2,3-Dimethyl-2-butanethiol	C$_6$H$_{14}$S	CH$_3$ CH$_3$ CH$_3$–CH–C–CH$_2$ S–H	39.3 ± 0.1	$n = 2$ 10.95	5.0	7.4;3	1.23

^{a}302 K

stabilization of the specific interactions in comparison with the *tert*-butyl mercaptan and is destabilized relative to the same interaction at the methyl-2-butanethyol. Therefore, the presence in the molecule of the isostructural methyl groups is accompanied by their participation in the shifting of the electron density in the molecule and the distribution of charges at the carbon and sulfur atoms, causing the formation of specific interactions with increased stability.

4.4.3   1,2–Ethanedithiol Series

The experimentally measured vaporization enthalpies of the series of compounds of 1,2-ethanedithiols for the four tioalcohols are described by the dependence $\Delta_{vap}H^0(298 \text{ K})$edto $= f(Cn)$ (Fig. 4.25). This allows the use of a similar dependence $\Delta_{vap}H^0(298 \text{ K})$pto for alcohols of the 1-propanethiol series and to use the more precise correlation $\Delta_{vap}H^0(298 \text{ K})$edto $= f(\Delta_{vap}H^0(298 \text{ K})pto)$, described by a straight line, to significantly expand the information on the vaporization enthalpies of the alcohol series 1,2-ethanedithiols (Table 4.19). Compounds of the 1,2-ethanedithiol series with two bond vacancies of the free electron pairs of the two sulfur atoms, two bond vacancies of the two hydrogen atoms, and two methylene groups form two specific interactions DS $\rightarrow$ CH$_2$- and two hydrogen bonds DS•••H–S with an energy of 5.0 kJ mol^{-1}, forming a network of these bonds (Fig. 4.26a). An increase in the number of methylene groups in the chain stabilizes the formed specific interactions (Fig. 4.26b). Using the known energy value of the hydrogen bond and Eq. (4.21), the energy of the specific interaction of the liquid 1,2-ethanedithiol is found to be equal to 6.17 kJ mol^{-1}:

$$DS \rightarrow CH_2 - S = \left(\Delta_{vap}H^0(298\,K)edto - 2DS \bullet \bullet \bullet H - S\right)/2 \qquad (4.21)$$

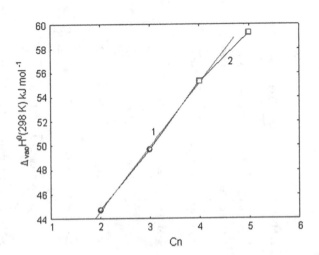

Fig. 4.25 Dependence of vaporization enthalpy versus the number of carbon atoms in the series of 1,2-ethanedithiols (*1,2*)

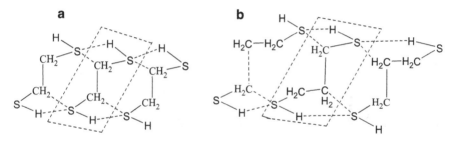

Fig. 4.26 Schematic of the network of specific interactions of liquid 1,2-ethanedithiol (**a**) and 1,3-propanedithiol (**b**)

This value is reduced in comparison with the interactions of methyl mercaptan formed by the methyl group (6.92 kJ mol^{-1}), due to the lower number of hydrogen atoms in the methylene group. Transforming Eq. (4.15) by replacement of the vaporization enthalpy of the corresponding tioalcohols, we determined the energies of the specific interactions of 1,2-ethanedithiol, which are presented in Table 4.19. The obtained energies of the specific interactions illustrate their stabilization with an increasing number of methylene groups in the hydrocarbon chain, with the maximum stability at 1,6-hexanedithiol:

DS → CH$_2$–(CH$_2$)$_n$: 1,2-ethanedithiol (6.17) < 1,3-propanedithiol (7.62) < 1,4-butanedithiol (8.66) < 1,5-pentanedithiol (9.75) < 1,6-hexanedithiol (10.78) < 1,7-heptanedithiol (10.78) < 1,8-heptanedithiol (10.78 kJ mol^{-1})

This is caused by the completion of the influence of the reverse dative bond on the specific interactions and draws attention to the following points:

First, the reduction in the differences in the energies of the specific interactions formed by the methylene and the methyl group (Table 4.16) following replacement of the methylene ligand by an ethyl ligand

Second, the complete coincidence of the energy value of the specific interaction of 1,4-butanediitiol, formed by the ethylene group (8.66 kJ mol^{-1}) and that of the ethyl group (8.65 kJ mol^{-1}) of the ethyl mercaptan molecule of the two differing series of compounds, which reflects the correctness of the selected energy estimation of the hydrogen bond in the calculation of the energy of specific interactions

4.4.4 Cyclopentanethiol and Cyclohexanethiol

The increased value of the vaporization enthalpy of the cyclopentyl methylsulfide (45.1 ± 0.1 kJ mol^{-1}) in comparison with those of cyclopentanethiol and cyclohexanethiol (44.6 ± 0.1 kJ mol^{-1}) indicates the increased stability of the two formed specific interactions of the methyl group DS → CH$_3$–S compared with the two

Table 4.19 Energies (kJ mol^{-1}) of hydrogen bonds and specific interactions of liquid compounds of series 1,2-ethanedithiols at 298 K

Compounds	Formula	Structure	$\Delta_{vap}H^0$ (298 K) [12]	DS $\rightarrow$ CH$_2$–(CH$_2$)$_n$	DS•••H-S	$\sum$DCH$_2$
1,2-Ethanedithiol	C$_2$H$_6$S$_2$	H-S–CH$_2$–CH$_2$–S–H	44.7	$n=0$ 6.17	5.0	–
1,3-Propanedithiol	C$_3$H$_9$S$_2$	H-S–H$_2$C–H$_2$C–H$_2$C–S–H	49.7	$n=0.5$ 7.62	5.0	–
1,4-Butanediol	C$_4$H$_{10}$S$_2$	H-S–H$_2$C–(CH$_2$)$_2$–CH$_2$–S-H	55.3	$n=1$ 8.66	5.0	–
1,5-Pentanedithiol	C$_5$H$_{12}$S$_2$	H-S–H$_2$C–(H$_2$C)$_3$–H$_2$C-S-H	59.3	$n=1.5$ 9.75	5.0	–
1,6-Hexanedithiol	C$_6$H$_{14}$S$_2$	H-S–H$_2$C–(H$_2$C)$_4$–H$_2$C-S-H	63.1[a]	$n=2$ 10.78	5.0	–
1,7-Heptanedithiol	C$_7$H$_{16}$S$_2$	H-S–CH$_2$–(CH$_3$)$_5$–CH$_2$-S-H	59.0[b] 66.9[a]	10.78	5.0	3.8
1,8-Octanedithiol	C$_8$H$_{18}$S$_2$	H-S–CH$_2$–(CH$_2$)$_6$–CH$_2$–S-H	60.9[c] 70.7[a]	10.78	5.0	7.6

[a]Elucidated at 298 K
[b]Corresponding to 407 K
[c]Corresponding to 420 K

Table 4.20 Energies (kJ mol^{-1}) of hydrogen bonds of liquid cyclopentanethiol and cyclohexanethiol

Compounds	Formula	Structure	$\Delta_{vap}H^0$ (298 K) [12]	T (K)	$\Delta_{vap}H^0(T)-$ DH	DS•••H–S
Cyclopentanethiol	$C_5H_{10}S$		38.4	363	26.7	5.85
Cyclopentane	C_5H_{10}	–	27.3	322	26.7	–
Cyclohexanethiol	$C_6H_{12}S$		44.6 ± 0.1	298	31.7	6.45
Cyclohexane	C_6H_{12}	–	32.3 ± 0.3	298	31.7	–

formed hydrogen bonds of cyclohexanethiol. This means that the cyclic thiols do not form associated molecules in vapors and, unlike alcohols, do not form specific interactions by the second free electron of the sulfur atom.

Therefore, the energies of the hydrogen bonds of these compounds can be determined from the difference between the vaporization enthalpy of the cyclopentanethiol and that of the cyclopentane minus the energy contribution of the substituted hydrogen atoms and, correspondingly, from the vaporization enthalpies of cyclohexanethiol and cyclohexane minus the same energy contribution of the saturated hydrogen atom:

$$DS \bullet \bullet \bullet H - S = \left(\Delta_{vap}H^0(298\,K)cpto - \left(\Delta_{vap}H^0(298\,K)cp - DH\right)\right)/2 \quad (4.22)$$

$$DS \bullet \bullet \bullet H - S = \left(\Delta_{vap}H^0(298\,K)chto - \left(\Delta_{vap}H^0(298\,K)ch - DH\right)\right)/2 \quad (4.22a)$$

In the energy calculation of the hydrogen bond of liquid cyclopentanethiol, we used the only available value for the vaporization enthalpy, which had been obtained at an increased temperature. As a result, the obtained energy value of the hydrogen bond has a reduced value. The increased stability of the energy of the specific interaction of DS → CH$_2$–S (8.6 kJ mol^{-1}) in cyclopentyl methylsulfide (Table 4.8) in comparison with the energy of the hydrogen bond of cyclopentanethiol and cyclohexanethiol DS•••H–S (Table 4.20) is caused by the reduced shifting of the electron density from the hydrogen atom of the thiol H–S group.

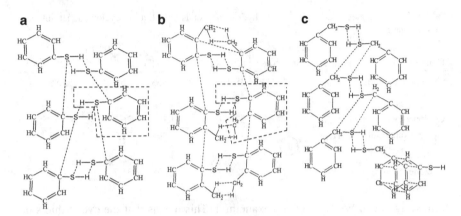

Fig. 4.27 Schematic of the network of specific interactions of liquid benzenethiol (**a**), 2-methylbenzenethiol (**b**), and benzenemethanethiol (**c**)

4.4.5 Benzenethiol(thiophenol), Methylbenzenethiols, and Benzenemethanethiol

Earlier in this chapter, in the Sect. 4.2.2 entitled "Phenyl Sulfides, Benzyl Methyl Sulfide, and (Thio)toluenes", it was shown that, in the three series of compounds of the phenyl sulfides, benzo methyl sulfides, and (thio)toluenes, only the sulfide molecule forms a specific interaction with the carbon atom of the benzene ring (with an energy of 6.17 kJ mol^{-1}) by the one free electron pair of the sulfur atom.

The benzenethiol molecule forms five specific interactions of the CH groups of the benzene ring DHC $\rightarrow$ CH (with an energy of 5.08 kJ mol^{-1} at 348–361K), two specific interactions of the carbon atom with the sulfur atom of DS $\rightarrow$ C=, and two hydrogen bonds DS•••H–S, forming a network of the specific interactions of the liquid condition (Fig. 4.27a). The energy of the hydrogen bond can be obtain from Eq. (4.23):

$$DS \bullet \bullet \bullet H - S = \left(\Delta_{vap}H^0(298\,K)btl - 5DHC \rightarrow CH - 2DS \rightarrow C\right)/2 \quad (4.23)$$

As shown above, the energies of the formed specific interactions of the sulfur with the carbon atom of the benzene ring exceeds the energy of the specific interaction of DC $\rightarrow$ C= by 1.0 kJ mol^{-1}; therefore, at the energy value of this specific interaction of 4.5 kJ mol^{-1}, the energy value of the DS $\rightarrow$ C= interaction should be taken to be equal to 5.5 kJ mol^{-1}. The calculated energy value of the hydrogen bond of DS•••H–S (Table 4.21) reflects its reduced value in comparison with the energy values of the same type of interaction of thiols with a linear saturated and allyl group or a saturated hydrocarbon ring:

DS•••H–S: benzenethiol (4.75) < methyl mercaptan (5.0) < cyclopentanethiol (5.85) < cyclohexanethiol (6.45 kJ mol^{-1}) < DS $\rightarrow$ CH$_3$ (6.92 kJ mol^{-1})

Table 4.21 Energies (kJ mol^{-1}) of hydrogen bonds and specific interactions of liquid benzenethiol, methylbenzenethiols, and benzenemethanethiol

Compounds	Formula	Structure	$\Delta_{vap}H^0$ (298 K) [12]	T (K)	DS → C=	DS•••H–S	DisoCH$_3$
DC → C = 4.5, DHC → CH = 5.08							
Benzenethiol (thiophenol)	C$_6$H$_6$S		45.9	348	5.5	4.75	–
Benzene	C$_6$H$_6$	–	30.5 / 33.8^a	361	–	–	–
2-Methylbenzene-thiol	C$_7$H$_8$S		48.1	361	5.65	4.75	2.10
3-Methylbenzene-thiol	C$_7$H$_8$S		48.7	368	5.95	4.75	2.10
4-Methylbenzene-thiol	C$_7$H$_8$S		48.1	361	5.65	4.75	2.10
					$\Delta_{vap}H^0$ (298 K)–DH		DS → CH$_2$–S
Benzenemethanethiol	C$_7$H$_8$S		56.6 ± 0.1	298	33.2	7.63	4.08

^{a}Corresponding to 320 K

The influence of a rigid benzene ring produced a value of the vaporization enthalpy at a temperature of 348 K that was significantly increased in comparison with the standard value. The molecule of methylbenzenethiol with the methyl group allows clarification of the influence of the location at the different carbon atoms of the benzene ring on the energy of the hydrogen bond. Figure 4.27b presents a schematic of the network of the specific interactions of liquid 2-methylbenzenethiol, illustrating the formation of four specific interactions of DHC → CH, two D=C → C= with the energies of 5.63 and 5.03 kJ mol^{-1}, and the two hydrogen bonds DS•••H–S. Additionally, the two specific interactions with low stability DH$_3$C → H–CH$_2$ form the isostructural methyl group, whose energy contribution to the vaporization enthalpy of 2-methylbenzenethiol is equal to the difference (4.20 kJ mol^{-1}) in the vaporization enthalpies of toluene and benzene (see the earlier Sect. 4.2.2 entitled "Phenyl Sulfides, Benzyl Methyl Sulfide, and (Thio)toluenes"). The energy of the hydrogen bond is determined from Eq. (4.23a):

$$DS \bullet \bullet \bullet H - S = \left(\Delta_{vap}H^0(298\ K) - 5DHC \to CH - 2D = C \to C \right.$$
$$\left. = -2\,DH_3C \to H - CH_2\right)/2 \qquad (4.23a)$$

This equation is used to obtain the hydrogen bonds energies of 3-methylbenzenethiol and 4-methylbenzenethiol. Given in Table 4.21 are the calculation results, which point to the small difference, which is within experimental error.

The molecules of benzenemethanethiol and benzo methyl sulfide are analogous in the number of specific interactions formed by the benzene ring 5DHC → CH, D=C → C= and the methylene group 2DS → CH$_2$–S (Figs. 4.27c and 4.15a), with energies of 5.63, 5.03, and 4.08 kJ mol^{-1}, respectively. The difference in the vaporization enthalpies of the compounds (56.6 ± 0.1 and 55.2 ± 2.1 kJ mol^{-1}) at the increased value of benzenemethanethiol, is an indication of the increased stability of the formed hydrogen bond in comparison with the energy of DS → CH$_3$–S (6.92 kJ mol^{-1}). The energy of the hydrogen bond can be obtained with the help of Eq. (4.24), taking into account the energies of the specific interactions of the benzene ring and the energy of the formed specific interaction of the methylene group:

$$DS \bullet \bullet \bullet H - S = \left(\Delta_{vap}H^0(298\,K)bmtl - 5DHC \to CH - D = C \to C \right.$$
$$\left. = -2DS \to CH_2 - S\right)/2 \qquad (4.24)$$

The obtained energy value of the hydrogen bond (Table 4.21) points to its increased value in comparison with the interactions in liquid thiols with linear and cyclic saturated hydrocarbon fragments. This stabilization effect of the hydrogen bond is caused by the location of the methylene CH$_2$ group between the sulfur atom, which is a strong electron acceptor, and the benzene ring, with a low ability to shift the electron density. The electron density thus shifts to the sulfur atom and provides an increase in the difference in charge between the sulfur and hydrogen atoms.

4.5 Energies of Specific Intermolecular Interactions and Hydrogen Bonds of Dithianes and 4,5,6,7-Tetrahydro-1,4-benzodithiol-2-tione and Its Derivatives

4.5.1 Dithianes, 1,3,5-Tritiane, and 1,3-Dithian-2-thyone

Presented in Table 4.22 are the sublimation enthalpies of dithianes and 1,3,5-tritiane with saturated methylene groups, indicating the influence on the energies of the specific interactions of the location of the sulfur atoms in the molecules of 1,4-dithiane and 1,3-dithiane, and its number in the molecule of 1,3,5-tritiane. The molecule of 1,4-dithiane with symmetrical location of the methylene groups provides the possibility for two free electron pairs for each sulfur atom in the formation of the specific interaction. As a result, the molecule forms eight specific interactions of $DS \rightarrow CH_2$, forming a network of these interactions in the crystal (Fig. 4.28a). The energies of this type of interaction are given in Table 4.22 and were obtained using Eq. (4.25):

$$DS \rightarrow CH_2 - S = \Delta_{vap}H^0(298\,K)/\,8 \qquad (4.25)$$

The location of the sulfur atoms in positions 1 and 3 of the ring of the 1,3-dithiane molecule (Fig. 4.28b) only provides the possibility for the sulfur atom of the contacting molecule to form, by the two electron pairs, the two specific interactions $DS \rightarrow CH_2$-S. In this situation, the sulfur atoms form a specific interaction of the second type with the third methylene group, and the methylene group is connected only with the sulfur atoms of $DS \rightarrow CH_2$-S. In this interaction, one free electron pair participates directly and the second electron pair increases the shifting of electron density to the carbon atom of the same methylene group. This shift increases the difference in the charges of the carbon and sulfur atoms, causing significant stabilization of the formed specific interaction $DS \rightarrow CH_2$-S. The energy value of this type of interaction significantly increases the energy value of the second type of $DS \rightarrow CH_2$ interaction. Taking the energy value of this type of interaction to be equal to the energy of the same type in crystal 1,4-dithiane (8.61 kJ mol^{-1}), the energy values of the second type of interaction $DS \rightarrow CH_2$-S of the 1,3-dithianes can be determined using Eq. (4.26) and are presented in Table 4.22:

$$DS \rightarrow CH_2 - S = \left(\Delta_{vap}H^0(298\,K) - 2DS \rightarrow CH_2\right)/\,4 \qquad (4.26)$$

The equal location of the sulfur atoms and the methylene groups in the molecule of 1,3,5-trithiane (Fig. 4.29a) causes the formation of six equal interactions $DS \rightarrow CH_2$-S of one type, forming a network of this specific interaction in the crystal. The energy of this interaction is obtained in a similar way to the energy of

Table 4.22 Energies (kJ mol^{-1}) of specific interactions and hydrogen bonds of dithianes, 1,3-dithian-2-thyone

Compounds	Formula	Structure	$\Delta_{sub}H^0$ (298 K) [18]	DS → CH$_2$	DS → CH$_2$-S	DC=S → C=S
1,4-Dithiane	C$_4$H$_8$S$_2$		68.9	8.61	—	—
1,3-Dithiane	C$_4$H$_8$S$_2$		69.9 ± 0.4	8.61	13.18	—
1,3,5-Trithiane	C$_3$H$_6$S$_3$		93.2 ± 0.2	—	15.53	—
1,3,5-Trioxane	C$_3$H$_6$O$_3$		56.20.2	DO → CH$_2$-O = 9.40	—	—
1,3-Dithian-2-thyone	C$_4$H$_6$S$_3$		91.4 ± 2.5	—	15.53	14.64
3,6-Diphenyl-1,2-dithiin	C$_{16}$H$_{12}$S$_2$	DHC → CH = 7.43	DC → C = 6.43 183.1 ± 2.9	DC → C 6.43	DS → CH 8.0	DS → C≡C 20.0
Vaporization [12]						
1,3-Dithiane	C$_4$H$_8$S$_2$		52.3 ± 0.8	6.9	9.62	—
1,4-Dithiane	C$_4$H$_8$S$_2$		55.2^a	6.9^a	—	—

^{a}Estimated

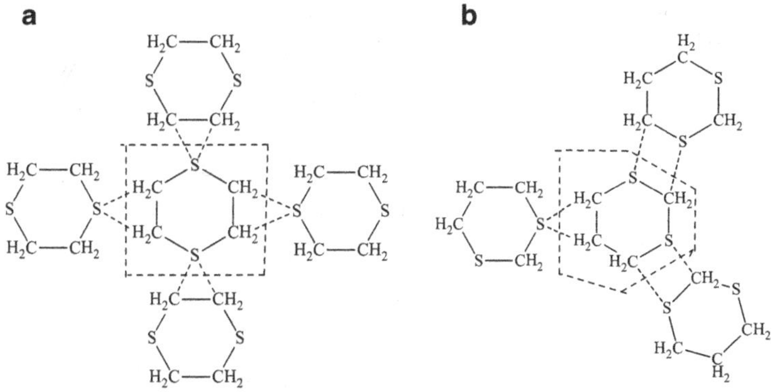

Fig. 4.28 Schematic of the network of specific interactions of crystalline 1,4-dithiane (**a**) and 1,3-dithiane (**b**)

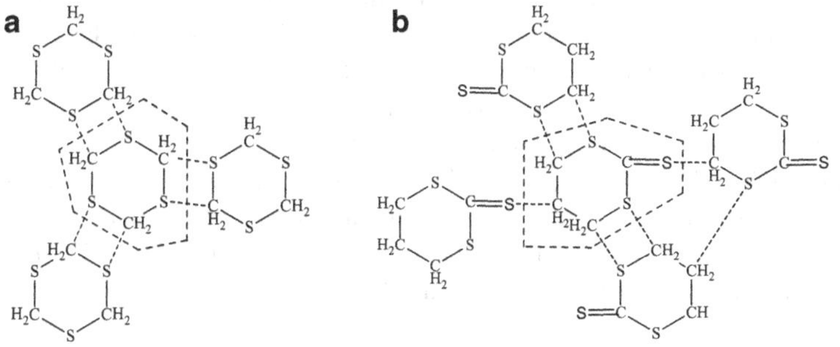

Fig. 4.29 Schematic of the network of specific interactions of crystalline 1,3,5-trithiane (**a**) and 1,3-dithian-2-thyone (**b**)

the DS $\rightarrow$ CH$_2$ interaction of the crystal 1,4-dithiane by dividing the enthalpy of sublimation by the number of specific interactions. The obtained energy values of the specific interactions (Table 4.22) illustrate their stabilization in the crystal compounds:

DS $\rightarrow$ CH$_2$: 1,4-dithiane (8.61) = 1,3-dithiane (8.61) < DS $\rightarrow$ CH$_2$–S: 1,3-dithiane (13.18) < 1,3,5-trithiane (15.53 kJ mol^{-1})

The calculated energy value of the six formed specific interactions of one type by the molecule of 1,3,5-trioxane, like 1,3,5-trithiane (Fig. 4.29a), as well as the value obtained by division of the sublimation enthalpy [11] by the number of interactions (Table 4.22), points to its low value in comparison with the formed interactions of 1,3,5-trithiane. This leads to the conclusion that the inequality in the energies of specific interactions DS $\rightarrow$ CH$_2$–S (15.53) > DO $\rightarrow$ CH$_2$–O (9.40 kJ mol^{-1}) is caused by the shifting of the electron density from one free

Fig. 4.30 Schematic of the network of specific interactions of crystalline 3,6-diphenyl-1,2-dithiin

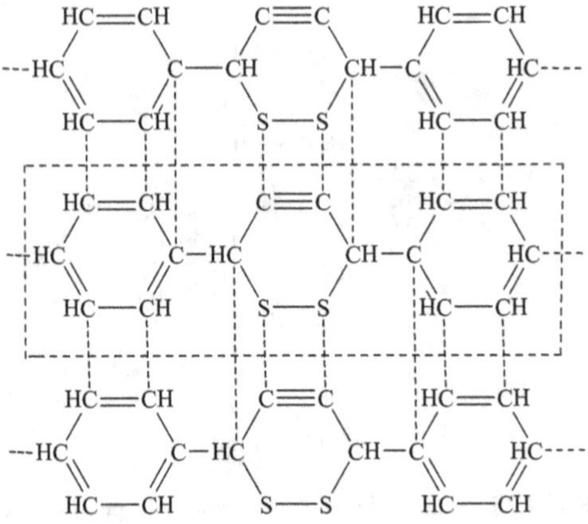

electron pair of the sulfur atom, with additional shifting from the second electron pair located at a higher energy level of the sulfur atom than at the oxygen atom.

The 1,3-dithian-2-thyone molecule with two sulfur atoms in the ring and a thyone group forms four specific interactions of the free electron pair of the sulfur atom and the significantly undivided $2s^2$ electron pair of the carbon atom of the methylene group, with an energy equal to the energy value of the same type of interaction DS → CH_2–S of 1,3,5-trithiane. Two interactions DC=S → CH_2–S are formed by the sulfur atom of the thyone group, jointly with four bonds of the second type, which form a network of the stable crystal bonds (Fig. 4.29b). The energy of this type of interaction is determined using Eq. (4.27):

$$DC = S \rightarrow CH_2 - S = \left(\Delta_{vap} H^0 (298\,K) dtti - 4DS \rightarrow CH_2 \right) / 2 \qquad (4.27)$$

The obtained energy value indicates the high stability of this type of specific interaction, which nevertheless have a reduced value compared with the second type of interaction DS → CH_2–S.

The energy value of the specific interaction DS → CH_2–S of the liquid 1,3-dithiane is equal to 6.9 kJ mol^{-1} and was obtained on the basis of the correlation of sublimation and vaporization enthalpies and the energies of the formed specific interactions of 1,3-dithiane and 1,4-dithiane. Taking into account the estimated energy value of the specific interaction of liquid 1,4-dithiane, the vaporization enthalpy of this compound can be obtained with the help of Eq. (4.25), and the results are given in Table 4.22.

The 3,6-diphenyl-1,2-dithiin compound with benzene rings in the molecule forms ten specific interactions of DHC → CH with an energy of 7.43 kJ mol^{-1} and two specific interactions D=C → C= with an energy of 6.43 kJ mol^{-1}. This gives a total energy contribution of 87.2 kJ mol^{-1} by the two benzene rings to the sublimation enthalpy (Fig. 4.30). The 1,2-dithiine fragment forms four specific interactions DS → C≡C with the carbon atom, forming a triple intermolecular bond and two specific interactions DS → CH.

The 18 formed specific interactions of four types with a planar molecular structure form a network of specific interactions that burst in the process of sublimation and provide the transition to vapor of the monomer molecules and a high value of the enthalpy of sublimation (Fig. 4.30). The formed specific interaction of DS → CH with the substituted hydrogen atom is similar to the interaction DS → CH$_2$–S, with an energy of 8.61 kJ mol^{-1}, formed by the crystal 1,3-dithiane. Taking the energy value of the specific interaction DS → CH to be equal to 8.0 kJ mol^{-1} minus the energy contribution of the substituted hydrogen atom, the energy of the specific interaction DS → C≡C can be calculated using Eq. (4.28) taking into account all the specific interactions in the crystal of this compound. The calculation results are given in Table 4.22.

$$DS \rightarrow C \equiv C$$
$$= \left(\Delta_{vap}H^0(298\,K) - 10DHC \rightarrow CH - 2DC \rightarrow C - 2DS \rightarrow CH\right)/4 \quad (4.28)$$

4.5.2 4,5-Benzo-dithiol-thiones, Tetramethylene-dithiole-thyone, and 4,5-Tetramethylene-1,3-dithiolan-2-thyone

The compounds of the three series with a seven-membered carbon ring and the four double bonds of 4,5-benzo-1,2-dithiol-3-thione and 4,5-benzo-1,3-dithiole-2-thione, two double bonds of 4,5-tetramethylene-1,3-dithiole-2-thyone and 4,5-tetramethylene-1,2-dithiole-3-thyone, a single double bond at the two thiol groups, and a single thyone group allow clarification of the influence of the location of these groups in the ring on the energy and stability of the formed hydrogen bond. The presence in the molecules of ring carbon atoms and functional groups indicates a similar number of bond vacancies of the formed specific interactions and hydrogen bonds. A change in the number of double bonds in the molecule of these compounds must be expressed by the types of specific interactions formed by the carbon atoms of the ring and their energies.

The 4,5-benzo-1,2-dithiol-3-thyone molecule with 12 bond vacancies forms four hydrogen DS•••H–S bonds by the two functional alcohol groups and the two specific interactions DC=S → C=S of the functional group of thione (Fig. 4.31a). The energy value of this specific interaction of the crystal 4,5-benzo-1,2-dithiol-3-thione should be taken to be equal to the value of the same type of specific interaction (14.65 kJ mol^{-1}) formed by the 1,3-dithian-2-thyone.

The estimation of the energy value is quite correct because the carbon atom of the functional C=S group forms single bonds with the contacting carbon atoms of the ring. The remaining six carbon atoms of the ring with double bonds (Fig. 4.31a) form one specific interaction DHC → CH, two DHC → C, and three D=C → C= interactions, whose energies are taken to be equal to the interactions in crystal benzene, 7.43, 6.92, and 6.43 kJ mol^{-1}, respectively. Such an assumption of the accepted energy estimation of the specific interactions is proved by the approximation of their energies to the maximum value, with a weakening of the reverse dative

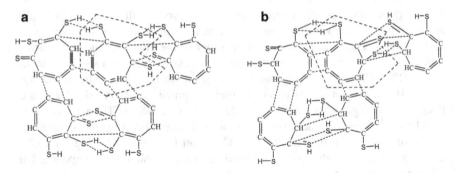

Fig. 4.31 Schematic of the network of specific interactions of crystalline 4,5-benzo-1,2-dithiol-3-thione (**a**) and 4,5-benzo-1,3-dithiole-2-thione (**b**)

bond influence with an increasing number of methylene groups up to 6–7 and, correspondingly, CH groups with double bonds in the benzene ring [11]. This allows the energy of the hydrogen bond of the crystal 4,5-benzo-1,2-dithiol-3-thione to be obtained using Eq. (4.29):

$$DS \bullet \bullet \bullet H - S = \left(\Delta_{vap}H^0(298\,K) - 2DC = S \to C = S - DHC \right.$$
$$\left. \to CH - 2DHC \to C = \ -3D = C \to C = \right)/4 \qquad (4.29)$$

The 4,5-benzo-1,3-dithiole-2-thione (Fig. 4.31b) molecule with 12 bond vacancies forms with the same value of the energy of the hydrogen bond, which is determined using Eq. (4.29):

DS•••H–S: 4,5-benzo-1,2-dithiol-3-thione (8.9) < 4,5-benzo-1,3-dithiole-2-thione
(11.8 kJ mol^{-1})

Presented in Table 4.23 are the calculation results, which allow a conclusion to be made on the stabilization of the hydrogen bond at the replacement location of the functional S-H groups at C(1) and C(2) for the positions C(1) and C(3). Such a significant stabilization of the hydrogen bond points to the fact that the transfer of the functional group=C=S from the location C(2) to the third carbon atom of the chain C(3) at the double bonds C(1) and C(4) does not express as significantly as the transfer of the alcohol S–H group, with the strong acceptor sulfur atom shifting the electron density from the CH group at location C(4) and changing the difference in charges at the sulfur and hydrogen atoms.

The 4,5-tetramethylene-1,3-dithiole-2-thyone molecule with the double bonds and two methylene groups in the ring forms one specific interaction D=HC → CH= and two D=C → C= with energies of 7.43 and 6.43 kJ mol^{-1}, respectively. The CH$_2$ methylene groups form two DH$_2$C → CH$_2$ and one DHC → CH interactions with energies of 6.27 and 5.77 kJ mol^{-1}, estimated from the sublimation enthalpies of the cyclohexane (37.6 kJ mol^{-1}). Additionally to these interactions, four

Table 4.23 Energies (kJ mol^{-1}) of specific interactions and hydrogen bonds of crystalline 4,5-benzo-1,2-dithiol-3-thione and its derivatives at 298 K

Compounds	Formula	Structure	$\Delta_{sub}H^0$ (298 K) [18]	DC=S → C=S	DS•••H–S
DHC → CH = 7.43, DHC → C= = 6.93, D=C → C= = 6.43					
4,5-Benzo-1,2-dithiol-3-thione	C$_7$H$_4$S$_3$		107 ± 0.4	14.64	8.9
4,5-Benzo-1,3-dithiole-2-thione	C$_7$H$_4$S$_3$		118.8 ± 0.4	14.64	11.8
4,5-Tetramethylene-1,3-dithiole-2-thyone	C$_7$H$_8$S$_3$		102.1 ± 2.9	14.64	11.0
4,5-Tetramethylene-1,2-dithiole-3-thyone	C$_7$H$_8$S$_3$		105.3	14.64	8.95
4,5-Tetramethylene-1,3-dithiolan-2-thyone	C$_7$H$_{10}$S$_3$		103.9 ± 2.9	14.64	8.8
4,5,6,7-Tetrahydro-1,4-benzodithiol-2-tione	C$_7$H$_8$S$_3$		99.0[a]	14.64	7.62

[a]346 K

hydrogen bonds and two specific interactions DC=S → C=S are formed. All of the 12 formed interactions of the six types participate in the formation of the network of the specific interactions of the crystal 4,5-tetramethylene-1,3-dithiole-2-thyone (Fig. 4.32a). The energy of the hydrogen bond can be determined using Eq. (4.30):

$$\begin{aligned} DS \bullet\bullet\bullet H - S &= \left(\Delta_{vap}H^0(298\,K) - 2DC = S \to C = S - D = HC \to CH\right. \\ &= -D - HC \to CH - 2D = C \to C = \left. -2DH_2C \to CH_2\right)/4 \end{aligned}$$

$$(4.30)$$

The 4,5-tetramethylene-1,2-dithiole-3-thyone molecule forms, by the carbon atoms of the ring with the double bonds, the two specific interactions D=HC → CH=, two D=C → C=, and two DH$_2$C → CH$_2$ with energies of 7.43, 6.43, and 6.27 kJ mol^{-1}, respectively, forming a network of interactions jointly with the two DC=S → C=S interactions and the four hydrogen bonds DS•••H–S (Fig. 4.32b). The energy of hydrogen bond can be determined from Eq. (4.31):

$$\begin{aligned} DS \bullet\bullet\bullet H - S &= \left(\Delta_{vap}H^0(298\,K) - 2DC = S \to C = S - 2D = HC \to CH\right. \\ &= \left. -2D = C \to C = -2DH_2C \to CH_2\right)/4 \end{aligned}$$

$$(4.31)$$

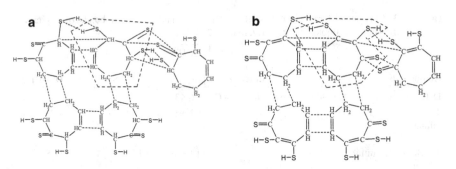

Fig. 4.32 Schematic of the network of specific interactions of crystalline 4,5-tetramethylene-1,3-dithiolan-2-thyone (**a**) and 4,5-tetramethylene-1,2-dithiole-3-thyone (**b**)

The presence of the ten hydrogen atoms in the molecule of 4,5-tetramethylene-1,3-dithiolan-2-thyone causes the formation, by the carbon atoms of the ring, of two of each specific interaction D=HC→CH= and DH$_2$C→CH$_2$ (Fig. 4.31b) and the interaction of the methylene groups with a saturated hydrogen atom D-HC→CH– with energies of 7.43, 6.27, and 5.77 kJ mol^{-1}, respectively. These interactions, together with the four hydrogen bonds DS•••H–S and the two DC=S→C=S interactions of increased stability form a network of the specific interactions in the crystal. The energies of the hydrogen bonds are determined using the Eq. (4.31a):

$$DS \bullet \bullet \bullet H - S = \left(\Delta_{vap}H^0(298\,K) - 2DC = S \rightarrow C = S - 2D = HC \rightarrow CH\right.$$
$$= \left. - 2DH_2C \rightarrow CH_2 - D - HC \rightarrow CH - \right)/4$$

$$(4.31a)$$

From the calculation results given in Table 4.23 for the energies of hydrogen bonds, it follows that their stabilization in the series of compounds is caused, first by the replacement of the location of the functional group at C(2) to C(3) in the first two compounds and, second, because there is no such increase in the shifting of electron density in the hydrocarbon ring with an increased number of methylene groups because of the difference in charges at the sulfur and hydrogen atoms under the influence of the strong acceptor of electron density, the=C=S group located at C (2):

DS•••H–S: 4,5-tetramethylene-1,3-dithiole-2-thyone (11.0) > 4,5-tetramethylene-1,2-dithiole-3-thyone   (8.95) > 4,5-tetramethylene-1,3-dithiolan-2-thyone (8.8 kJ mol^{-1})

The molecules of 4,5,6,7-tetrahydro-1,4-benzodithiol-2-tione and 4,5-tetramethylene-1,2-dithiole-3-thyone have a similar number of bond vacancies and a similar number of each type of specific interactions and can use an analog of Eq. (4.31) for calculation of the energy of the hydrogen bonds. The energy of the hydrogen bonds at 346 K, presented in Table 4.23, has a reduced value in comparison with the values obtained at standard conditions.

4.5.3 1,7,7-Trimethylbicyclo[2,2,1]heptane-2-thione and Trithiocarbonic Acid (Cyclic Ethylene Ester)

The two given compounds of the differing series are combined by the presence of a thioketone group=C=S, in which the carbon atom is located in the ring of the molecule and forms the stable specific interaction D=C=S → C=S with an energy of 14.64 kJ mol^{-1}; for example, in the compound 4,5-benzo-1,2-dithiol-3-thione. On the basis of the photoelectron spectra in the work by Nefedov and Vovna [21], it is noted that the electron configuration of $H_2C=S$ coincides with the configuration of the $H_2C=O$ molecule. The first band basically reflects only the population with the lower vibration state of the ions, and the vibrational structure of the second and the third bands is characterized as the binding MO. In this connection, the value of the sublimation enthalpy of 40.3 ± 0.32 kJ mol^{-1} for bicyclo[2,2,1]heptane, obtained with a low error, seems low at 12.5 kJ mol^{-1} in comparison with the value derived from the sublimation enthalpy of cyclopropane minus the contributions of the two saturated hydrogen atoms (2.00 kJ mol^{-1}) and excluding the energy contribution of the methylene group (Table 4.24). It leads to the conclusion that the methylene group located between the two rings of the cyclopropane has a particular impact on the shifting of the electron density in the hydrocarbon ring, providing it with a reduction in charge differences at the carbon and hydrogen atoms responsible for the energies of the formed specific interactions. The energy contributions of the methylene groups of cyclopropane and bicyclo[2,2,1] are 9.13 and 6.73 kJ mol^{-1}, respectively. The given feature is kept in the molecule of 1,7,7-trimethylbicyclo [2,2,1]heptane-2-thione. In this connection, we can estimate the value of the sublimation enthalpy of bicyclo[2,2,1]heptane-2-thione, based on the value of sublimation enthalpy of 1,7,7-trimethylbicyclo[2,2,1]heptane-2-thione, taking the energy contribution of the isostructural methyl group at location C(1) as being equal to the 4.9 kJ mol^{-1} of the same 2-methylthiirane group and that of the two methyl groups at location C(7) as having a value of 4.25 kJ mol^{-1} for each molecule of 2,2-dimethylthiirane (Table 4.11). The obtained value of the sublimation enthalpy for bicyclo[2,2,1]heptane-2-thione is given in Table 4.24. The difference between this value and that of the sublimation enthalpy of bicyclo[2,2,1]heptane minus the energy contribution of the three methylene groups (20.15 kJ mol^{-1}) is equal to the sum of the energies of the two formed two specific interactions:

Table 4.24 Energies (kJ mol^{-1}) of specific interactions of crystalline compounds with cyclopropane at 298 K

Compounds	Formula	Structure	$\Delta_{sub}H^0$ (298 K) [18]	DisoCH$_3$	DCH$_2$	DC=S → CH$_2$-C
1,7,7-Trimethylbicyclo[2,2,1]heptane-2-thione	C$_9$H$_{16}$S		61.7 ± 0.9	4.9 / 4.25 × 2/ DCH$_2$ = 6.43 / DCH$_2$	1.7	14.64
Cyclopropane	C$_3$H$_6$		27.4	9.13	–	–
Ethylene sulfide thiirane	C$_2$H$_4$S		30.5	–	DS → CH$_2$ = 7.62	–
Bicyclo[2,2,1]heptane	C$_7$H$_{14}$		40.3 ± 0.32	6.43	1.7	–
Bicyclo[2,2,1]heptane-2-thione	C$_7$H$_{10}$		51.0	6.43	1.7	14.64

$$DC = S \rightarrow CH_2$$

$$= \left(\Delta_{sub}H^0(298 \text{ K})bht - \Delta_{sub}H^0(298 \text{ K})bh - 3DCH_2 \right)/2 \qquad (4.32)$$

The obtained energy value (15.4 kJ mol^{-1}) exceeds the correctly determined energy value of the same type of specific interaction of 1,3-dithian-2-thyone by 0.85 kJ mol^{-1}, which is justified by the uncounted contribution of the methylene group to the sublimation enthalpy of bicyclo[2,2,1]heptane-2-thione, equal to 1.7 kJ mol^{-1}.

4.6 Energies of Specific Interactions of Alkylthioboranes

The compounds of the tris(methylthio) borane series should be considered as complexes where the boron atom acts as the central atom and the intermolecular interactions are carried out by the S–R fragment between the sulfur atom and the carbon atom of the end methyl group of the alkyl ligand, whose electron density shifts from the essentially undivided $2s^2$ electron pair. The shifting of the electron density from the boron atom enriches the negative charge of the sulfur atom by partly returning to the p_z orbital of the boron atom in the form of a reverse dative bond.

The weakening of the influence of the reverse dative bond is expressed by the vaporization enthalpies of the first representatives of the series of compounds, ending at the tri(propylthio)borane with the propyl ligand (Fig. 4.33). The further enthalpy increase is caused by the energy contribution of an increasing number of methylene groups of the alkyl chain . The molecules of the tris(methylthio) boranes with six bond vacancies form six specific interactions of one type DS → CH$_3$–S, differing for the compounds of the given series by the number of methylene groups of the alkyl chain, DS → CH$_3$–CH$_2$–S and DS → CH$_3$–CH$_2$–CH$_2$–S, forming the network of specific interactions (Fig. 4.34a). The energy of the specific interaction formed by the alky ligand DS → R–S is determined by division of the enthalpy value by the number of interactions:

$$DS \rightarrow R - S = \Delta_{vap}H^0(298 \text{ K})/6 \qquad (4.33)$$

where R is CH$_3$, CH$_2$–CH$_3$, or CH$_2$–CH$_2$–CH$_3$.

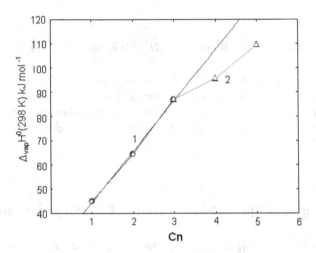

Fig. 4.33 Dependence of vaporization enthalpy on the number of carbon atoms in the alkyl group of liquid alkylthioboranes: decreasing influence of reverse dative bond (*1*) and contribution energy of CH_2 groups to the enthalpy property (*2*)

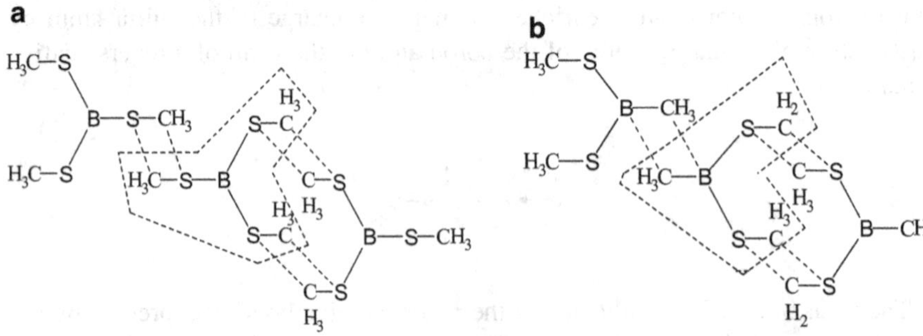

Fig. 4.34 Schematic of the network of specific interactions of liquid tris(methylthio)borane (**a**) and methylbis(methylthio)borane (**b**)

As presented in Table 4.25, the calculation results are described by the natural sequence of stabilization of the specific interaction, with maximum stability at the tri(propylthio)borane and remaining stable at all further compounds of the series:

DS → CH_3–S, tris(methylthio)borane (7.48) < DS → CH_3–CH_2–S, tri(ethelthio) borane (tri(ethelthio)borane) < DS → CH_3–CH_2–CH_2–S, tri(propylthio)borane (14.5 kJ mol^{-1})

The inequality of the specific interaction energies of the liquid tris(methylthio) borane and trimethylborate points to the increased shifting of the electron density to the more distant energy level of the sulfur atom than to the oxygen atom of the corresponding molecules:

Table 4.25 Energies (kJ mol^{-1}) of specific interactions of liquid alkylthioboranes

Compounds	Formula	Structure	$\Delta_{vap}H^0$ (298 K) [12]	T (K)	DS → CH$_3$–(CH$_2$)$_n$–S	DO → CH$_3$–O	DCH$_2$
Tris(methylthio) borane	C$_3$H$_9$BS$_3$	H$_3$C–S, H$_3$C–S \ B–S–CH$_3$	44.9	394	$n=0$ 7.48	–	–
Trimethylborate	C$_3$H$_9$BO$_3$	H$_3$C–O, H$_3$C–O \ B–O–CH$_3$	41.9	317	–	7.0	–
Tri(ethelthio)borane	C$_6$H$_{15}$BS$_3$	H$_3$C–H$_2$C–S, H$_3$C–H$_2$C–S \ B–S–CH$_2$–CH$_3$	61.5 ± 2.1		$n=1$ 10.25	–	–
Tri(propylthio) borane	C$_9$H$_{21}$BS$_3$	H$_3$C–(H$_2$C)$_2$–S, H$_3$C–(H$_2$C)$_2$–S \ B–S–(CH$_2$)$_2$–CH$_3$	87. ±2.1	298	$n=2$ 14.5	–	–
Tri(butylthio)borane	C$_{12}$H$_{27}$BS$_3$	H$_3$C–(H$_2$C)$_3$–S, H$_3$C–(H$_2$C)$_3$–S \ B–S–(CH$_2$)$_3$–CH$_3$	95.8 ± 2.1	298	$n=2$ 14.5	–	2.93 × 3
Tri(pentylthio)borane	C$_{15}$H$_{33}$BS$_3$	H$_3$C–(H$_2$C)$_4$–S, H$_3$C–(H$_2$C)$_4$–S \ B–S–(CH$_2$)$_4$–CH$_3$	109.6 ± 2.1	298	$n=2$ 14.5	–	3.43 × 6
					DN → CH$_3$–N	DB → CH$_3$–B	
Dimethyl (methylthio)borane	C$_3$H$_9$BS	H$_3$C, H$_3$C \ B–S–CH$_3$	31.6	269	7.48	4.15	–
Methylbis (methylthio)borane	C$_3$H$_9$BS$_2$	H$_3$C–S, H$_3$C \ B–S–CH$_3$	44.7	315	7.48	7.35	7.35
					DN → CH$_3$–N	DO → CH$_3$–O	DB → CH$_3$–(CH$_2$)$_3$–B
Butyl(diethyl-amino) methoxyborane	BNOC$_9$H$_{22}$	C$_2$H$_5$, C$_2$H$_5$ \ N–B / O–CH$_3$, C$_4$H$_9$	58.2 ± 2.5	298	5.85	7.0	10.2

DS $\rightarrow$ CH$_3$–S,   tris(methylthio)   borane   (7.48,   394   K) > DO $\rightarrow$ CH$_3$–O, trimethylborate (7.0 kJ mol^{-1}, 317 K)

The molecules of methylbis(methylthio)borane (Fig. 4.34b) and dimethyl (methylthio)borane, with one and two binding methyl ligands with the boron atom, form two and four specific interactions DH$_3$C $\rightarrow$ B–CH$_3$ and, additionally, the two interactions DS $\rightarrow$ CH$_3$–S with known energy, which allows the energies of two specific interactions formed by the boron atom to be obtained:

$$DB \rightarrow CH_3 - B = \left(\Delta_{vap}H^0(298\,K) - nDS \rightarrow CH_3 - S\right)/q \qquad (4.34)$$

The obtained calculation results (Table 4.25) indicate the significant stabilization of the specific interaction DB $\rightarrow$ CH$_3$-B (7.35 kJ mol^{-1}), located in the *trans* position to the two interactions DS $\rightarrow$ CH$_3$–S in the molecule of methylbis(methylthio) borane.

The molecule of butyl(diethylamino)methoxyborane with eight bond vacancies forms eight specific interactions of three types, including 4DN $\rightarrow$ CH$_3$–CH$_2$–N and 2DO $\rightarrow$ CH$_3$–O with energies of 5.85 and 7.0 kJ mol^{-1}, and the specific interaction 2D(–CH$_2$)$_3$–CH$_3$ $\rightarrow$ B of unknown energy value, which can be determined using Eq. (4.35):

$$2D(B \rightarrow CH_3 - CH_2)_2 \rightarrow B = \left(\Delta_{vap}H^0(298\,K) - 4DN\right.$$
$$\left. \rightarrow CH_3 - CH_2 - N - 2DO \rightarrow CH_3 - O\right)/2 \qquad (4.35)$$

The obtained energy contribution of the butyl group with four carbon atoms to the vaporization enthalpy is equal to 10.2 kJ mol^{-1}(Table 4.25). Taking into account that the maximum energy value of the formed specific interaction is reached at the propyl ligand (7.0 kJ mol^{-1}; see the section in Chap. 1 entitled "Specific Interactions of Boron Alkyls Subgroup Elements"), the difference between these two values (3.2 kJ mol^{-1}) corresponds to the energy contribution of the third methylene group of the butyl fragment. Therefore, there is a good reason to estimate the standard vaporization enthalpy of tributylborane to be equal to 61.2 kJ mol^{-1}.

References

1. Chang FC, Young VY, Prather JW, Cheng KL (1986) Study of methyl chalcogen compounds with ultraviolet photoelectron spectroscopy. J Electron Spectrosc 40(4):363–383
2. Aue DH, Webb HM, Davidson WR et al (1980) Proton affinities and photoelectron spectra of three-membered-ring heterocycles. J Am Chem Soc 102(16):5151–5157
3. Bock H, Wagner G, Wittel K et al (1974) n/π-Konjugation in heterosusubsttituierten athylene. Photoelectronespectren und photoelectronspectro molekuleigenschutten. XXXII. Chem Ber Bd 107(6):S.1869–S.1881
4. Mueller C, Schaefer W, Schweig A et al (1976) Detection of rotational isomers by variable temperature photoelectron spectroscopy. A new technique in the realm of molecular conformational analysis. J Am Chem Soc 98(18):5440–5443

5. Kamphuis J, Bos HJT, Worrell CW, Runge W (1986) The molecular structure of allenes and ketenes. Part 19. Photoelectron spectra and conformations of donor-substituted allenes. J Chem Soc Perkin Trans 2 10:1509–1516.
6. Sarneel R, Worrell CW, Pasman P et al (1980) The photoelectron spectra of 4-methylene thiacyclohexane derivatives through-bond interaction. Tetrahedron 36(22):3241–3248
7. Batich C, Heilbronner E, Quinn CB, Wiseman JR (1976) The electronic structure of vinyl ethers and sulfides with interrupted conjugation examined by photoelectron spectroscopy. Helv Chim Acta 59(2):512–522
8. Baev AK (1969) Problems of chemical nature of phase transformation, vol 1, General and applied chemistry. Vaisheshika Educating, Minsk, pp 197–206
9. Baev AK (1969) Phase condition and complex formation ability of galogenide metals, vol 1, General and applied chemistry. Vaisheshika Educating, Minsk, pp 207–218
10. Baev AK (1972) Complex formation ability halogenides of second–six groups periodical system, vol 5, General and applied chemistry. Vaisheshika Educating, Minsk, pp 35–51
11. Baev AK (2012) Specific intermolecular interactions of organic compounds. Springer, Heidelberg/Dordrecht/London/New York, 434 p
12. Chickos JS, Acree WE (2003) Enthalpies of vaporization of organic and organometallic compounds, 1880–2002. J Phys Chem Rev 32(2):519
13. Hargittai I (1987) Structural chemistry of sulfur compounds. Nauka, Moscow, p 264
14. Wagner G, Bock H (1974) Die Delokalisation von schwefel-elektronenpaaren in alkylsulfiden und -disulfiden. Chem Ber 107(1):68–77
15. Solouki B, Bock H (1977) Photoelectron spectra and molecular properties. 59. Ionization energies of disulfur dihalides and isomerization surfaces $XSSX \leftrightarrow SSX_2$. Inorg Chem 16 (3):665–669
16. Guimon MF, Guimon G, Pfister-Guillouzo G (1975) Application of photoelectron spectroscopy to conformational analysis of 1,2,4-trithiolanes. Tetrahedron Lett 16(7):441–444
17. Bock H, Stein U. Semkow A (1980) 1,2-Dithiolan – Bindungsmodell fur α-Liponsaure. Chem Ber 113(10): 3208–3220
18. Chickos JS, Acree WE (2002) Enthalpies of sublimation of organic and organometallic compounds, 1910–2001. J Phys Chem Rev 31(2):537
19. Baev AK (1987) Chemistry of gas-heterogenic systems of elementorganic compounds. Science and Technology, Minsk, 175 p
20. Pignataro S, Distefano G (1974) n-σ Mixing in pentatomic heterocyclic compounds of sixth group by photoelectron spectroscopy. Chem Phys Lett 26(3):356–360
21. Nefedov VI, Vovna VIJ (1989) Electronic structure of organic and elementorganic compounds. Nauka, Moscow, 198 p
22. Nohre CL, Reineck L, Veenhuizen H et al (1984) High resolution He(I) photoelectron spectra of 2-halothiophenes. Prepr Univ No UUIP-111, Uppsala
23. Von Niessen W, Kraemer WP, Cederbaum LS (1976) The electronic structure of molecules by a many-body approach III. Ionization potentials and one-electron properties of furan and thiophene. J Electron Spectrosc 8(3):179–197
24. Adam MY, Cauletti C, Piancastelli MN (1987) Final state multiple structures in photoionization of the inner valence shell of H2S: angular distribution in the 40–70 eV photon energy range. J Electron Spectrosc 42(1):1–10
25. Cradock S, Whiteford RA (1972) Photoelectron spectra of the methyl, silyl and germyl derivatives of the group VI elements. J Chem Soc Faraday Trans 2 68(2):281–288
26. Baev AA, Baev AK (2011) Structural-energetic analysis of liquid isostructural propanol-2, 2-methylpropanol-1 and their solutions with network of specific interactions: challenges and insigne. In: Abstracts of the XI international conference "Problems of salvations and complexing at solutions". Ivanovo, Russia, p 6

Chapter 5
Specific Intermolecular Interactions of Sulfur, Oxygen, and Nitrogenated Organic Compounds

5.1 Specific Interactions of Sulfur and Oxygenated Organic Compounds

5.1.1 Carbonyl Sulfide, Carbon Disulfide, Dimethylsulfoxide, Divinylsulfoxide, and Diacetylsulfide BY-Oxythiane

The vaporization enthalpies of the sulfide compounds of carbon and carbon dioxide are often measured at temperatures that differ from the standard temperature by 10–14 K, which causes certain errors in the obtained energy characteristics. Nevertheless, they are of special interest in the thermodynamic analysis of specific interactions of the various oxygen- and sulfur-containing compounds. The inherent similarity of the C=O and C=S groups in the compounds' molecular structure, along with significant differences in the donor–acceptor properties of oxygen and sulfur atoms, allow their influence on the energies of the specific interactions formed to be traced. The difference of 1.5 kJ mol^{-1} between the vaporization enthalpies of carbon dioxide and carbonyl sulfide points to the unequal influence of the oxygen and sulfur atoms on the stabilization of a specific interaction. In turn, the difference of 11.8 kJ mol^{-1} between the vaporization enthalpies of carbon dioxide and carbon disulfide indicates the formation of more stable specific interactions by the sulfur atoms. The deficit of electron acceptors in the carbon atoms in the formation of specific interactions provides the sulfur atoms with the participation of one free electron pair.

© Springer International Publishing Switzerland 2015

A.K. Baev, *Specific Intermolecular Interactions of Element-Organic Compounds*,
DOI 10.1007/978-3-319-08563-0_5

The energies of the four specific interactions $DO \rightarrow C$ formed by the molecules of carbon dioxide and $DS \rightarrow C$ by carbon dioxide can be determined directly from the vaporization enthalpy by dividing it by the number of interactions. The energy of specific interactions of $DS \rightarrow C$ carbonyl sulfide is determined using Eq. (5.1), considering the energy contribution of the two specific interactions of $DO \rightarrow C$ to the enthalpy characteristic:

$$DS \rightarrow C = \left(\Delta_{vap}H^0(T) - 2DO \rightarrow C\right)/2 \qquad (5.1)$$

$DO \rightarrow C$, Carbon dioxide $(4.18) < DS \rightarrow C$, Carbonyl sulfide $(4.97) < DS \rightarrow C$, Carbon disulfide $(7.12$ kJ mol$^{-1})$

The obtained energy values of the specific interactions (Table 5.1) illustrate its stabilization following replacement of the oxygen atom by the sulfur atom, which is caused by shifting of the electron density from the carbon atom to a more distant energy level of the sulfur atom than the oxygen atom. Maximum stabilization of the same type of interaction of $DS \rightarrow C$ is reached in the presence of the two sulfur atoms in the carbon disulfide molecule, providing the increased shifting of the electron density, accompanied by the significant difference in its negative charges by the increased positive charge of the carbon atom. An even more significant shifting of the electron density following replacement of the oxygen atom in the molecule of O=C=O with a selenium atom is expressed in the energy value of $O=C \rightarrow S < O=C \rightarrow Se$ interactions located at the more distant energy level of the selenium atom:

O=C=O: $DO \rightarrow C$ $(4.18) < O=C=S$: $DS \rightarrow C$ $(4.97) < O=C=Se$: $DSe \rightarrow C$ $(6.7$ kJ mol$^{-1})$

O=C=O: $DO \rightarrow C$ $(4.18) < S=C=S$: $DS \rightarrow C$ $(7.12) < Se=C=Se$; $DSe \rightarrow C$ $(9.3) < S=C=Se$; $DSe \rightarrow C$ $(9.7$ kJ mol$^{-1})$

The stabilization of the specific interactions in the compounds achieving maximum stability of the $DSe \rightarrow C$ interaction of carbon selenide sulfide allow a conclusion to be made regarding the increased shifting of the electron density from the sulfur atom in the molecule and with its location at a distant energy level of the selenium atom.

A distinctive feature of the dimethylsulfoxide molecule is the presence of two methyl groups and two free electron pairs of the oxygen atom enriched by the shifted electron density from the sulfur atom, providing it with high negative charges. The four formed specific interactions $DS=O \rightarrow CH_3$ invoke the formation of chains crosslinked by van der Waals interactions in the united network of specific interactions (Fig. 5.1a). The energy of the specific interaction $DS=O \rightarrow CH_3–C$ can be obtained from Eq. (5.2):

Table 5.1 Energies (kJ mol^{-1}) of hydrogen bonds and specific interactions of liquid carbonyl sulfide, carbonyl selenides, dimethylsulfoxide, divinylsulfoxide, and diacetylsulfide

Compounds	Formula	Structure	$\Delta_{vap}H^0$ (298 K) [1]	T (K)	DO → C	DSe → C	DS → C
Carbon dioxide	CO_2	O=C=O	16.7	288	4.18	–	–
Carbonyl sulfide	COS	O=C=S	18.3	299	4.18	–	4.97
Carbon disulfide	CS_2	S=C=S	28.5	275	–	–	7.12
Carbon oxyselenide	COSe	O=C=Se	21.7	236	4.18	6.7	–
Carbon selenide sulfide	CSSe	S=C=Se	33.6	288	–	9.7	7.12
Carbon diselenide	CSe_2	Se=C=Se	37.2 ± 0.8	298	–	9.3	–
Crystalline							
Carbon diselenide	CSe_2	Se=C=Se	46.3	224	–	11.6	–
Dimethylsulfoxide	C_2H_6OS	$(H_3C)_2$S=O	52.3	308	–	DS=O → CH$_3$ 13.1	DS=O → CH$_2$=CH
Divinylsulfoxide	C_4H_6OS	$(H_2C$=CH$)_2$S=O	51.2 ± 0.9	298	–	–	12.8
Diacetylsulfide	$C_4H_6O_2S$	$(H_3C$–C(=O)$)_2$S	50.9	340	DS → CH$_2$–C 6.92	DO → CH$_2$–C –	DC=O → C=O 5.80
Acetyl anhydride	$C_4H_6O_3$	$(H_3C$–C(=O)$)_2$O	45.3	335	–	5.63	5.70

a   **b**

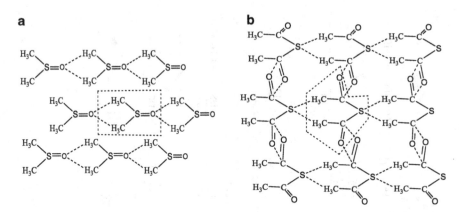

Fig. 5.1 Schematic of the network of specific interactions of liquid dimethylsulfoxide (**a**) and diacetylsulfide (**b**)

$$DS = O \rightarrow CH_3 - C = \left(\Delta_{vap}H^0(T)\text{dms}\right)/4 \qquad (5.2)$$

The molecule of divinylsulfoxide with unsaturated ligands and similar molecular structure forms four specific interactions $DS=O \rightarrow CH_2=CH-$ of one type, forming a network of these interactions in the liquid condition. The energy of this interaction is determined using Eq. (5.2a):

$$DS = O \rightarrow CH_2 = CH- = \left(\Delta_{vap}H^0(T)\text{dvs}\right)/4 \qquad (5.2a)$$

As a result of the implemented calculations (Table 5.1), the high stability of the specific interaction of $DS=O \rightarrow CH_3$, obtained from the vaporization enthalpy, is given without errors [1]:

$$DS \rightarrow CH_2 - S, \text{Dimethyl sulfide}(6.92) < DS \rightarrow CH_2 - S, \text{Dimethyldisulfide}(9.62)$$
$$< DS = O \rightarrow CH_3, \text{Dimethylsulfoxide}\left(13.1\,\text{kJmol}^{-1}\right)$$

Nevertheless, it leads to the formation of the S=O group with the more stable specific interaction $DS=O \rightarrow CH_3$ in comparison with the formed dimethyl-disulfide. The natural sequence in the specific interaction energies in the series of compounds formed by the ethyl and vinyl functional groups is the result of the increased negative charge at the oxygen atom of divinylsulfoxide in comparison with the charges of the sulfur atoms in the diethyl sulfide molecule, and is smaller than that of diethyl sulfide:

Diethyl   sulfide $(9.05) <$ Divinylsulfoxide $(12.8) <$ Diethyl   disulfide
$(11.35\,\text{kJ mol}^{-1})$

The analogy of the molecules and structures of diacetylsulfide and acetyl anhydride with an equal number of bond vacancies and the similarity of the two types of four interactions $DS \rightarrow CH_3$, $DC=O \rightarrow C=O$, $DO \rightarrow CH_3$, and $DC=O \rightarrow C=O$ is manifested in the energies of the formed interactions. In particular, work by Baev [2] showed the small difference in the energies of the specific interaction of dimethyl ether $DO \rightarrow CH_3$ and realizing in liquid acetyl anhydride. This allows an assumption to be made regarding the similarity of the energies of $DS \rightarrow CH_3$ interactions (6.92 kJ mol^{-1}) of the liquid diacetylsulfide and dimethyl sulfide, and the energy value of the specific interaction $DC=O \rightarrow C=O$ to be determined using Eq. (5.3):

$$DC = O \rightarrow C = O = \left(\Delta_{vap}H^0(T)aa - 4DS \rightarrow CH_3 - C \right)/4 \qquad (5.3)$$

$DC=O \rightarrow C=O$: Acetyl anhydride (5.70) < Diacetyl sulfide (5.92 kJ mol^{-1})

The difference of 0.98 kJ mol^{-1} in the energies of the specific interaction indicates the increased shifting of the electron density from the sulfur atom to the carbonyl oxygen atoms of the diacetylsulfide acid molecule compared with the shifting to similar oxygen atoms in the molecule of acetyl anhydride.

5.1.2 1-(Methylthio)-2-(Vinyloxy)ethane, Thioacetyl Acid, and Acetic Acid

The fragment of 2-(vinyloxy)ethane located in the molecule of 1-(methylthio)-2-(vinyloxy)ethane (Fig. 5.2) between the sulfur atom and the hydroxyl group implements, on the one hand, by the vinyl group, the property of the functional isostructural group and, on the other hand, by the ethane group, the specific interaction $DO \rightarrow CH_2$–CH– with an energy of 2.1 kJ mol^{-1}. This value is practically identical to the energy contribution of the single methylene group to the vaporization enthalpy of 1-(methylthio)-2-(vinyloxy)ethane and should not be equal to the energy contribution D–$HC=CH_2 \rightarrow CH=CH_2$ (4.4 kJ mol^{-1}) of the similar group to the vaporization enthalpy of 2-vinylthiirane (Table 4.11). The given assumption is proved by the examples of the sharp reduction in the energy contribution of the isostructural methyl group to the enthalpy characteristic

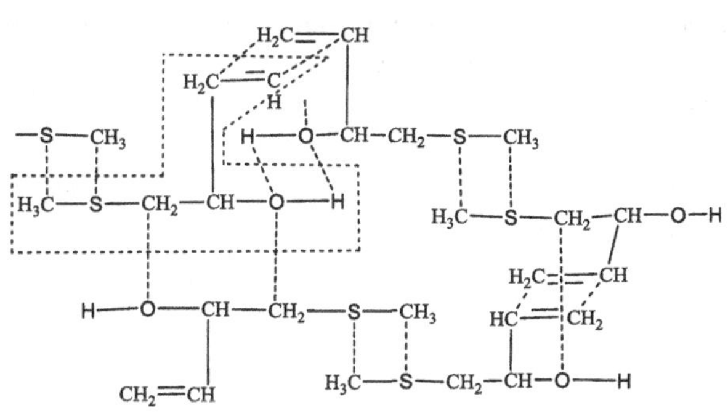

Fig. 5.2 Schematic of the network of specific interactions of liquid 1-(methylthio)-2-(vinyloxy) ethane

[2]. Significant reduction in the energy contribution of the 2-(vinyloxy)ethane fragment to the vaporization enthalpy of 1-(methylthio)-2-(vinyloxy)ethane allows an assumption to be made regarding the conservation of the charge difference between the sulfur atom and carbon atom of the methyl group and, on the other hand, between the oxygen atom and the hydrogen atom of the hydroxyl group. This assumption provides the energy conservation of the formed specific interaction DS → CH$_3$ (6.92 kJ mol^{-1}) and the hydrogen bond DH–O•••H–O (14.4 kJ mol^{-1}) [2] of the liquid dimethyl sulfide and ethanol, respectively. Thus, the molecule of 1-(methylthio)-2-(vinyloxy)ethane forms six specific interactions of three types; the pentacoordinated carbon atom and the two hydrogen bonds of DH–O•••H–O forming a network of specific interactions in the liquid condition. Taking into account the definite correctness of the assumption, it seems realistic to estimate the energies of these interactions and the formed specific interaction of the vinyl group D–HC=CH$_2$→CH=CH$_2$ of reduced stability using Eq. (5.3):

$$D - HC = CH_2 \rightarrow CH = CH_2$$
$$= (\Delta_{vap}H^0(T)\text{mtve} - 2DS \rightarrow CH_3 - 2DH - O \bullet\bullet\bullet H - O - DO \quad (5.3a)$$
$$\rightarrow CH_2 - CH-)/2$$

The calculated energy value (Table 5.2) of the energy contribution of the isostructural vinyl group to the enthalpy vaporization of 1-(methylthio)-2-(vinyloxy)ethane is equal to 0.66 kJ mol^{-1}. According to the data given in Table 5.2 on the enthalpy characteristics of thioacetyl acid and acetic acid, it follows that the reduced value of the vaporization enthalpy of thioacetyl acid by 16.4 kJ mol^{-1} compared with that of acetic acid is caused by the ability of the sulfur atom to form the stable interaction DS•••H–S (Fig. 5.3a) by one free electron pair. The second electron pair is expressed by the significant stabilization of this interaction. At the same time, the oxygen atom of acetic acid (Fig. 5.3b) forms, by the two free electron pairs, a single hydrogen bond DH–O•••H–O and one specific interaction DO→CH$_3$–C– (7.7 kJ mol^{-1}) with the essentially undivided 2s^2 electron pair of the carbon atom of the methyl group of the acetic fragment [2]. The lack of electron acceptor atoms allows the hydrogen atom in the *trans* condition of the methyl group in relation to the oxygen atom of the hydroxyl group to participate in the formation of the hydrogen bond DC=O•••H–CH$_2$ (0.60 kJ mol^{-1}) with the oxygen atom of the carbonyl group [2]. Therefore, the molecule of acetic acid forms with the molecules of the near environment two stable hydrogen bonds of the hydroxyl group with unknown energy, two hydrogen bonds of low stability by the carbonyl oxygen atom and hydrogen atom of the methyl group, and two specific interactions DO→CH$_3$–C with an energy of 7.7 kJ mol^{-1} [2]. The energy of the hydrogen bond is determined using Eq. (5.4):

Table 5.2 Energies (kJ mol^{-1}) of hydrogen bonds and specific interactions of liquid 1-(methylthio)-2-(vinyloxy)ethane and thioacetyl acid

Compounds	Formula	Structure	$\Delta_{vap}H^0$ (298 K) [1]	T (K)	DO → CH$_2$–CH–CH$_2$ DisoCH=CH$_2$	DS → CH$_3$–S	DH–O•••H–O
1-(Methylthio)-2-(vinyloxy)ethane	C$_5$H$_{10}$OS	OH over CH$_2$=CH–CH–CH$_2$–S–CH$_3$	47.5	330	2.1/0.66	6.92	14.4
					DC=O•••H–CH$_2$	DO → CH$_3$–C=	
Acetic acid	C2H4O2	CH$_3$–C(=O)–O–H	51.6 ± 1.8	298	0.60 × 2	7.7	17.5 [2]
Thioacetyl acid	C$_2$H$_4$OS	CH$_3$–C(=O)–S–H	35.2	333	0.60 × 2	D=C–CH$_3$ → C–CH$_3$ 7.25	DS•••H–S = 9.75

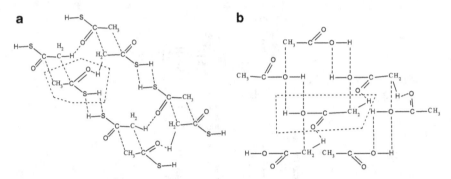

Fig. 5.3 Schematic of the network of specific interactions of liquid thioacetyl acid (**a**) and acetic acid (**b**)

$$DH-O\bullet\bullet\bullet H-O = \left(\Delta_{vap}H^{0}(T)aa - 2DO \rightarrow CH_3 - C = -2DC = O\bullet\bullet\bullet H - CH_3\right)/2 \tag{5.4}$$

The molecule of thioacetyl acid forms two hydrogen bonds DS•••H–S, two hydrogen bonds DC=O•••H–CH$_3$ of low stability, and two specific interactions D=C–CH$_3 \rightarrow$C–CH$_3$ formed by the acetyl group.

$$
\begin{array}{c}
\text{H} \\
\diagdown \\
\text{H}_2\text{C}-\text{C} \overset{\text{O}}{\diagup}\diagdown \text{S}- \\
-\text{S} \overset{}{\diagdown}\text{C}\overset{\diagup}{\diagdown}-\text{CH}_2\text{-H} \\
\text{O}
\end{array}
$$

These interactions have an estimated energy value of 7.85 kJ mol^{-1} (Table 5.3) minus the energy contribution of the hydrogen atom (0.60 kJ mol^{-1}), forming a hydrogen bond of low stability with the carbonyl oxygen atom. The energy of the hydrogen bond is obtained from Eq. (5.4a):

$$DS \bullet\bullet\bullet H - S = \left(\Delta_{vap}H^{0}(T)taa - 2DS \rightarrow CH_3 - C - 2DC = O\bullet\bullet\bullet H - CH_3\right)/2 \tag{5.4a}$$

Presented in Table 5.2 are the energy values of the hydrogen bond, which illustrate its stabilization in comparison with the thioalcohols, whose energies have significantly reduced values compared to acetic acid:

Ethyl mercaptan (5.0, 298 K) < thioacetyl acid (9.75, 333 K) < acetic acid (17.5 kJ mol^{-1}, 298 K)

Table 5.3 Energies (kJ mol^{-1}) of specific interactions and hydrogen bonds of liquid alkyl thiolactates at 298 K

Compounds	Formula	Structure	$\Delta_{vap}H^0$ (298 K) [1]	T (K)	DisoCH$_3$	DS $\rightarrow$ CH$_3$–(CH$_2$)$_n$–S
DC=O••H–CH$_2$ = 0.60×2; DC–CH$_3$ $\rightarrow$ C–CH$_3$ = 7.85						
S-Ethyl thiolactate	C$_4$H$_8$OS		40.0 ± 0.2	298	–	$n=1$ 11.55
S-Isopropyl thiolactate	C$_5$H$_{10}$OS		49.1 ± 0.2	298	9.1	$n=1$ 11.55
S-Propyl thiolactate	C$_5$H$_{10}$OS		42.3 ± 0.2	298	–	$n=2$ 12.70
S-Butyl thiolactate	C$_6$H$_{12}$OS		48.1 ± 0.2	298	DCH$_2$ = 5.8	$n=2$ 12.70
S-tert-Butyl thiolactate	C$_6$H$_{12}$OS		42.9 ± 0.2	298	1.45 × 2	$n=1$ 11.55

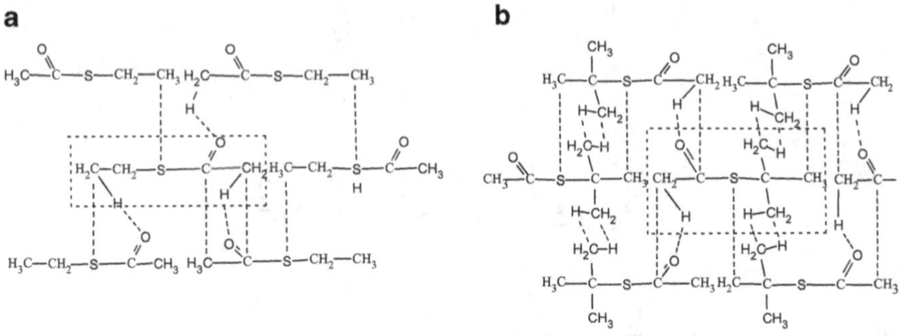

a   **b**

Fig. 5.4 Schematic of the network of specific interactions of liquid S-ethyl thiolactate (**a**) and S-*tert*-butyl thiolactate (**b**)

5.1.3   Alkyl Thiolactates

The information on the energies of the thioacetyl acid interactions presented in the previous section allows the energies of the specific interactions of a number of compounds of the ethyl thiolactates series to be obtained. These compounds are characterized by the invariability of the energies of the DC=O•••H–CH$_2$ (0.60×2) and DC–CH$_3 \to$ C–CH$_3$ (7.85 kJ mol^{-1}) interactions. The unknown energy value is the specific interaction formed by the changing alcohol fragment. Therefore, the six formed interactions of three types form a network crosslinked by the stable specific interactions and the weak hydrogen bonds (Fig. 5.4a) and functional isostructural methyl groups of the liquid S-isopropyl thiolactate and S-*tert*-butyl thiolactate (Fig. 5.4b).

The energy value of the formed specific interaction of the alcohol ethyl, propyl, and butyl ligands can be determined using Eq. (5.4b):

$$DS \to CH_3 - (CH_2)n - S$$
$$= \left(\Delta_{vap}H^0(T) - 2DS \to CH_3 - C - 2DC = O \bullet\bullet\bullet H - CH_2\right)/2 \quad (5.4b)$$

where n is the number of methylene groups in the alkyl ligand.

The obtained energy values of the specific interactions increase in the following sequence, with the maximum value for the propyl thiolactate due to completion of the influence of the intermolecular reverse dative bond:

DS→CH$_3$–CH$_2$–S (11.55), ethyl thiolactate < DS→CH$_3$–CH$_2$–CH$_2$–S (12.70), propyl thiolactate = DS→CH$_3$–CH$_2$–CH$_2$–S (12.70 kJ mol^{-1}), butyl thiolactate

Taking into account that the energy of the specific interaction is caused by the number of carbon atoms of the alkyl chain, the energy contribution of the isostructural methyl group to the enthalpy characteristic can be obtained from the difference between the vaporization enthalpies of isopropyl thiolactate and *tert*-butyl thiolactate, respectively, and the ethyl thiolactate with the same ethyl ligand. The obtained data (Table 5.3) indicate the extremely high value of the energy contribution of the isostructural methyl group of isopropyl thiolactate, exceeding the energy contribution of the vinyl group to the enthalpy of vaporization of 2-vinylthiirane (8.8 kJ mol^{-1}). Nevertheless, the specific interaction DH$_3$C $\rightarrow$ H–CH$_2$ with an energy of 4.55 kJ mol^{-1} formed by the isostructural methyl group of isopropyl thiolactate does not exceed the energy contribution of the methylene group (5.8 kJ mol^{-1}) to the vaporization enthalpy of butyl thiolactate. Such a high value for the energy contribution of the single CH$_3$ group is doubtful, as the second isostructural methyl group of the *tert*-butyl thiolactate molecule reduces the total contribution to the vaporization enthalpy down to 2.9 kJ mol^{-1}.

5.1.4 BY-Oxythiane and Thiopirans

The molecule of BY-oxythiane with one oxygen and one sulfur atom at locations C(1) and C(4) of the six-membered ring with four methylene groups repeats the property of 1,4-dioxane and, in the liquid and crystal conditions, forms a network of alternating bond types by four specific interactions of two types, DO $\rightarrow$ CH$_2$ and DS $\rightarrow$ CH$_2$ (Fig. 5.5a). The energies of the formed specific interactions of the two electron pairs of oxygen atoms should correspond due to their stability, formed by the 1,4-dioxane, and can be obtained by the division of the vaporization enthalpy or sublimation enthalpy of the liquid or crystal compound by the eight specific interactions (Fig. 5.5b). The obtained energy value of the specific interaction DO $\rightarrow$ CH$_2$ (4.56 kJ mol^{-1}) is used for obtaining the energy of the specific interaction of DS $\rightarrow$ CH$_2$, formed by the free electron pairs of the sulfur atom of the same BY-oxythiane, using Eq. (5.5):

$$DS \rightarrow CH_2 = \left(\Delta_{vap}H^0(350\,K)ot - 4DO \rightarrow CH_2\right)/4 \qquad (5.5)$$

The obtained energy value of the specific interaction (6.64 kJ mol^{-1}) at a temperature of 350 K (Table 5.4) is in good compliance with the data presented in

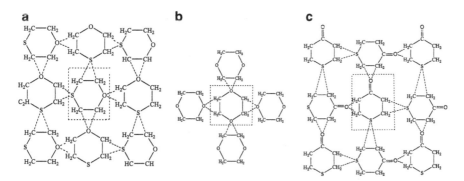

Fig. 5.5 Schematic of the network of specific interactions of liquid and crystalline BY-oxythiane (**a**), 1,4-dioxane (**b**), and tetrahydro-4*H*-thioperan-4-one (**c**)

Table 5.4 Energies (kJ mol^{-1}) of specific interactions of BY-oxythiane, 1,4-oxaselenane, and thiopirans

Compounds	Formula	Structure	$\Delta_{vap}H^0$ (298 K) [1]	T (K)	DO→CH$_2$	DS→CH$_2$	DC=O→CH$_2$
BY-Oxythiane	C$_4$H$_8$OS		44.8	352	4.56	6.64	–
1,4-Dioxane	C$_4$H$_8$O$_2$		38.6 ± 0.1	298	4.82	–	–
			36.5	350	4.56		
1,4-Oxaselenane	C$_4$H$_8$OSe		46.6	367	4.56	7.1	–
Crystalline [3]							
Tetrahydro-4H-thioperan-4-one	C$_5$H$_8$OS		72.6 ± 1.7	298	–	8.61	9.53
1,4-Dithiane	C$_4$H$_8$S$_2$		68.9	298	–	8.5	–

Table 4.21 (6.9 kJ mol^{-1}, 298 K) at standard conditions of the liquid 1,4-dithiane and 1,3-dithiane. The stabilization of the specific interaction DSe → CH$_2$ (7.1) > DS → CH$_2$ (6.64 kJ mol^{-1}), formed by the sulfur atom of the 1,4-oxaselenane molecule of the analog of BY-oxythiane, reflects the stabilizing influence of the shifting electron density from the carbon atom of the methylene group located at the more distant energy level of the selenium atom than the sulfur atom.

The molecule of tetrahydro-4H-thioperan-4-one with the sulfur atom, C=O group, and four symmetrically located methylene groups in the ring forms four specific interactions DS → CH$_2$ (Fig. 5.5c). By analogy with BY-oxythiane, the energy of this type of interaction is taken to be equal to the energy value of the same bond type in crystal 1,4-dithiane (8.5 kJ mol^{-1}), forming a similar network of specific interactions to 1,4-dioxane (Fig. 5.5b). The second type of four specific interactions DC=O → CH$_2$ is formed by the carbonyl groups with methylene groups of the molecules of the near environment (Fig. 5.5c). Within the framework of the considered procedures, the energy calculations for this type of specific interaction are obtained from Eq. (5.5a):

$$DC = O \rightarrow CH_2 = \left(\Delta_{vap}H^0(350K)ttp - 4DS \rightarrow CH_2 \right)/4 \qquad (5.5a)$$

Taking into account the energies of the specific interactions DC=O → CH$_2$ and DS → CH$_2$ of the crystal tetrahydro-4H-thioperan-4-one (Table 5.4), it must be stated that the ketone group forms a more stable specific interaction.

5.1.5 1,3-Dithiol-2-one, 1,3-Dithiole-2-thione, and 1,3-Dithiolan-2-one

It should be noted that 1,3-dithiol-2-one, 1,3-dithiole-2-thione, and 1,3-dithiolan-2-one are the only representatives of the crystalline thiols with known sublimation enthalpy. However, the compounds with eight bond vacancies form eight specific interactions of two types, DS=O → C–C= and DS•••H–S for 1,3-dithiol-2-one, and two other compounds by four specific interactions DS=O → CH–CH and DC=S=O → C–C= and the four hydrogen bonds DS•••H–S correspondingly (Fig. 5.6). The invariable values of the energies of the formed hydrogen bonds in the liquid thiols allows an assumption to be made regarding its immutability in the crystal condition too. There is a lack of information in the literature regarding the energies of specific interactions of cyclopropanol and its analog with the double bond, which determines the necessity to choose the energy value of the specific interaction of the crystal of 1,4-cyclohexanedione (9.45 kJ mol^{-1}) [2]. The presence of the two carbonyl groups provides a definite analogy to the shift of the electron density in the molecules of the considered compounds with the sulfur atom. Taking the value equal to the energy value of the specific interaction DC=S=O → C–C=of

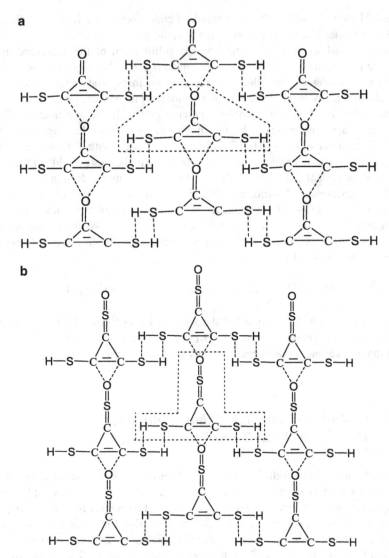

Fig. 5.6 Schematic of the network of specific interactions of crystalline 1,3-dithiol-2-one (**a**) and 1,3-dithiole-2-thione (**b**)

the crystalline 1,3-dithiol-2-one, the energies of the hydrogen bond can be obtained using Eq. (5.6):

$$DS \bullet\bullet\bullet H - S = \left(\Delta_{sub}H^{0}(350\,K) - 4\,DC = S = O \rightarrow C - C = \right)/4 \qquad (5.6)$$

The results of the calculations are presented in Table 5.5. Using the energy value of the hydrogen bond and the vaporization enthalpy of the corresponding thiols

Table 5.5 Energies (kJ mol^{-1}) of specific interactions of crystalline 1,3-dithiolan-2-one, 1,3-dithiolan-2-one, 1,3-dithiolan-2-one, and 1,3-dithiole-2-thione at 298 K

Compounds	Formula	Structure	$\Delta_{sub}H^0$ (298 K) [3]	DC=O → C-C=	DS•••H-S	DC=S=O → C-C=
1,3-Dithiol-2-one	$C_3H_2OS_2$		73.6 ± 0.0.8	9.45	8.95	–
1,3-Dithiolan-2-one	$C_3H_4OS_2$		80.3 ± 0.4^a	DC=O → C-CH 10.6	8.95	–
1,3-Dithiole-2-thione	$C_3H_2OS_3$		75.4 ± 0.4	–	8.95	9,9

^{a}Temperature was not indicated

(alcohols), the energy values of the specific interactions are determined. It follows that, from the data in Table 5.5, the specific interactions of the derivatives of 1,3-dithiol-2-one reaches increased stability in the compound with the cyclopropyl fragment:

$$DC = O \; \rightarrow \; C - C \; = (9.45) < DC = S = O \; \rightarrow \; C - C = (9.9) < DC = O$$
$$\rightarrow \; C - CH(10.6 \, kJ \, mol^{-1})$$

5.2 Energies of the Specific Interactions of Sulfones

5.2.1 Alkyl Sulfones and Divinylsulfone

In the sulfones group $> SO_2$, the four upper orbitals a_1, a_2, b_1, and b_2 are the combination of the O $2p_\pi$ orbital with the binding of sulfur orbitals. Interactions of $\pi(b_1, a_2)$ and $\sigma(b_2, a_1)$ orbitals of the sulfone group with the saturated and unsaturated substituents at the sulfur atom [4–8] grounded the sequence of the vertical ionizing potential (Iv) for the upper levels of $b_1 < b_2$, and $a_1 < a_2$; however, the strained cyclic sulfones a_2 and b_2 tend to be converted [6]. According to the data in Solouki et al. [5], the vinyl π orbital "intervenes" between the levels a_1 and a_2. In accordance with X-ray spectroscopy data [9], the high sulfur positive charge (~0.7 e) of the sulfone group stabilizes the π orbitals of the phenyl groups in the CH_3PhSO_2 molecules and $PhSO_2$ at 0.5 eV. The binding character of the O $2p_\pi$ orbital of the sulfone oxygen atoms and the sulfoxide oxygen atom is proved by the wide oscillating contours of the photoelectron strip and the values of the oxygen $2p_\pi$ electrons [10, 11]. Therefore, the presence of the binding orbitals of the oxygen and sulfur atoms determines the molecular character of this group in the formation of the specific interaction by the oxygen atom.

The experimentally measured sublimation enthalpies of alkyl sulfones, including dipropyl sulfone, are described by the linear dependence log $\Delta_{sub}H^0(298 \text{ K}) = f$ (Cn) (Fig. 5.7). It follows that the influence of the intramolecular reverse dative bond of the given series of compounds completes at the di-*n*-butyl sulfone and, therefore, this correlation can be used for the estimation of the sublimation enthalpy of the unstudied dipropyl sulfone (Table 5.6).

In the molecule of dimethyl sulfone are two oxygen atoms with two free electron pairs and two methyl groups, where the essentially undivided $2s^2$ electron pair of the carbon atom acts as the bond vacancy and provide the deficit of acceptors of the electron density for the formation of specific interactions. In this connection, the molecule forms four specific interactions $DS=O \rightarrow CH_3$ of one type, forming a network of these bonds in the crystal of the compound (Fig. 5.8a). The compounds of the homologous series with a normal structure and four specific interactions of one type are characterized by the possibility to obtain the energies of interactions by the division of sublimation enthalpies by the number of bonds. The obtained energy

Fig. 5.7 Dependence of sublimation enthalpy on the number of carbon atoms (Cn) in the alkyl sulfones

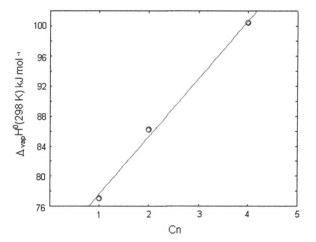

values of specific interactions reflect their stabilization with an increasing number of methylene groups at the maximum value of the propyl ligand:

DS=O → CH_2, dimethyl sulfone $(19.25) <$ DS=O → CH_3–CH_2, diethyl sulfone $(21.55) <$ DS=O → CH_3–CH_2–CH_2, dipropyl sulfone $(23.32) =$ DS=O → CH_3–CH_2–CH_2, dibutyl sulfone $(23.32$ kJ mol$^{-1})$

The presence of the two ligands in the ethyl methyl sulfone molecule allows the energy value of the specific interaction formed by the methyl ligand to be taken as being equal to 19.25 kJ mol^{-1}, and the energy of the second type of interaction can be calculated using Eq. (5.7):

$$DS = O \rightarrow CH_3 - CH_2 = \left(\Delta_{vap}H^0(298\,K)des - 2DS \rightarrow CH_3 \right)/2 \qquad (5.7)$$

The obtained energy value (Table 5.6) for DS=O → CH_3 has a reduced value because of the structural tension caused by the difference in the lengths of the methyl and ethyl ligands (Fig. 5.8b):

DS=O → CH_3, ethyl methyl sulfone $(19.65) <$ DS=O→CH_3–CH_2, diethyl sulfone $(21.55$ kJ mol$^{-1})$

The obtained high energy value of the specific interaction DS=O → CH_3, formed by the methyl group of the dimethyl sulfone molecule, reflects the interaction's specificity and is the natural energy value of the interaction formed by the open-structure molecules of the ethyl and propyl ligands, with further stabilization at 2.30 and 4.07 kJ mol^{-1} (Table 5.6).

The energy contribution of the isostructural methyl groups to the sublimation enthalpy of *tert*-butyl sulfone, *tert*-butyl ethyl sulfone, and di-*tert*-butyl sulfone can be obtained from ethyl methyl sulfone as the difference between the sublimation enthalpies of these compounds and that of diethyl sulfone . As presented in

Table 5.6 Energies (kJ mol^{-1}) of specific interactions of alkyl sulfones at 298 K

Compounds	Formula	Structure	$\Delta_{sub}H^{o}$ (298 K) [3]	DS=O → CH$_2$	DS=O → CH$_3$–CH$_2$	DS=O → CH$_3$– (CH$_2$)$_2$–
Dimethyl sulfone	C$_2$H$_6$O$_2$S		77.0 ± 2.9	19.25	–	–
Ethyl methyl sulfone	C$_3$H$_4$O$_2$S		77.8 ± 2.9	19.25	19.65	–
Diethyl sulfone	C$_4$H$_{10}$O$_2$S		86.2 ± 2.5	–	21.55	–
Dipropyl sulfone	C$_6$H$_{14}$O$_2$S		93.3[a]	–	–	$n = 2$ 23.32
Di-n-butyl sulfone	C$_8$H$_{18}$O$_2$S		100.4 ± 2.3	–	DCH$_2$ = 7.1	$n = 3$ 23.32
$tert$-Butyl sulfone	C$_5$H$_{12}$O$_2$S		82.4 ± 2.5	19.25	19.65	DisoCH$_3$ = 4.6:2
$tert$-Butyl ethyl sulfone	C$_6$H$_{14}$O$_2$S		86.6 ± 2.5	–	21.55	DisoCH$_3$ = 0.4
Di-$tert$-butyl sulfone	C$_8$H$_{18}$O$_2$S		94.1 ± 2.9	–	21.55	DisoCH$_3$ = 7.3:4
					Liquid [1]	
					DS=O → CH$_2$=CH–	
Divinylsulfone	C$_4$H$_6$O$_2$S	H$_2$C=CH–S(O$_2$)–HC=CH$_2$	56.4[b]	–	14.1	–

[a]Estimated
[b]298 K

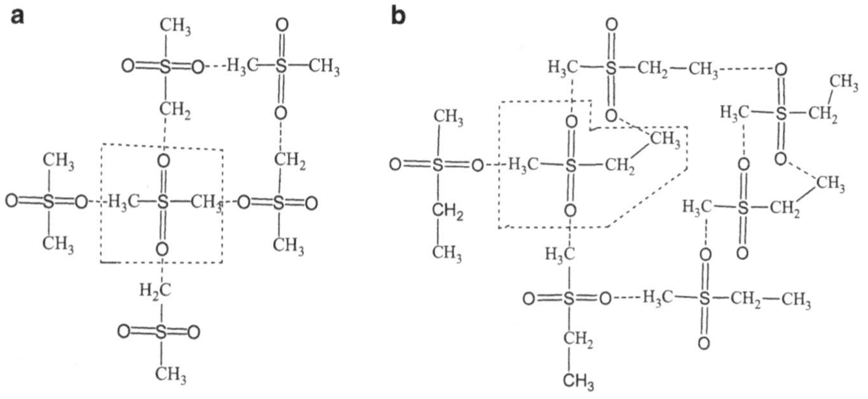

a
$$O{=}S{=}O{-}{-}H_3C{-}\overset{\displaystyle CH_3}{\underset{\displaystyle CH_2}{\underset{|}{\overset{|}{S}}}}{-}CH_3$$

$$O{=}S{=}O{-}\!\cdot\!\cdot\!H_3C{-}\underset{\displaystyle CH_3}{\overset{\displaystyle CH_3}{\underset{|}{\overset{|}{S}}}}{-}CH_3\!\cdot\!\cdot\!{-}O{=}S{=}O$$

b

Fig. 5.8 Schematic of the network of specific interactions of liquid and crystalline dimethyl sulfone (**a**) and ethyl methyl sulfone (**b**)

Table 5.6, the energy values contributed by the isostructural methyl groups are within range of experimental error.

5.2.2 Tetrahydrothiophene-1,1-dioxide(sulfolane), Thiete sulfone(2H-thiete)-1,1-dioxide, and 2.5-Dihydro-2-methyl-thiophene-1,1-dioxide

The high acceptor properties of the sulfur atom of the sulfone group $=S=O$, strengthened by the oxygen atom, provides a significant shift in the electron density from the carbon atoms of the methylene groups in the ring of tetrahydrothiophene-1,1-dioxide. The partial transfer of the electron density from the sulfur atom to the contacting carbon atoms of the two methylene groups as a reverse dative bond and the further shift of electron density to the oxygen atom reduces the negative charge of the sulfur atom and provides the increased negative charge of the oxygen atom.

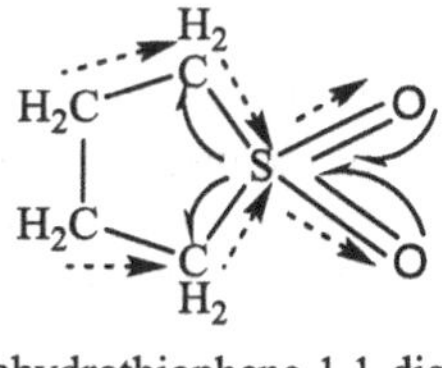

Tetrahydrothiophene-1,1-dioxyde. 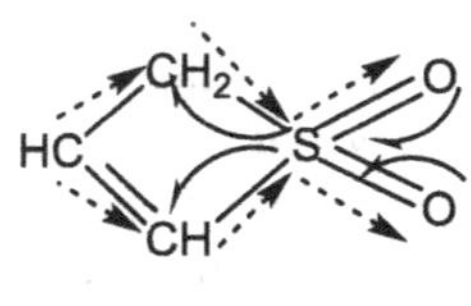   Thiete sulfone (2H-thiete)-1,1-dioxyde

According to these data, high energy values of the specific interactions of the different series of compounds with the sulfone group are expected, but with significantly reduced stability because the rigidity of the ring shifts the reduced electron density to the sulfone groups.

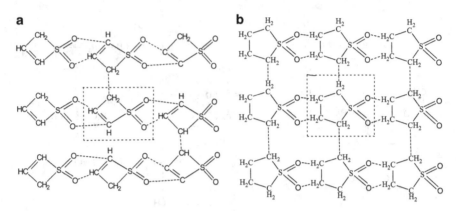

Fig. 5.9 Schematic of the network of specific interactions of crystalline thiete sulfone(2*H*-thiete)-1,1-dioxide (**a**) and tetrahydrothiophene-1,1-dioxide (**b**)

The molecules of thiete sulfone (2*H*-thiete)-1,1-dioxide forms, by the two sulfone groups, two specific interactions DS=O→CH= with CH groups and one specific interaction of the methylene group $DH_2C \rightarrow CH_2$ (Fig. 5.9a), crosslinking the chains to the networks of formed interactions.

The molecule of tetrahydrothiophene-1,1-dioxide, with four ring methylene groups and differing charges of the carbon atoms, forms the two most stable interactions DS=O → CH_2 with the sulfone groups. The two remaining CH_2 groups form two specific interactions $DH_2C \rightarrow CH_2$ with the molecules of the near environment (Fig. 5.9b), forming a network of these interactions.

The energy of this type of interaction of thiete sulfone (2*H*-thiete)-1,1-dioxide and tetrahydrothiophene-1,1-dioxide can be estimated from the sublimation enthalpy of cyclopropane and cyclopentane (42.6 kJ mol^{-1},122 K). The calculation results of the formed specific interactions $DH_2C \rightarrow CH_2$ reduce naturally with an increasing number of methylene groups and CH groups in the ring:

$$\text{Cyclopropane} \ (9.1) \ = \ \text{Cyclobutane} \ (9.1) \ < \ \text{Cyclopentane} \ (8.52)$$
$$< \ \text{Benzene} \ \left(7.43 \ \text{kJ mol}^{-1}\right)$$

The obtained energy value of the specific interaction of cyclopentane corresponds to 122 K; however, it is within the experimental error of the measured sublimation enthalpy. The estimated energies of the specific interactions $DH_2C \rightarrow CH_2$ were used in order to determine the energies of the second type of interaction using Eqs. (5.8a) and (5.8b):

Thiete sulfone (2*H*-thiete)-1,1-dioxide:

$$DS = O \rightarrow CH = \ = \left(\Delta_{vap}H^0(298\,\text{K})\text{tsdo} - DH\,C_2 \rightarrow CH_2\right)/4 \qquad (5.8a)$$

Tetrahydrothiophene-1,1-dioxide:

$$DS = O \rightarrow CH = \ = \left(\Delta_{vap}H^0(298\,K)thtdo - 2DHC_2 \rightarrow CH_2\right)/4 \qquad (5.8b)$$

The obtained energy value of the thiete sulfone(2H-thiete)-1,1-dioxide interaction $DS{=}O \rightarrow CH{=}$ has significantly increased stability in comparison with tetrahydrothio-phene-1,1-dioxide(sulfolane) (12.65 kJ mol^{-1}):

$DS{=}O \rightarrow CH{=}$, thiete sulfone(2H-thiete)-1,1-dioxide $(18.6) > DS{=}O \rightarrow CH_2$, tetrahydrothio-phene-1,1-dioxde(sulfolane) (12.65 kJ mol^{-1}) $< DS{=}O \rightarrow CH_2$, dimethyl sulfone (19.25 kJ mol^{-1})

The increased stability is caused by the double bond of the CH= group, providing a reduced shifting of the electron density and an increased difference between the charges of the carbon atoms of the group and the oxygen atoms of the sulfone group, compared with the difference between the charges of the carbon atom of the ethylene group of tetrahydrothio-phene-1,1-dioxide(sulfolane) and the oxygen atom of the same sulfone group. At the same time, the higher stability of the same type of specific interaction $DS{=}O \rightarrow CH_2$, formed by dimethyl sulfone with the open methyl group, illustrates the shifting of the electron density to the oxygen atom of the sulfone group, which provides the increased difference in the charges of the same carbon and oxygen atoms.

The inequality of the sublimation enthalpies of tetrahydrothiophene-1,1-dioxide (sulfolane), 2.5-dihydro-3-methyl-thiophene-1,1-dioxide and 2.5-dihydro-2-methyl-thiophene-1,1-dioxide points to the significant influence of the location of the isostructural methyl group because the influence of the double bond on the reduction of the enthalpy characteristic does not exceed 1.0 kJ mol^{-1}:

Tetrahydrothiophene-1,1-dioxide(sulfolane) $(67.8) < 2.5$-dihydro-3-methyl-thio-phene-1,1-dioxide $(64.0) < 2.5$-dihydro-2-methyl-thiophene-1,1-dioxide $(60.7$ kJ mol$^{-1})$

This means that the reduction by 7.1 kJ mol^{-1} of the enthalpy characteristic of the crystal compound, without changing the number of ring carbon atoms at the location of the isostructural methyl group at C(2), is caused by a significant reduction in the shifting of electron density in the molecule and a reduction in the charge difference at the carbon atoms and the oxygen atom of the C=S group. The removal of the isostructural group location at C(3) provides the reduction in the shifted electron density, mainly from the given carbon atom of the ring. Taking the energy contribution of the isostructural methyl group to the sublimation enthalpy as the reduced value of a similar contribution to the vaporization enthalpy of 3-methyltetrahydrothiophene (3.9 kJ mol^{-1}), it is possible to estimate the somewhat high energy value of the specific interaction $DS{=}O \rightarrow CH_2{-}CH{=}$ with the help of Eq. (5.9):

$$DS = O \rightarrow CH =$$
$$= \left(\Delta_{vap}H^0(298\,K)thmtp - 2DHC_2 \rightarrow CH_2 - DisoCH_3\right)/4 \qquad (5.9)$$

Table 5.7 Energies (kJ mol^{-1}) of specific interactions of tetrahydrothiophene-1,1-dioxide(sulfolane), thiete sulfone(2H-thiete)-1,1-dioxide, and 2.5-dihydro-2-methyl-thiophene-1,1-dioxide

Compounds	Formula	Structure	$\Delta_{sub}H^0$ (298 K) [3]	$DH_2C \to CH_2$	$DS=O \to CH_2$	$DS=O \to CH=$
Thiete sulfone(2H-thiete)-1,1-dioxide	$C_3H_4O_2S$		83.7 ± 2.5 (208 K)	9.1	–	18.6
Tetrahydrothio-phene-1,1-dioxide(sulfolane)	$C_4H_8O_2S$ vapor		67.8 ± 0.8	8.52 (122 K)	12.65	–
2.5-Dihydro-2-methyl-thiophene-1,1-dioxide	$C_5H_8O_2S$		60.7 ± 2.5	8.52	9.95	DisoCH$_3$ = 3.9
2.5-Dihydro-3-methyl-thiophene-1,1-dioxide	$C_5H_8O_2S$		64.0 ± 2.5	8.52	10.5	DisoCH$_3$ = 3.9

The obtained energy values of the specific interactions (Table 5.7) of two compounds with the isostructural methyl group located at carbon atoms C(2) and C(3), respectively, are the main indication expressed in the shape of the energy property:

DS=O → CH=CH$_2$: 2.5-dihydro-2-methyl-thiophene-1,1-dioxide (9.95) < 2.5-dihydro-3-methyl-thiophene-1,1-dioxide (10.5 kJ mol^{-1})

These groups play a special role in the shifting of electron density in the molecules and are extremely important in protein chemistry. This problem was discussed in the work by Baev [12].

5.2.3 Methylbenzylsulfone and Phenyl Sulfones with Unsaturated Ligands

The results of the implemented thermodynamic analysis of sulfones with alkyl ligands and with the saturated and unsaturated hydrocarbon ring allow investigation of the compounds of the sulfones series with the benzene ring and alkene ligand. The rigid benzene ring with a low ability to shift the electron density to the acceptor assumes the formation by the CH group of an interaction of reduced stability compared with the formed alkene ligand of phenyl sulfones. The molecule of phenyl vinylsulfone or phenyl propadienyl sulfone, with the bond vacancies of the essentially undivided 2s^2 electron pair of the carbon atoms of the end group of the alkene ligand and CH group of the benzene ring, as well as the bond vacancies of the free electron pairs of the oxygen atoms, forms (like alkylsulfones) four specific interactions of two types, DS=O → CH$_2$=CH– (or CH$_2$=C=CH– correspondingly) and DS=O → CH. The benzene ring with the substituted hydrogen atom forms two specific interactions DHC → C and three DHC → CH interactions with energies of 6.92 and 7.43 kJ mol^{-1}, respectively, and one interaction DS=O → CH with an unknown energy value (Fig. 5.10). The contacting benzene rings of the molecules of the near environment and the sulfone fragment form a network of four types of specific interactions. Taking the energy value of the formed interaction of the carbon atom of the vinyl group of DS=O → CH$_2$=CH to be equal to the energy value of the same type for crystal divinylsulfone (14.1 kJ mol^{-1}), the opportunity arises to estimate the energy value of the specific interaction DS=O → CH of the crystal phenyl vinylsulfone using Eq. (5.10):

$$DS=O \rightarrow CH$$
$$= \left(\Delta_{vap}H^0(298K)phvs - 2DS=O \rightarrow CH_2=CH - 2DHC \rightarrow C - 3DHC \rightarrow CH \right)/2$$
$$(5.10)$$

The obtained energy value of the specific interaction (Table 5.8) depends little on the influence of the unsaturated ligand and remains practically stable. The value can

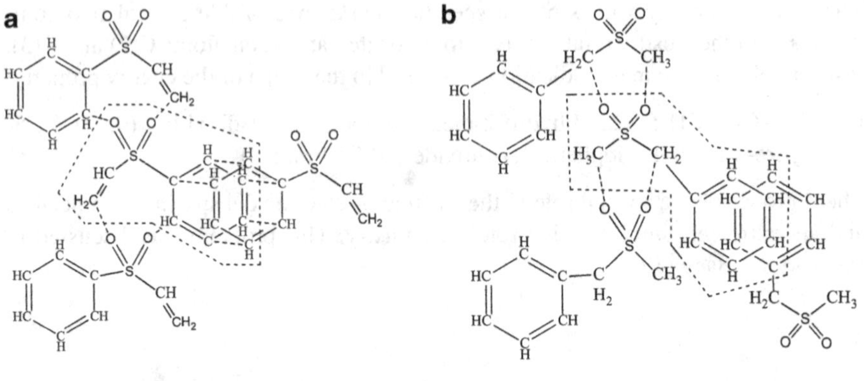

Fig. 5.10 Schematic of the network of specific interactions of crystalline phenyl vinylsulfone (**a**) and methylbenzylsulfone (**b**)

be used for the estimation of the energy of specific interactions formed by alkene ligands of phenyl sulfones using the modified Eq. (5.10a), replacing the value of the vaporization enthalpy of the compounds with the enthalpy of sublimation:

$$DS = O \rightarrow R$$
$$= \left(\Delta_{vap}H^0(298 \text{ K})\text{phvs} - 2DS = O \rightarrow CH - 2DHC \rightarrow C - 3DHC \rightarrow CH\right)/4$$
$$(5.10a)$$

The results of the implemented calculations of the energies of specific interactions are given in Table 5.8. The sequence in the energies of the specific interactions at the triple bond of the end carbon atom and the two double bonds in the ligand are as follows:

DS=O → CH$_2$=CH– (14.1), phenyl vinylsulfone < DS=O → H$_3$C–C≡C– (20.8), phenyl prop-1-ynyl sulfone < DS=O → HC≡C–CH$_2$– (25.6), phenyl prop-2-ynyl sulfone < DS=O → CH$_2$=CH=CH– (25.8 kJ mol^{-1}), phenyl propadienyl sulfone

The molecule of methylbenzylsulfone with the bond vacancy of the essentially undivided 2s^2 electron pair of the carbon atom of the methylene group forms the specific interaction DS=O → CH$_2$ of one type with the methyl group of the sulfone fragment, with reduced stability. In this situation, the benzene ring participates in the formation of the network of specific interactions of CH groups, DHC → CH and DHC → C, with an energy contribution equal to the sublimation enthalpy minus the energy contribution of the saturated hydrogen atom (Fig. 5.10b). The total energy of the two specific interactions DS=O → CH$_2$ and two DS=O → CH$_3$ interactions can be determined from the difference between the sublimation enthalpy of methylbenzylsulfone and that of benzene minus the contribution of the saturated hydrogen

Table 5.8 Energies (kJ mol⁻¹) of specific interactions of crystalline methylbenzylsulfone and phenyl sulfones with unsaturated ligand at 298 K

Compounds	Formula	Structure	$\Delta_{sub}H^0$ (298 K) [3]	DS=O→CH	Ligand	DS=O→L
DHC→CH = 7.6; DHC→C = 6.6						
Phenyl vinylsulfone	$C_8H_8O_2S$		82 ± 9.5	8.9	$CH_2=CH-$	14.1
Phenyl propadienyl sulfone	$C_9H_8O_2S$		105.4 ± 2.5	8.9	$CH_2=CH=CH-$	25.8
Phenyl prop-1-ynyl sulfone	$C_9H_8O_2S$		95.4 ± 2.5	8.9	$H_3C-C\equiv C-$	20.8
Phenyl prop-2-ynyl sulfone	$C_9H_8O_2S$		105 ± 2.5	5.2	$HC\equiv C-CH_2-$	25.6
Methylbenzylsulfone	$C_8H_{10}O_2S$		99.23	DS=O→CH_3 13.35	–	–

atom. Consequently, the average energy value of the interaction practically of one type is given by Eq. (5.11):

$$DS = O \rightarrow CH_3 = \left(\Delta_{sub}H^0(298 \ K)mbs - \Delta_{sub}H^0(298 \ K)bz - DH \right)/4 \quad (5.11)$$

The obtained energy value of the specific interaction in comparison with the energy of the same type of interaction of dimethyl sulfone is caused by the low ability of the benzene ring to shift the electron density, providing a small fluctuation in the charge of the carbon atom of the methylene group and a small difference in the charge of the oxygen atom:

$$DS = O \rightarrow CH_3 : \text{Methylbenzyl sulfone}(13.35)$$
$$< \text{Dimethyl sulfone}(19.25 \ kJ \, mol^{-1})$$

5.2.4 Methyl-p-*tolyl Sulfone and* p-*Tolyl Sulfone with Unsaturated Ligands*

The obtained conclusions and development of the thermodynamic analysis draw attention to the lack of reliable experimental data on the vaporization enthalpies of toluene, for which the values are given without error indications exceeding its sublimation enthalpy. The sublimation enthalpy ($43.1 \ kJ \, mol^{-1}$) of toluene with the functional methyl group has a reduced value in comparison with the value for benzene ($44.4 \ kJ \, mol^{-1}$), which indicates the influence of this group on the changes in the shift of electron density in the molecule and the reduction in the charge difference of the near carbon atoms of the ring. At the same time, the high values of the sublimation enthalpies of 1,2-dimethylbenzene ($60.1 \ kJ \, mol^{-1}$, 248 K) and 1,4-dimethylbenzene ($60.8 \ kJ \, mol^{-1}$, 286 K) may result from the large difference in the charges of ring carbon atoms that are not present. Taking the first assumption, it is possible to estimate the energies of the specific interactions formed by p-tolyl prop-1-sulfones and p-tolyl bat-1-sulfones.

The difference in the sublimation enthalpies of benzene and toluene is the energy contribution of the substituted hydrogen atom ($1.0 \ kJ \, mol^{-1}$) and the low energy contribution ($0.3 \ kJ \, mol^{-1}$) of the isostructural methyl group to the sublimation enthalpy of toluene. Taking the contribution of the isostructural methyl group to the sublimation enthalpy to be equal to $0.60 \ kJ \, mol^{-1}$, the calculated energy contributions of C_6H_4 and the specific interactions of the benzene ring are presented in Table 5.9. The formed specific interactions of the sulfone fragment of methyl-p-tolyl sulfone or its series of compounds: $DS=O \rightarrow C$ (two) and $DS=O \rightarrow CH_3$ with the carbon atom of the benzene ring and methyl group, and $DS=O \rightarrow CH_3$ (two) with the molecules of the near environment and the carbon atoms of the benzene ring form the network of specific interactions (Fig. 5.11).

Taking the energy of the $DS=O \rightarrow CH_3$ interaction to be equal to the energy of the same type of interaction of the crystal dimethyl sulfone ($19.25 \ kJ \, mol^{-1}$) and

Table 5.9 Energies (kJ mol^{-1}) of specific interactions of crystalline methyl-p-tolyl sulfone and p-tolyl prop sulfones at 298 K

Compounds	Formula	Structure	$\Delta_{sub}H^0$ (298 K) [3]	DS=O → C=	Ligand	DS=O → L
DC$_6$H$_4$ = 39.9; DHC → CH = 7.42; DHC → C = 6.92; DisoCH$_3$ = 0.60						
Methyl-p-tolyl sulfone	C$_8$H$_{10}$O$_2$S		100 ± 3.3	12.15	CH$_3$	19.25
p-Tolyl propadienyl sulfone	C$_{10}$H$_{10}$O$_2$S		113 ± 2.5	12.15	CH$_2$=CH=CH–	25.75
p-Tolyl prop-1-ynyl sulfone	C$_{10}$H$_{10}$O$_2$S		103.3 ± 2.5	12.25	H$_3$C-C≡C-	20.9
p-Tolyl prop-2-ynyl sulfone	C$_{10}$H$_{10}$O$_2$S		107.5 ± 2.5	12.15	HC≡C-CH$_2$–	23.0
p-Tolyl trans-prop-1-enyl sulfone	C$_{10}$H$_{12}$O$_2$S		83.7 ± 2.1	12.15	–	11.1
p-Tolyl prop-2-enyl sulfone	C$_{10}$H$_{12}$O$_2$S		95.8 ± 2.9	12.15	–	17.15
p-Tolyl isopropenyl sulfone	C$_{10}$H$_{12}$O$_2$S		88.7 ± 2.5	12.15	DisoCH$_3$ = 1.8	12.45

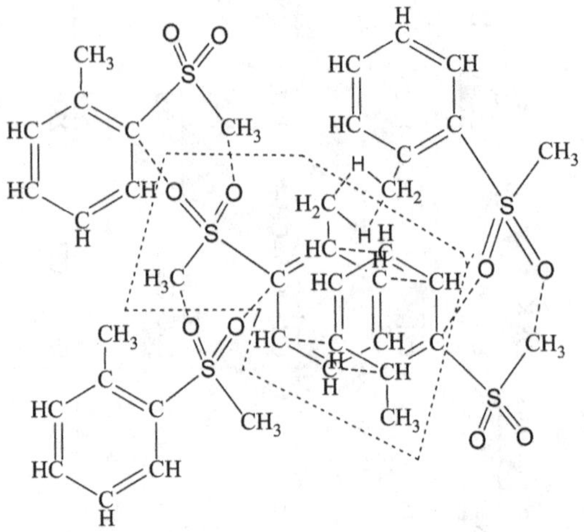

Fig. 5.11 Schematic of the network of specific interactions of crystalline methyl-*p*-tolyl sulfone

obtaining energy values for the interactions in crystalline methyl-*p*-tolyl sulfone, the unknown energy value of DS=O → C can be determined using Eq. 5.12):

$$DS = O \rightarrow C$$
$$= \left(\Delta_{sub}H^0(298\,K) - 2DS = O \rightarrow CH_3 - 4DHC \rightarrow CH - DC \rightarrow C - DisoCH_3 \right)/2$$
$$(5.12)$$

The calculated energy value used to determine the energies of the formed specific interaction by the unsaturated ligand DS=O → L of the *p*-tolyl prop-sulfone series of compounds can be obtained using Eq. (5.12a):

$$DS = O \rightarrow L$$
$$= \left(\Delta_{sub}H^0(298\,K) - 2DS = O \rightarrow C - 4DHC \rightarrow CH - DC \rightarrow C - DisoCH_3 \right)/2$$
$$(5.12a)$$

From the data in Table 5.9, it follows that the energies of the specific interactions of the given series of compounds are described by the sequence of its stabilization:

$$DC=O \rightarrow C = (12.25) < DC=O \rightarrow CH_3 \ (19.25) < DC=O \rightarrow H_3C–C\equiv C– \ (20.9) <$$
$$DC=O \rightarrow HC\equiv C–CH_2 \ (23.0) < DC=O \rightarrow CH_2=CH=CH– \ (25.75 \text{ kJ mol}^{-1})$$

The maximum energy value of the crystal *p*-tolyl propadienyl sulfone with the propadienyl ligand exceeds the energy value of the interaction of the propyl ligand of dipropyl sulfone (23.32 kJ mol^{-1}). The sequence in the stabilization of specific interactions is observed for compounds with three carbon atoms of the chain with a move in the double bond location and with two double bonds:

$$-CH = CH_2 - CH_3 \ (11.1) < -CH_2 - CH = CH_2 \ (17.15) < -CH = C$$
$$= CH_2 \ (25.75\,kJ\,mol^{-1}).$$

The analog of the specificity of interactions of phenol fragments in the liquid condition and in the crystals of p-tolyl ligand-1-ynyl sulfone and p-tolyl ligand-1-enyl sulfone allow calculations to be made of the energies of specific interactions of p-tolyl bat-1-enyl sulfones using Eq. (5.12a). The calculation results are presented in Table 5.10, from which it follows that the energy of the formed specific interaction is increased with the removal of the location of the double bond from the sulfur atom in the molecule of p-tolyl bat-3-enyl sulfone:

DC=O $\rightarrow$ CH$_2$=CH–CH$_2$–CH$_2$–   (22.4) < DC=O $\rightarrow$ CH$_3$–CH=CH–CH$_2$–
(23.0) < DC=O $\rightarrow$ CH$_3$–CH$_2$–CH=CH– (25.95 kJ mol^{-1})

A similar sequence in the stabilization of the specific interaction is expressed in the alkyl ligands with three carbon atoms in the chain $\begin{smallmatrix}-HC \equiv C-CH_3\\ H_3C\end{smallmatrix}$ $\begin{smallmatrix}-CH_2-C \equiv CH_2\\ CH_3\end{smallmatrix}$ and a double bond.

5.2.5 Diphenyl Sulfone, Dibenzyl Sulfone, and Di-p-tolyl Sulfone

The sulfone groups of dibenzyl sulfone molecules form the specific interactions of DC=O $\rightarrow$ CH$_2$ by the free electron pairs of the oxygen atoms with the essentially undivided $2s^2$ electron pair of the carbon atom of the methylene group, providing it with the pentacoordinated condition. Benzene rings form the same specific interactions as in crystalline benzene with substituted hydrogen atoms (Fig. 5.12a). Therefore, the energy of the specific interaction DC=O $\rightarrow$ CH$_2$ can be obtained from the difference between the sublimation enthalpy of dibenzyl sulfone and the doubled value of the benzene sublimation enthalpy minus the substituted hydrogen atoms, according to Eq. (5.13):

$$DS = O \rightarrow CH_2 = \left(\Delta_{sub}H^0(298\,K)dbs - 2\Delta_{sub}H^0(298\,K)b - 2DH\right)/4 \quad (5.13)$$

The obtained energy value of the specific interaction DS=O $\rightarrow$ CH$_2$ (10.42 kJ mol^{-1}) allows calculation of the average energy values of two types of interaction (DS=O $\rightarrow$ CH$_2$ + DS=O $\rightarrow$ CH$_3$)/2 (13.3 kJ mol^{-1}) of crystal methylbenzylsulfone as being equal to 10.4 and 16.2 kJ mol^{-1}, respectively.

The molecule of diphenyl sulfone forms a network of specific interactions in the crystal by the four bond vacancies of the sulfone fragment with the carbon atoms of the benzene rings DS=O $\rightarrow$ C with unknown energy and the five specific

Table 5.10 Energies (kJ mol^{-1}) of specific interactions of crystalline p-tolyl bat-enyl sulfones at 298 K

Compounds	Formula	Structure	$\Delta_{sub}H^0$ (298 K) [3]	DC=O → C	Ligand	DS=O → L
DC$_6$H$_4$ = 41.5; DHC → CH = 7.42; DC → C = 5.92; DisoCH$_3$ = 0.60						
p-Tolyl bat-1-enyl sulfone	C$_{11}$H$_{14}$O$_2$S		106.3 ± 2.5	12.85	–CH=CH–CH$_2$–CH$_3$	22.4
p-Tolyl bat-2-enyl sulfone	C$_{11}$H$_{14}$O$_2$S		107.5 ± 2.5	12.85	–CH$_2$–CH=CH–CH$_3$	23.0
p-Tolyl bat-3-enyl sulfone	C$_{11}$H$_{14}$O$_2$S		113.4 ± 2.5	12.85	–CH$_2$–CH$_2$–CH=CH$_2$	25.95
p-Tolyl isobutenyl sulfone	C$_{11}$H$_{14}$O$_2$S		102.1 ± 2.5	12.85	DisoCH$_3$ = 1.8	19.4
p-Tolyl 2-methylprop-2-enyl sulfone	C$_{11}$H$_{14}$O$_2$S		106.7 ± 2.9	12.85	DisoCH$_3$ = 1.8	21.2

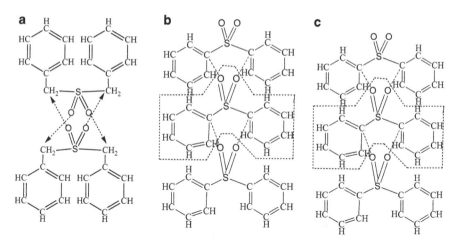

Fig. 5.12 Schematic of the network of specific interactions of crystalline dibenzyl sulfone (**a**), diphenyl sulfone (**b**) and di-*p*-tolyl sulfone (**c**)

interactions of the CH groups of each benzene ring (Fig. 5.12b). Thus, the calculation of the energy of the specific interaction is conducted using Eq. (5.14):

$$DS = O \rightarrow C = \left(\Delta_{sub}H^0(298\,K)dphs - 10DHC \rightarrow CH\right)/4 \qquad (5.14)$$

The molecule of diphenyldisulfone with two sulfone fragments forms eight specific interactions $DS=O \rightarrow C$ and eight interactions of CH groups of the two benzene rings. The energy of the first type of interaction is calculated using Eq. (5.14a):

$$DS = O \rightarrow C = \left(\Delta_{sub}H^0(298\,K)dphs - 8DHC \rightarrow CH\right)/4 \qquad (5.14a)$$

Di-*p*-tolyl sulfone forms, by the two isostructural methyl groups, the specific interactions of low stability $DH_3C \rightarrow H–CH_2$ with an energy of 0.60 kJ mol^{-1}, six interactions $DHC \rightarrow CH$ and four interactions $DHC \rightarrow C$ with known energy values, and four specific interactions of the sulfone fragment $DS=O \rightarrow C$, forming a network of these types of specific interactions (Fig. 5.12c). The energies of this type of interaction are calculated using Eq. (5.14b):

$$DS = O \rightarrow C$$
$$= \left(\Delta_{sub}H^0(298\,K)dts - 6DHC \rightarrow CH - 4DHC \rightarrow C - DH_3C \rightarrow H - CH_2\right)/4 \qquad (5.14b)$$

According to the results of the implemented calculations (Table 5.11), the energies of the specific interaction $DS=O \rightarrow C$ indicate its stabilization in the range of

Table 5.11 Energies (kJ mol^{-1}) of specific interactions of crystalline diphenyl sulfone, dibenzyl sulfone, dibenzyl sulfone and di-p-tolyl sulfone at 298 K

Compounds	Formula	Structure	$\Delta_{sub}H^0$ (298 K) [3]	T (K)	DS=O $\rightarrow$ CH$_2$	DS=O $\rightarrow$ C
Dibenzyl sulfone	C$_{14}$H$_{14}$O$_2$S		128.5 ± 2.9	–	10.42	–
Diphenyl sulfone	C$_{12}$H$_{10}$O$_2$S		106.3 ± 2.9	–	–	7.6
Diphenyldisulfone	C$_{12}$H$_{10}$O$_4$S$_2$ vapor		149.0 ± 4.9	298	–	11.1
DHC $\rightarrow$ CH = 7.42; DHC $\rightarrow$ C = 6.92; DisoCH$_3$ = 1.20						
Di-p-tolyl sulfone	C$_{14}$H$_{14}$O$_2$S		109.6 ± 2.9	–	–	8.8

compounds caused by the increased shift of the electron density to the oxygen atom of the S=O group:

DS=O → C: diphenyl sulfone (7.6) < di-*p*-tolyl sulfone (8.8) < diphenyldisulfone (11.1 kJ mol^{-1})

5.2.6   2-Thiophene Carboxylic Acid

The crystalline 2-thiophene carboxylic acid is the only representative of the derivatives of carboxylic acids for which there is a reliable measurement of the sublimation enthalpy at 319 K, which allows the thermodynamic analysis and the energy of the hydrogen bonds to be obtained. The molecule of the given compound with four bond vacancies of carbon atoms and one sulfur atom, one oxygen atom, one hydrogen atom of the hydroxyl group, and one oxygen atom of the carbonyl group forms the following interactions in the crystal with the molecules of the near environment: two specific interactions DS → CH, two specific interactions DHC → CH, two specific interactions DO → C=, and two specific interactions DC=O → H–C with the corresponding energies of 12.25 kJ mol^{-1} (Table 4.15), 7.43 kJ mol^{-1}, 5.8 kJ mol^{-1} (1.0 kJ mol^{-1} lower than the energy value of the contribution of the saturated hydrogen atom of the CH group), and DC=O → H–C with an energy of 1.0 kJ mol^{-1}. The given energies of the specific interactions allow an estimation of the energy value of the stable hydrogen bond using Eq. (5.15):

$$DO \bullet \bullet \bullet H - O = (\Delta_{sub}H^0(298\,K) - 2DS \rightarrow CH - 2DHC \rightarrow CH$$
$$- 2DO \rightarrow C = -DC = O \rightarrow H - C)/2 \tag{5.15}$$

The obtained energy value of the hydrogen bond of 2-thiophene carboxylic acid (22.0 kJ mol^{-1}) points to its increased stability in comparison with benzoic acid (19.45 kJ mol^{-1}) [2]. This stability is caused by the influence of the sulfur atom of the ring on the shifting of the electron density in the carboxyl fragment and the change in the charges of the oxygen and hydrogen atoms of the hydroxyl group (Fig. 5.13).

5.3   Specific Intermolecular Interactions of Sulfur and Nitrogenated Organic Compounds

5.3.1   Thiocyanate, Isothiocyanates, and Phenyl Isothiocyanate

Presented in Fig. 5.14a is a schematic of the network of specific interactions, illustrating the formation in the molecule of thiocyanic acid of the four bond

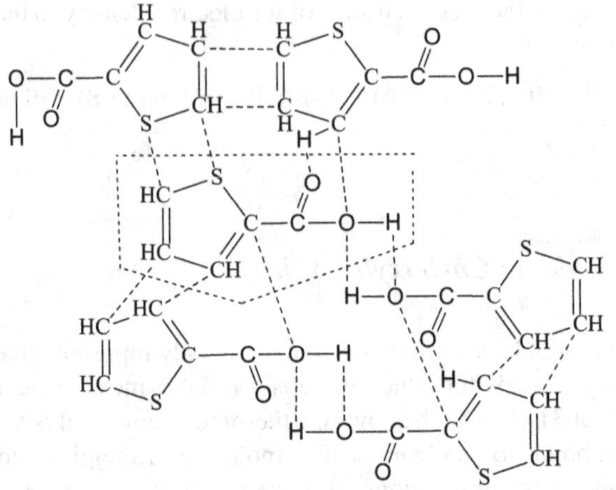

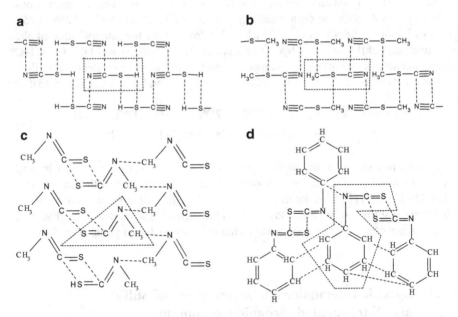

Fig. 5.13 Schematic of the network of specific interactions of crystalline 2-thiophene carboxylic acid

Fig. 5.14 Schematic of the network of specific interactions of liquid thiocyanic acid (**a**), methyl thiocyanate (**b**), methyl isothiocyanate (**c**), and phenyl isothiocyanate (**d**)

vacancies of the two hydrogen bonds DS•••H–S and two stable interactions D≡N → C≡, formed by the C≡N group with the triple bond. The molecule of thiocyanic acid not containing the carbonyl and C=S group, with a linear structure and low acid properties, allows the energy of the hydrogen bond to be accepted as being equal to the energy value of the same bond (5.0 kJ mol^{-1}) of liquid methyl mercaptan with a linear structure (see Chap. 4). This allows the energy value of specific interactions to be obtained from the vaporization enthalpies of the acid and the total value of the two hydrogen bonds using Eq. (5.16a):

$$D \equiv N \rightarrow C \equiv \ = \left(\Delta_{vap}H^0(298\,K)tca - 2DS \bullet \bullet \bullet H - S\right)/2 \qquad (5.16a)$$

The molecules of methyl thiocyanate and ethylthiocyanate with the four bond vacancies of the essentially undivided $2s^2$ electron pair of the carbon atom of the methyl group of the alkyl ligand, the C≡N group, and the free electron pairs of the sulfur and nitrogen atoms form the four specific interactions of the two types, forming a network of these bonds in the liquid compounds (Fig. 5.14b). The energies of the formed specific interactions of the methyl DS $\rightarrow$ CH$_3$–S and ethyl group DS $\rightarrow$ CH$_3$–CH$_2$–S with the sulfur atom correspond to the energy values of the same interactions of liquid dimethyl sulfide (6.92 kJ mol^{-1}) and diethyl sulfide (9.05 kJ mol^{-1}). The energy of the second type of specific interaction is determined from Eq. (5.16b):

$$D \equiv N \rightarrow C \equiv\equiv \left(\Delta_{vap}H^0(298\,K) - 2DS \rightarrow CH_3 - (CH_2)n - S\right)/2 \qquad (5.16b)$$

Presented in Table 5.12 are the calculation results of the energies of specific interactions of D≡N→C≡: thiocyanic acid (9.0) < methyl thiocyanate (13.45, 274 K) < ethylthiocyanate (13.45, 373 K, kJ mol^{-1}), reflecting its stabilization with an increasing number of carbon atoms in the alkyl chain with the increasing difference in charges at the carbon and nitrogen atoms of the same cyanide group. A characteristic feature of methyl isothiocyanate, ethyl isothiocyanate, and allyl isothiocyanate is the presence in their molecules of four bond vacancies forming the four specific interactions of two types DN → CH$_3$–N, DN → CH$_3$–CH$_2$–N, and DN → CH$_2$=CH–CH$_2$–N and the second type DC=S → C=S, forming the networks in the liquid compounds crosslinked by the nonspecific interactions (Fig. 5.14c). Here, attention should be directed to the fact that the electron configuration of the H$_2$C=S molecule with the carbonyl group coincides with the configuration of H$_2$C=O. The first band reflects only the lower vibrational state of ions, and the second ($\nu = 840$ cm^{-1}) and third bands ($\nu = 930$ cm^{-1}) correspond to the binding MO. The significant destabilization during the transition from H$_2$C=O to H$_2$C=S π orbitals (27 eV) in comparison with $\Delta I(n)$ (1.5 eV) (tenfold reduction) assumes the reduced stability of the formed specific interaction and, as a result, the participation of one free electron pair of the sulfur atom in one specific interaction DC=S → C=S. The energies of the formed specific interactions of methyl, ethyl, and allyl ligands are equal to 4.8, 5.8, and 8.1 kJ mol^{-1}, respectively [2]. The energy of the second type of specific interaction was calculated using Eq. (5.16c):

$$DC = S \rightarrow C = S = \left(\Delta_{vap}H^0(298\,K) - 2DN \rightarrow R - N\right)/2 \qquad (5.16c)$$

Table 5.12 Energies (kJ mol^{-1}) of specific interactions of liquid thiocyanates and isothiocyanates

Compounds	Formula	Structure	$\Delta_{vap}H^0$ (298 K) [1]	T (K)	DS•••H-S	D≡C→N≡
Thiocyanic acid	CHNS	N≡C–S–H	28.0	293	5.0	9.0
					DS→CH$_3$–(CH$_2$)$_n$-S	
Methyl thiocyanate	C$_2$H$_3$NS	H$_3$C–S–C≡N	40.7	274	$n = 0$; 6.92	13.45
Ethylthiocyanate	C$_3$H$_5$NS	CH$_3$–CH$_2$–S–C≡N	44.2	373	$n = 1$; 9.05	13.05
					DN→CH$_3$–(CH$_2$)$_m$-N	DS→C=
Methyl isothiocyanate	C$_2$H$_3$NS	CH$_3$–N=C=S	37.3	298	$n = 0$; 4.8	13.75
Ethyl isothiocyanate	C$_3$H$_5$NS	CH$_3$–CH$_2$–N=C=S	40.2; 39.8	298	$n = 1$; 5.8	14.3; 14.1
Allyl isothiocyanate	C$_4$H$_5$NS	H$_2$C=HC–H$_2$C–N=C=S	42.1	298	$n = 2$; 8.1	12.95
DHC→CH =5.76					DN→C=	–
Phenyl isothiocyanate	C$_7$H$_5$NS		52.6	335	5.8	9.0?

where R is CH_3, CH_3-CH_2-, or $CH_2=CH-CH_2-$. The calculation results presented in Table 5.12 reflect the natural character of their stabilization with an increasing number of carbon atoms in the alky ligand:

$$DC = S \rightarrow C = S : \text{Methyl isothiocyanate}(13.75)$$

$$< \text{Ethylisothiocyanate}\left(14.3\,\text{kJ}\,\text{mol}^{-1}\right)$$

The molecule of phenyl isothiocyanate forms two similar specific interactions DC=S $\rightarrow$ C=S and five specific interactions of CH groups of the benzene ring DHC $\rightarrow$ CH and DN $\rightarrow$ C= with energies of 5.63 and 5.8 kJ mol^{-1}, respectively, forming a network of these interactions in the liquid condition of the compound (Fig. 5.14d). The energy value of the DC=S $\rightarrow$ C=S interaction is determined from Eq. (5.16d), taking into account the energy contributions to the enthalpy characteristic of phenyl isothiocyanate:

$$DC = S \rightarrow C = S = H_2C$$

$$= S \left(\Delta_{vap}H^0(298\,\text{K}) - 4DHC \rightarrow CH - DN \rightarrow C =\right)/2 \qquad (5.16d)$$

The obtained reduced energy value of the given specific interaction is caused by the low ability of the rigid benzene ring to shift the electron density, changing the charge differences at the sulfur and carbon atoms of the chain.

5.3.2 Sulfur Diimide, Thioacetamide, and Its Derivatives

The photoelectron spectra of thiooxamides point to the fact that the first ionizing potential is determined by n_S electrons, and the upper π orbital is localized mainly at the group C=S [13, 14], providing it with the formation of the specific interaction C=S $\rightarrow$ C=S.

The series of compounds allows the discovery of the mutual influence of sulfur atoms and nitrogen atoms with high acceptor properties on the stability of the specific interaction formed by the C=S group. The schematic showing the network of specific interactions of liquid ammonium carbamate presented in Fig. 5.15a illustrates the seven formed specific interactions DN•••H–N of hydrogen atoms and nitrogen atoms, and the two hydrogen atoms with the sulfur atoms of DS•••H–N of the molecules of the near environment, with energies of 3.86 kJ mol^{-1} [2] and 5.0 kJ mol^{-1}, respectively (Table 5.12). The energies of the formed specific interactions of the C=S group can be obtained using Eq. (5.17):

$$DC = S \rightarrow C = S$$

$$= \left(\Delta_{vap}H^0(298\,\text{K}) - 7DN \bullet \bullet \bullet H - N - 2DS \bullet \bullet \bullet H - S/2 \qquad (5.17)\right)$$

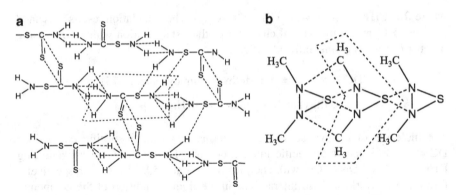

Fig. 5.15 Schematic of the network of specific interactions of liquid ammonium carbamate (**a**) and sulfur diimide (**b**)

The molecule of sulfur diimide forms a network of specific interactions in the liquid condition by two types of interactions: two interactions $N \to CH_3–N$ of the methyl group of methylamine with an energy of 4.8 kJ mol^{-1} and four specific interactions of the second type of interaction $DS \to N$, whose energy is determined from Eq. (5.17a):

$$DS \to N - N = \left(\Delta_{vap}H^0(298\,K) - 2N \to CH_3 - N\right)/4 \qquad (5.17a)$$

Given in Table 5.13 are the results of the energy calculations of specific interactions directly influenced by the increased stability of $DS \to C = \,> DS \to N–N$ formed by the sulfur atom with the carbon atom. The increased stability is caused by the increased shifting of electron density with the essentially undivided $2s^2$ electron pair of the carbon atom compared with the nitrogen atom. The compound dithiooxamide of the ammonium carbamate series with substituted sulfur atom at the C=S group forms eight hydrogen bonds of the amino group with an energy of 6.34 kJ mol^{-1} and four specific interactions $DC=S \to C=S$, forming a network in the crystal (Fig. 5.16a). The energies of these specific interactions are obtained using Eq. (5.17b):

$$DC = S \to C = S = \left(\Delta_{sub}H^0(298\ K) - 8DN \bullet \bullet \bullet H - N\right)/4 \qquad (5.17b)$$

The obtained energy value of 14.8 kJ mol^{-1} has an importance in determining the energies of the same type of specific interaction in combination with the sublimation enthalpies of dithiooxamide analogs. First, the total energy contribution of the formed hydrogen bonds of amine groups is 50.7 kJ mol^{-1}. Following the replacement of one C=S with the energy of two specific interactions (29.6 kJ mol^{-1}) by the imide =N–H group, the sublimation enthalpy increases to 22 kJ mol^{-1}. This means that the energy of the two hydrogen bonds D=N•••H–N= and the two specific interactions $DC=S \to C=S$ is equal to 75.1 kJ mol^{-1}. Such a high value points to

Table 5.13 Energies (kJ mol^{-1}) of specific interactions of liquid ammonium carbamate and sulfur diimide

Compounds	Formula	Structure	$\Delta_{vap}H^0$ (298 K) [1]	T (K)	DN•••H–N	DS•••H–N	DC=S → C=S
Ammonium carbamate	$CH_4N_2S_2$		54.1	247–331	3.86	5.0	8.55
Sulfur diimide	$C_2H_6N_2S$		37.2	263	DN → CH_3–N = 4.8	DS → N–N	–
						6.9	

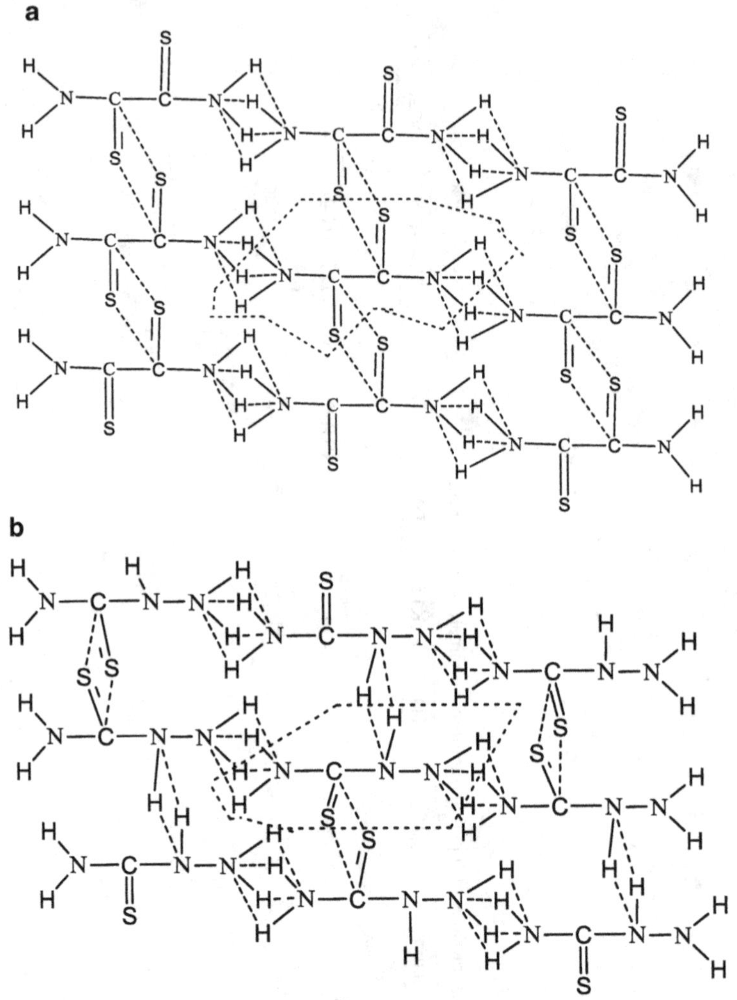

Fig. 5.16 Schematic of the network of specific interactions of crystalline dithiooxamide (**a**) and thiosemicarbazide (**b**)

the stabilization of the remaining two specific interactions DC=S → C=S in the molecule of thiosemicarbazide under the influence of the contributed imide group. Second, the contribution of the second imide group to the molecule of thiocarbohydrazide is accompanied by an increase in the sublimation enthalpy of 26.3 kJ mol^{-1} to reach 152.1 ± 3.0 kJ mol^{-1}. Precisely the given difference in the sublimation enthalpies indicates that the stabilization of specific interactions DC=S → C=S by the contribution of the first imide group reached its maximum value and that the value of 26.3 kJ mol^{-1} represents the energy contribution of two additional hydrogen bonds of the second imide group. Therefore, the obtained energy value of the formed hydrogen bond of the imide group D=N•••H–N= is equal to 13.15 kJ mol^{-1} and that of the specific interaction DC=S → C=S is equal to 18.8 kJ mol^{-1}. In the crystal thiosemicarbazide, the eight formed hydrogen bonds, two more stable hydrogen bonds of the imide group, and two specific interactions DC=S → C=S form a network of these interactions (Fig. 5.16b).

Like amides, the molecule of thioacetamide possesses an increased energy value of the hydrogen bond (9.7 kJ mol^{-1}) formed by the amine group and an increased contribution of the methyl and ethyl groups to the vaporization enthalpy [2]. Therefore, the energy contribution of the methyl group was taken to be equal to 7.7 kJ mol^{-1}. The obtained energy value of the specific interaction is equal to 14.55 kJ mol^{-1} using Eq. (5.17c):

$$DC = S \rightarrow C = S = \left(\Delta_{sub}H^0(298\,\text{K}) - 8DN \bullet \bullet \bullet H - N - 2DCH_3 \right)/2 \quad (5.17c)$$

The energies of the specific interactions given in Table 5.14 are described by the natural sequence of stabilization, reflecting the influence of the two C=S groups to the shift in the electron density and its distribution at the sulfur atoms, and the further participation of the imide group in the formation of charges at the sulfur atom and the imide nitrogen atom and the two imide groups:

DC=S → C=S: Thioacetamide   (14.55) ≤ Dithiooxamide   (14.8) < Thiosemicarbazide (18.8) = Thiocarbohydrazide (18.8 kJ mol^{-1})

5.3.3   Thiobenzamide and Its Derivatives

The compounds of thiobenzamide and its derivatives allow the clarification of the influence of sulfur atoms on the energies of hydrogen bonds and specific interactions formed by the amine group with the methyl and ethyl groups. Presented in Fig. 5.17 is a schematic of the network of the specific interactions of liquid thiobenzamide with the 12 bond vacancies: five interactions DHC → CH and one interaction D=C → C= formed by the benzene ring, with energies of 7.43 and 6.43 kJ mol^{-1}, respectively, and the four hydrogen bonds with energies of 6.34 kJ mol^{-1} formed by the amino group, and additionally two specific interactions DC=S → C=S formed by the sulfur atom of C=S group. The energy of this type of interaction is determined from the sublimation enthalpy and the total energy values of each type of specific interaction and hydrogen bond using Eq. (5.18):

$$DC = S \rightarrow C = S$$
$$= \left(\Delta_{sub}H^0(298\,\text{K}) - 4DN \bullet \bullet \bullet H - N - 5DHC \rightarrow CH - D = C \rightarrow C = \right)/2$$
$$(5.18)$$

Table 5.14 Energies (kJ mol^{-1}) of specific interactions of crystalline thioacetamides, thiosemicarbazide, and thiocarbohydrazide at 298 K

Compounds	Formula	Structure	$\Delta_{sub}H^0$ (298 K) [3]	DN$\cdots$H–N	DC=S $\rightarrow$ C=S	DH–N$\cdots$H–N
Dithiooxamide	$C_2H_4N_2S_2$	$H_2N-C-C-NH_2$ (=S =S)	103.8	6.34	14.8	–
Thiosemicarbazide	CH_5N_3S	$NH_2-C-NH-NH_2$ (=S)	125.8 ± 1.5	6.34	18.8	13.15
Thiocarbohydrazide	CH_6N_4S	$H_2N-NH-C-NH-NH_2$ (=S)	152.1 ± 3.0	6.34	18.8	13.15
Thioacetamide	C_2H_5NS	$H_3C-C-NH_2$ (=S)	83.3 ± 0.3	9.7	14.55	DCH$_3$ = 7.7

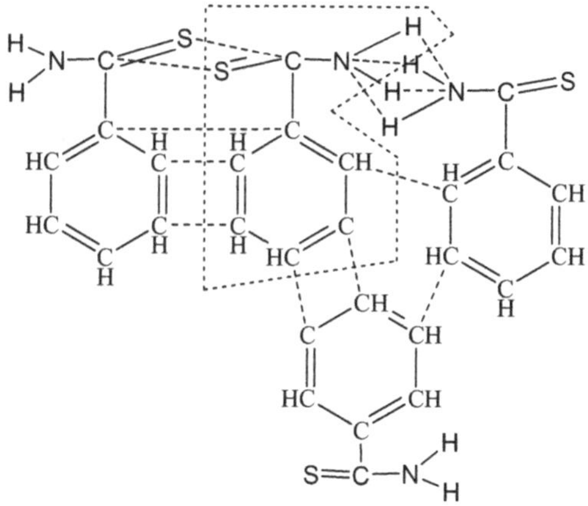

Fig. 5.17 Schematic of the network of specific interactions of crystalline thiobenzamide

A distinctive feature of *N,N*-dimethylthiobenzamide and *N,N*-diethylthiobenzamide compounds is the formation of the four specific interactions DN → CH$_3$–N and DN → CH$_3$–CH$_2$–N, whose energies are obtained from the transformed Eq. (5.18a):

$$DN \rightarrow R - N$$
$$= \left(\Delta_{sub}H^0(298\,K) - 2DC = S \rightarrow C = S - 5DHC \rightarrow CH - D = C \rightarrow C =\right)/2$$

$$(5.18a)$$

Given in Table 5.15 are the calculation results of the energies, which have increased errors. Nevertheless, they reflect the increased stability in comparison with the hydrogen bond.

5.3.4 2,3,4,5-Tetrahydro-(methylthio)pyridine and 4-(Methylthio)pyridine

For compounds of 2,3,4,5-tetrahydro-(methylthio)pyridine and 4-(methylthio) pyridine with a nitrogen atom in the six-membered ring, with the four methylene groups and the four CH groups of the benzene ring, there is a small difference of 8.45 and 8.80 kJ mol^{-1} in the energies of the specific interactions DC=N → C=N [2] formed in the liquid condition, which allow the influence of the sulfide group location at the carbon atom of the ring to be revealed. The molecules of the considered compounds with a similar number of bond vacancies form the same number of specific interactions of differing types. The molecule of 2,3,4,5-tetrahydro-(methylthio)pyridine forms the four specific interactions DH$_2$C → CH$_2$ (Fig. 5.18) with an energy of 5.53 kJ mol^{-1}, which is little different from the interaction with the double bond DHC → CH (5.63 kJ mol^{-1}), formed by the benzene ring of the liquid 4-(methylthio)pyridine. Of interest is the energy contribution formed by the S–CH$_3$ groups of these compounds, located at C(2) and C(4).

Table 5.15 Energies (kJ mol^{-1}) of specific interactions of crystalline thiobenzamides at 298 K

Compounds	Formula	Structure	$\Delta_{sub}H^0$ (298 K) [3]	DN••H-N	DC=S → C=S	DN → CH$_3$-N
DHC → CH = 7.43, D = C → C = =6.43						
Thiobenzamide	C$_7$H$_7$NS		103.4 ± 2.2	6.34	9.1	–
			97.2 ± 0.6			
N,N-Dimethylthiobenzamide	C$_9$H$_{11}$NS		94.4 ± 2.0	–	9.1	8.15
N,N-Diethylthiobenzamide	C$_{11}$H$_{15}$NS		91.4 ± 3.2	–	9.1	7.4

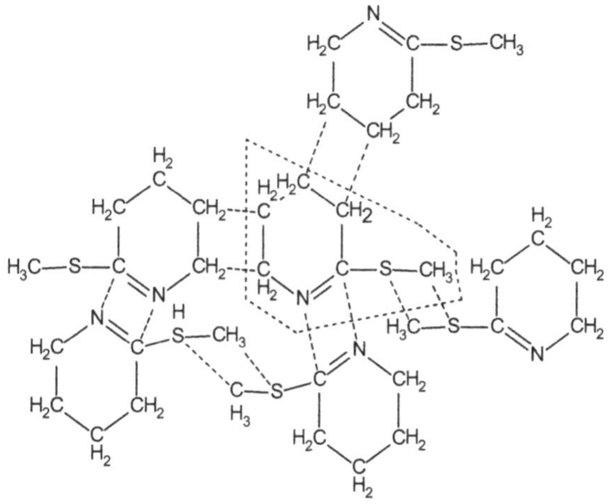

Fig. 5.18 Schematic of the network of specific interactions of liquid 2,3,4,5-tetrahydro-(methylthio)pyridine

The energy values of specific interactions of $DC=N \rightarrow C=N$ can be obtained using Eqs. (5.19) and (5.19a):

2,3,4,5-Tetrahydro-(methylthio)pyridine 4-(methylthio)pyridine:

$$DS \rightarrow CH_3 - S-$$

$$= \left(\Delta_{vap}H^0(298\,K) - 2DC = N \rightarrow C = N - 4DH_2C \rightarrow CH_2\right)/2 \quad (5.19)$$

4-(Methylthio)pyridine:

$$DS \rightarrow CH_3 - S-$$

$$= \left(\Delta_{vap}H^0(298\,K) - 2DC = N \rightarrow C = N - 4DHC \rightarrow CH\right)/2 \quad (5.19a)$$

The increased energy value of specific interactions $DS \rightarrow CH_3$–S of liquid 4-(methylthio)pyridine with the sulfide group at C(4) reflects the increase in the shift of the electron density to the sulfur atom, providing an increased difference in charge between the sulfur atom and the carbon atom of the methyl group.

5.3.5 Specific Interactions of Sulfur and Nitrogen in Cyclic Thiazol and Its Derivatives

The heterocycle of thiazol contains nitrogen and sulfur atoms with strong acceptor properties of electron density, CH groups located between them and, on the other hand, HC=CH– groups with the double bond of carbon atoms. This allow an assumption to be made regarding the increased difference between the charges of the carbon atoms (Fig. 5.19). However, the low value of vaporization enthalpy of

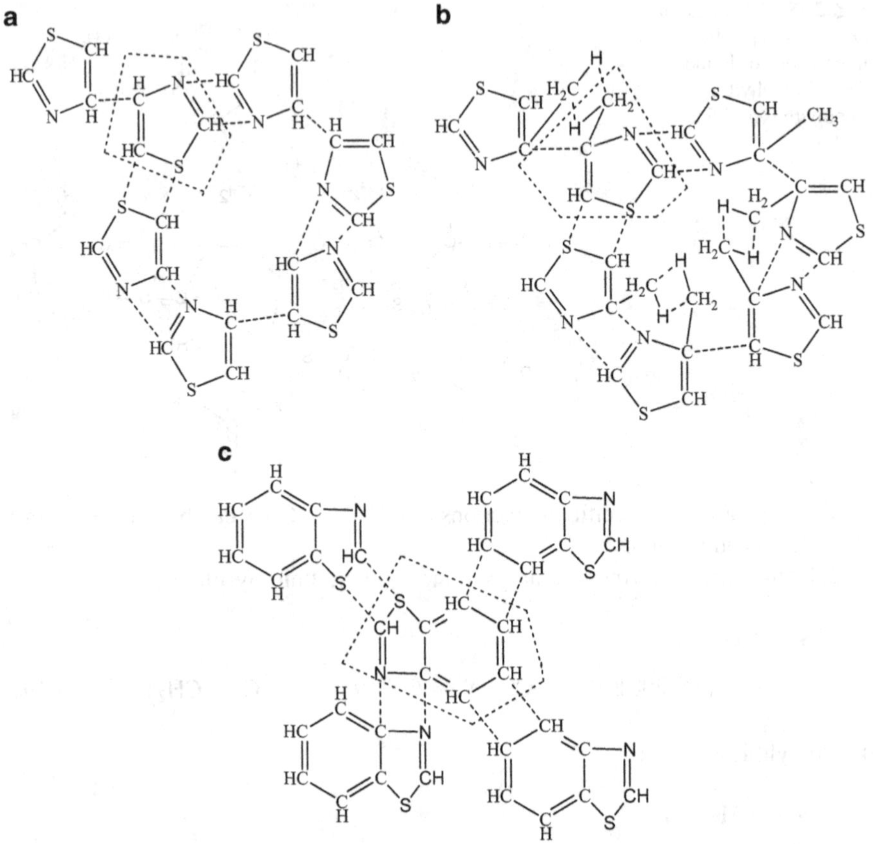

Fig. 5.19 Schematic of the network of specific interactions of liquid thiazol (**a**), 4-methylthiazol (**b**), and benzothiazole (**c**)

the compound (Table 5.16) allows the conclusion to be made that the significant obstacle to the shifting of the electron density is the small difference in the acceptor properties of the nitrogen and sulfur atoms. The molecules of thiazol and the molecules of the near environment form two specific interactions DS → CH and one DHC → CH interaction with energies of 8.85 and 5.63 kJ mol^{-1}, respectively, and two interactions of DN → CH (Fig. 5.19a). These allow the construction of an equation with the vaporization enthalpy of the compound being equal to the sum of the energy contributions of the specific interactions forming the network of these interactions in the liquid and crystal conditions:

$$DN \rightarrow CH = \left(\Delta_{vap}H^0(298\,K) - 2DS \rightarrow CH - DHC \rightarrow CH\right)/2 \qquad (5.20)$$

The obtained energy value of the specific interaction DN → CH illustrates its reduced stability DN → CH (8.25) < DS → CH (8.85 kJ mol^{-1}). The energy

Table 5.16 Energies (kJ mol^{-1}) of specific interactions of pyridine derivatives

Compounds	Formula	Structure	$\Delta_{vap}H°$(298 K) [1]	T (K)	$DH_2C \rightarrow CH_2$/ $DHC \rightarrow CH$	$DC=N \rightarrow C=N$	$DS \rightarrow CH_3-S$
2,3,4,5-Tetrahydro-(methylthio)pyridine	C$_6$H$_{11}$NS		52.6	328	5.53	8.45	6.2
4-(Methylthio)pyridine	C$_6$H$_7$NS		55.8	361	5.63	8.80	7.8

contribution of the isostructural methyl group to the vaporization enthalpy of 2-methylthiazol and 4-methylthiazol, forming two specific interactions of low stability (Fig. 5.19b), is equal to the difference in their enthalpy characteristics, pointing to the stabilization of the formed specific interaction following the replacement of location C(2) with C(4) for the carbon atom with the double bond:

$$DH_3CH - CH_2; \text{ Methylthiazol } (0.15) < \text{4-Methylthiazol } (1.75 \, kJ \, mol^{-1})$$

Presented in Fig. 5.19c is a schematic of the network of specific interactions of liquid benzothiazole, which indicates the formation of the two specific interactions $DN \rightarrow C=$ with an energy of 7.75 kJ mol^{-1}, four interactions $DHC \rightarrow CH$ with an energy value of 5.63 kJ mol^{-1}, and two interactions $DS \rightarrow CH=$ with unknown energy, which is determined using Eq. (5.20a):

$$DS \rightarrow CH == (\Delta_{vap}H^0(298\,K) - 2DN \rightarrow CH - 4DHC \rightarrow CH)/2 \qquad (5.20a)$$

The calculation results (Table 5.17) described the similar regularity of the stabilization of the specific interactions $DS \rightarrow CH$ (10.25) > $DN \rightarrow C$ (7.75 kJ mol^{-1}), caused by the placement of the increased electron density from CH groups at the more distant energy level of the sulfur atom.

5.3.6 Energies of Specific Interactions of Trithiocarbonic Acid (Cyclic Ethylene Ester) and 4,5,6,7-Tetrahydro-1,4-benzodithiol-2-tione

The obtained information on the energy properties of compounds with the $=C=S$ fragment allow a thermodynamic analysis to be carried out on the trithiocarbonic acid (cyclic ethylene ester) and 4,5,6,7-tetrahydro-1,4-benzodithiol-2-tione compounds. The molecule of trithiocarbonic acid with two methylene groups, two sulfur atoms, and the C=S analog of a carbonyl group forms four specific interactions $DS \rightarrow CH_2$ (Fig. 5.20) and two interactions $DC=S \rightarrow C=S$, forming a network of these interactions. The energy of the formed specific interaction $DC=S \rightarrow C=S$ can be estimated based on the analogy with ammonium carbamate, with the sulfur and nitrogen atoms (strong acceptors of electron density) providing the maximum approximation of the real energy value, equal to 8.55 kJ mol^{-1} (Table 5.18). The energy value of the second type of specific interaction $DS \rightarrow CH_2$ can be determined from Eq. (5.21):

$$DS \rightarrow CH_2 = (\Delta_{vap}H^0(298\,K)\text{ttca} - 2DC = S \rightarrow C = S)/4 \qquad (5.21)$$

The obtained energy value of the specific interaction $DS \rightarrow CH_2$ formed by methylene group has an increased stability (16.45 kJ mol^{-1}) caused by the high

Table 5.17 Energies (kJ mol^{-1}) of specific interactions of liquid thiazol and its derivatives

Compounds	Formula	Structure	$\Delta_{vap}H^0$ (298 K) [1]	T (K)	DS → CH=	DN → CH=	DisoCH$_3$
DHC → CH = 5.63							
Thiazol	C$_3$H$_2$NS		39.7	348	8.85	8.25	–
2-Methylthiazol	C$_4$H$_5$NS		40.0	352	8.85	8.25	0.30
4-Methylthiazol	C$_4$H$_5$NS		43.2 ± 0.2	298	8.85	8.25	3.5
Benzothiazole	C$_7$H$_5$NS		58.7	320	10.25	DN–C= 7.75	–

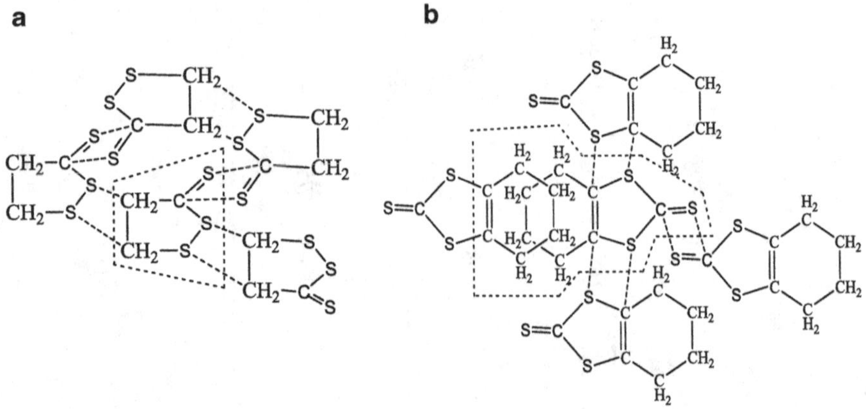

a

b

Fig. 5.20 Schematic of the network of specific interactions of liquid trithiocarbonic acid (cyclic ethylene ester) (**a**) and 4,5,6,7-tetrahydro-1,4-benzodithiol-2-tione (**b**)

difference between the charges of the sulfur atom and the carbon atom of the CH_2 groups of the heterocycle.

The molecule of 4,5,6,7-tetrahydro-1,4-benzodithiol-2-tione with ten bond vacancies forms the same number of specific interactions, forming a network of these interactions. The structure of the liquid condition is schematically reflected in Fig. 5.20b. Four specific interactions $DH_2C \rightarrow CH_2$ are formed by the methylene groups of the cyclohexene ring with an energy of 5.58 kJ mol^{-1}. Two specific interactions of the second type $DC=S \rightarrow C=S$ are formed by the $=C=S$ group, with an energy of 8.55 kJ mol^{-1}, and four interactions of the third type of specific interaction $DS \rightarrow C=$ involve two sulfur atoms of one electron pair, while the second electron pair provides additional stabilization. The energy of this type of interaction is determined using Eq. (5.21a):

$$DS \rightarrow C =$$

$$= \left(\Delta_{vap} H^0 (298 \text{ K}) \text{thbt} - 4DH_2C \rightarrow CH_2 - 2DC = S \rightarrow C = S \right) / 4 \quad (5.21a)$$

The obtained energy value of this type of interaction $DS \rightarrow C=$ has reduced stability $DS \rightarrow C$ (14.9) $< DS \rightarrow CH_2$ (16.45 kJ mol^{-1}) in comparison with the value in liquid trithiocarbonic acid.

5.3.7   *2,1,3-Benzothiodiazole, Phenyl-1,2,3-selenadiazole, and 2-Methylthio-4-ethylamine-6-isopropylamino-1,3,5-triazine*

The two nitrogen atoms of the 2,1,3-benzothiodiazole molecule connected to the carbon atoms of the benzene ring have small differences in charge. This is due to

Table 5.18 Energies (kJ mol^{-1}) of specific interactions of liquid ammonium carbamate and sulfur diimide

Compounds	Formula	Structure	$\Delta_{vap}H^0$ (298 K) [1]	T (K)	$DH_2C \rightarrow CH_2$	$DC{=}S \rightarrow C{=}S$	$DS \rightarrow CH_2$
Trithiocarbonic acid (cyclic ethylene ester)	$C_3H_4S_3$		82.9	298	-	8.55	16.45
4,5,6,7-Tetrahydro-1,4-benzodithiol-2-tione	$C_7H_8S_3$		99.0	346	5.58	8.55	$DS \rightarrow C{=}14.9$

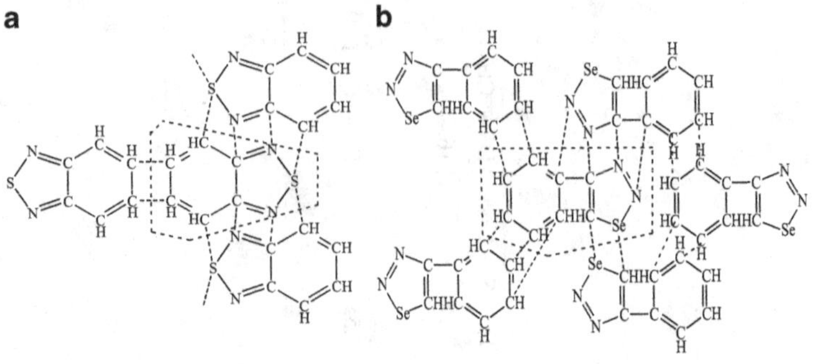

a   b

Fig. 5.21 Schematic of the network of specific interactions of crystalline 2,1,3-benzothiodiazole (a) and phenyl-1,2,3-selenadiazole (b)

the fact that the nitrogen atoms are connected directly to the CH groups and the electron density from the carbon atoms shifts to the nitrogen atom connected to the sulfur atom. The high acceptor ability and the location of the sulfur atom with the nitrogen atoms provide the molecule with the possibility to shift the electron density from the carbon and nitrogen atoms, reducing its negative charge. The rigid benzene ring remains little exposed to the changes in the charges at the carbon atoms and, therefore, its participation in the formation of specific interactions $DN \rightarrow HC=$ with the nitrogen atom provides reduced stability in comparison to the formed interaction of the sulfur and carbon atoms $DS \rightarrow C=$. The presence of the ten bond vacancies in the molecule of 2,1,3-benzothiodiazole causes the formation of the same number of specific interactions. Presented in Fig. 5.21a is the network of specific interactions of crystalline 2,1,3-benzothiodiazole, which includes three types of interactions: two interactions $DHC \rightarrow CH$ with an energy of 7.43 kJ mol^{-1}, four interactions $DN \rightarrow C=$, and four interactions of the third type $DS \rightarrow CH=$.

The energy value of the $DN \rightarrow C=$ interaction of the crystalline 2,1,3-benzothiodiazole can be estimated as being equal to the energy value of the same type of interaction (12.9 kJ mol^{-1}) in crystalline 2-hydroxypyridine and uracil, 2,4-dihydroxypirimidine [12]. The energy of the $DS \rightarrow C=$ specific interaction is determined with the help of Eq. (5.22):

$$DS \rightarrow CH = \left(\Delta_{vap}H^0(298\,K) - 4DN \rightarrow C = -2DHC \rightarrow CH\right)/4 \qquad (5.22)$$

The obtained energy values (Table 5.19) are described by the natural sequence of stabilization:

$$DHC \rightarrow CH(7.43) < DN \rightarrow C = (12.9) < DS \rightarrow CH\left(14.7\,kJ\,mol^{-1}\right)$$

The location of the nitrogen and selenium atoms directly with the carbon atoms of the CH groups in the molecule of 4-phenyl-1,2,3-selenadiazole provides the difference in the charges of nitrogen atoms, which should form specific interactions of differing energies. This creates difficulties in the estimation of the energy values of the bonds, so the problem is simplified and the average value of its energies can be estimated. The molecule of 4-phenyl-1,2,3-selenadiazole with the 12 bond vacancies forms 11 specific interactions of three types (Fig. 5.21b): five DHC $\rightarrow$ CH interactions with an energy of 7.43 kJ mol^{-1}, four interactions DN $\rightarrow$ C=, and two DSe $\rightarrow$ CH interactions, formed by the free electron pair of the selenium atom stabilized by the participation of the second electron pair of the same selenium atom. Therefore, the energy of this type of specific interaction can be accepted to be equal to the energy value (10.6 kJ mol^{-1}) of the same type in the crystalline diphenyl diselenide, with the shifting of electron density from the two free electron pairs of the selenium atom. The energy of the specific interaction DN $\rightarrow$ C= is determined using Eq. (5.22a):

$$DN \rightarrow C = \left(\Delta_{vap}H^0(298\,K) - 2DS \rightarrow CH = -5DHC \rightarrow CH \right)/4 \qquad (5.22a)$$

Given in Table 5.19 are the calculation results describing the stabilization of the energies of specific interactions in the crystalline 4-phenyl-1,2,3-selenadiazole:

$$DHC \rightarrow CH(7.43) < DN \rightarrow C = (8.94) < DN \rightarrow C = \ \left(10.6\,kJ\,mol^{-1}\right)$$

5.4 Energies of Specific Interactions of Sulfur, Oxygen, and Nitrogenated Compounds

The thermodynamic characteristics of the phase transformations of the sulfur, oxygen, and nitrogen compounds obtained at standard conditions, or close to them, are known for an insignificant number of compounds with the three differing heteroatoms of the representatives of this class. In this connection, we are limited in the implementation of thermodynamic analysis to a single compound. The molecule of N-sulfinylmethanamine with an anglular structure has four bond vacancies, including an essentially undivided $2s^2$ electron pair of the carbon atom and a free electron pair of the sulfur atom.

Oxygen atoms and nitrogen atoms form four specific interactions of two types: 2DS=O $\rightarrow$ S=O with unknown energy value and the second type 2DN $\rightarrow$ CH$_3$ with an energy of 4.8 kJ mol^{-1} [2]. The formed chains are connected by the nonspecific interactions in the network in the condensed state. Transition to vapor of the molecules of this compound is determined by the energy of the more stable

Table 5.19 Energies (kJ mol^{-1}) of specific interactions of crystalline 2,1,3-benzothiodiazole at 298 K

Compounds	Formula	Structure	$\Delta_{sub}H^0$ (298 K) [3]	T (K)	DHC → CH	DN → CH=	DS → C=
DHC → CH = 7.43; DC → C= = 6.43							
2,1,3-Benzothiodiazole	C$_6$H$_4$N$_2$S		125.3 ± 0.8	298	7.43	12.9	14.7
4-Phenyl-1,2,3-selenadiazole	C$_8$H$_6$N$_2$Se		94.1 ± 0.8	298	7.43	DN → C = 8.94	10.6

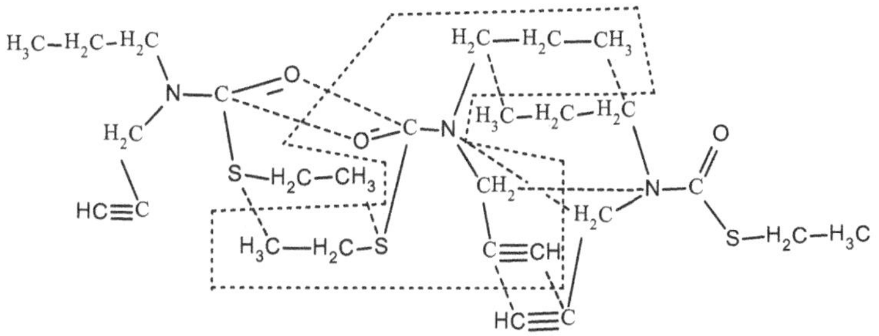

Fig. 5.22 Schematic of the network of specific interactions of liquid carbamothiole acid, propyl-2-propynyl-S-ethyl ester

DS$=$O$\rightarrow$S$=$O specific interaction, whose energy value is given in Table 5.19, as determined using Eq. (5.23):

$$DS = O \rightarrow S = O = \left(\Delta_{vap}H^0(298\,K) - 2DN \rightarrow CH_3\right)/2 \qquad (5.23)$$

The molecule of carbamothiole acid; propyl-2-propynyl-S-ethyl ester (Fig. 5.22), forms ten specific interactions of five types: $2DN \rightarrow CH_3$–CH_2–CH_2–N and $2DN \rightarrow CH_3$–CH_2–N with energy values of 8.1 and 6.0 kJ mol^{-1}; $2DN \rightarrow CH_2$–N and $2D{\equiv}HC \rightarrow C{\equiv}$ with energies of 4.2 and 4.07 kJ mol^{-1}, obtained from the energy of the specific interaction of $N \rightarrow CH_3$–N minus the contribution of the substituted hydrogen atom (0.60 kJ mol^{-1}) and from the vaporization enthalpy of acetylene[2]. This allows the energy of the fifth type of specific interaction, DC$=$O$\rightarrow$C$=$O, to be obtained using Eq. (5.23a):

$$
\begin{aligned}
DC = O &\rightarrow C = O \\
&= \left(\Delta_{vap}H^0(298\,K) - 2DN \rightarrow CH_3 - CH_2 - CH_2 - N - 2DS\right. \qquad (5.23a) \\
&\quad \left. \rightarrow CH_3 - CH_2 - S - 2DN \rightarrow CH_2 - - 2D \equiv HC \rightarrow C \equiv \right)/2
\end{aligned}
$$

The compound carbamothiole acid; (1-methylethyl)-2-propynyl-S-ethyl ester is obtained from carbamothiole acid; propyl-2-propynyl-S-ethyl ester by replacement of the propyl ligand with the isopropyl group. The ethyl ligand forms with the molecules of the near environment two specific interactions $DN \rightarrow CH_3$–CH_2–N with a total energy contribution of 12.0 kJ mol^{-1} and, additionally, there is a contribution of 1.65 kJ mol^{-1} of the isostructural methyl group to the vaporization enthalpy. At the same time, the two formed specific interactions of the substituted propyl ligand contribute to the slightly increased total energy value of 16.2 kJ mol^{-1}. It follows that the values cited in the literature [1] for the vaporization enthalpies without experimental error, or at least for propyl-2-propynyl-S-ethyl ester, are not correct.

Table 5.20 Energies (kJ mol^{-1}) of specific interactions of liquid N-sulfinylmethanamine, carbamothiole acids, and (methylethyl)-2-propynyl-S-ethyl ester

Compounds	Formula	Structure	$\Delta_{vap}H^0$ (298 K) [1]	T (K)	DN $\rightarrow$ CH$_3$	D$\equiv$HC $\rightarrow$ C$\equiv$	DC=O $\rightarrow$ C=O	DS=O $\rightarrow$ C=S
N-Sulfinylmethanamine	CH$_3$NOS		31.8	264	4.8	–		11.1
DN $\rightarrow$ CH$_3$–CH$_2$–CH$_2$–N = 8.1; D$_N$ $\rightarrow$ CH$_3$–CH$_2$–CH$_2$–N = 6.0; DisoCH$_3$ = 1.65; DS $\rightarrow$ CH$_3$–CH$_2$–S = 9.05								
Carbamothiole acid; propyl-2-propynyl-S-ethyl ester	C$_9$H$_{15}$NOS		64.6	305	4.2	4.07	6.9	
Carbamothiole acid; (1-methylethyl)-2-propynyl-S-ethyl ester	C$_9$H$_{15}$NOS		72.8	305	4.2	4.07	12.25	

$$DC = O \rightarrow C = O$$
$$= (\Delta_{vap}H^0(298\,K) - 2DCH_3 - CH_2 - N \rightarrow CH_3 - CH_2 - N$$
$$- 2DCH_3 - CH_2 - S \rightarrow CH_3 - CH_2 - S - 2DN \rightarrow CH_2 - \qquad (5.23b)$$
$$- 2D \equiv HC \rightarrow C \equiv -DisoCH_3)/2$$

Using the modified Eqs. (5.23a) and (5.23b), the calculations gave an essentially increased energy value for the DC=O → C=O specific interaction, as presented in Table 5.20.

References

1. Chickos JS, Acree WE (2003) Enthalpies of vaporization of organic and organometallic compounds, 1880–2002. J Phys Chem Rev 32(2):519
2. Baev AK (2012) Specific intermolecular interactions of organic compounds. Springer, Heidelberg/Dordrecht/London/New York, 434 p
3. Chickos JS, Acree WE (2002) Enthalpies of sublimation of organic and organometallic compounds, 1910–2001. J Phys Chem Rev 31(2):537
4. Bohm MC, Gleiter R (1979) Electronic structure and reactivity of propellanes. Tetrahedron 35 (5):675–679
5. Solouki B, Bock H, Appel R (1975) Schwefelsaure-derivate X2S=Y2: alkyl, vinyl- und arylsulfone, alkylsulfoimide und sulfurylhalogenide. Chem Ber 108(3):897–913
6. Martin HD, Iden R, Landen H, Mayer B, Distefano G, Modelli A, Gleiter R (1986) Photoelectron and electron transmission spectra of thiete 1,1-dioxide, thietane 1,1-dioxide and related compounds. The sulfone effect on a highly react. J Electron Spectrosc 41(2):385–397
7. Aitken RA, Gosney I, Farries H et al (1984) Chemical repercussions of orbital interactions through bond and through space. The reactivity of the double bond in unsaturated cyclic sulphones towards aziridine formation and epoxidation. Tetrahedron 40(13):2487–2503
8. Mueller C, Schweig A, Vermeer H (1975) Thiirene dioxides. Electronic structure. J Am Chem Soc 97(5):982–987
9. Dolenko GN, Kondratenko NV, Popov VN, Iagupolkiy LM (1987) Investigation of the values of charges on sulfur atoms aminalkyls, arylperftoralkyl sulphides of fluorescent spectroscopy. Izvestya AN SSSR Ser Chem 2:336–340
10. Nefedov VI, Vovna VI (1989) Electronic structure of organic and elementorganic compounds. Nauka, Moscow, 198 p
11. Vovna VI, Vilesov FI (1974) Photoelectron spectra the structure of molecular orbitals of methyl amines. Opt Spectrosc 36(2):251
12. Baev AK (2014) Specific intermolecular interactions of nitrogenated and bioorganic compounds. Springer, Heidelberg/Dordrecht/London/New York, p 579
13. Henriksen L, Isaksson R, Liljefors T, Sandstrom J (1981) Ultraviolet absorption and photoelectron spectra of some cyclic and open-chain mono- and dithiooxamides. Acta Chem Scand B 35(7):489–495
14. Guimon C, Pfister-Guillouzo G, Arbelot M (1975) Spectres photoelectroniques d'heterocycles carbonyles et thiocarbonyles. Tetrahedron 31(22):2769–2774

Index

© Springer International Publishing Switzerland 2015

A.K. Baev, *Specific Intermolecular Interactions of Element-Organic Compounds,*
DOI 10.1007/978-3-319-08563-0

Printed in the United States
By Bookmasters